ACS SYMPOSIUM SERIES **419**

Downstream Processing and Bioseparation

Recovery and Purification of Biological Products

Jean-François P. Hamel, EDITOR
Massachusetts Institute of Technology

Jean B. Hunter, EDITOR
Cornell University

Subhas K. Sikdar, EDITOR
National Institute of Standards and Technology

Developed from a symposium sponsored
by the Division of Industrial and Engineering Chemistry, Inc.,
at the Third Chemical Congress of North America
(195th National Meeting
of the American Chemical Society),
Toronto, Ontario, Canada,
June 5–11, 1988

American Chemical Society, Washington, DC 1990

Library of Congress Cataloging-in-Publication Data

Downstream processing and bioseparation: recovery and purification of biological products
  Jean-François P. Hamel, editor, Jean B. Hunter, editor, Subhas K. Sikdar, editor.

  p.   cm.—(ACS Symposium Series, 0097–6156; 419).

  "Developed from a symposium sponsored by the Division of Industrial and Engineering Chemistry, Inc., at the Third Chemical Congress of North America (195th National Meeting of the American Chemical Society), Toronto, Ontario, Canada, June 5–11, 1988."

  Includes bibliographical references.

  ISBN 0–8412–1738–6

  1. Separation (Technology)—Congresses.
  2. Biotechnology—Technique—Congresses.

  I. Hamel, Jean-François P., 1958– .   II. Hunter, Jean B., 1955– .   III. Sikdar, Subhas K.   IV. American Chemical Society. Division of Industrial and Engineering Chemistry.   V. Chemical Congress of North America (3rd: 1988: Toronto, Ont.)   VI. Series.

TP248.25.S47D68   1990
660'.2842—dc20   89–49336
                                                      CIP

The paper used in this publication meets the minimum requirements of American National Standard for Information Sciences—Permanence of Paper for Printed Library Materials, ANSI Z39.48–1984.   ∞

Copyright © 1990

American Chemical Society

All Rights Reserved. The appearance of the code at the bottom of the first page of each chapter in this volume indicates the copyright owner's consent that reprographic copies of the chapter may be made for personal or internal use or for the personal or internal use of specific clients. This consent is given on the condition, however, that the copier pay the stated per-copy fee through the Copyright Clearance Center, Inc., 27 Congress Street, Salem, MA 01970, for copying beyond that permitted by Sections 107 or 108 of the U.S. Copyright Law. This consent does not extend to copying or transmission by any means—graphic or electronic—for any other purpose, such as for general distribution, for advertising or promotional purposes, for creating a new collective work, for resale, or for information storage and retrieval systems. The copying fee for each chapter is indicated in the code at the bottom of the first page of the chapter.

The citation of trade names and/or names of manufacturers in this publication is not to be construed as an endorsement or as approval by ACS of the commercial products or services referenced herein; nor should the mere reference herein to any drawing, specification, chemical process, or other data be regarded as a license or as a conveyance of any right or permission to the holder, reader, or any other person or corporation, to manufacture, reproduce, use, or sell any patented invention or copyrighted work that may in any way be related thereto. Registered names, trademarks, etc., used in this publication, even without specific indication thereof, are not to be considered unprotected by law.

PRINTED IN THE UNITED STATES OF AMERICA

ACS Symposium Series

M. Joan Comstock, *Series Editor*

1990 ACS Books Advisory Board

Paul S. Anderson
Merck Sharp & Dohme Research Laboratories

V. Dean Adams
Tennessee Technological University

Alexis T. Bell
University of California—Berkeley

Malcolm H. Chisholm
Indiana University

Natalie Foster
Lehigh University

G. Wayne Ivie
U.S. Department of Agriculture, Agricultural Research Service

Mary A. Kaiser
E. I. du Pont de Nemours and Company

Michael R. Ladisch
Purdue University

John L. Massingill
Dow Chemical Company

Robert McGorrin
Kraft General Foods

Daniel M. Quinn
University of Iowa

Elsa Reichmanis
AT&T Bell Laboratories

C. M. Roland
U.S. Naval Research Laboratory

Stephen A. Szabo
Conoco Inc.

Wendy A. Warr
Imperial Chemical Industries

Robert A. Weiss
University of Connecticut

Foreword

The ACS SYMPOSIUM SERIES was founded in 1974 to provide a medium for publishing symposia quickly in book form. The format of the Series parallels that of the continuing ADVANCES IN CHEMISTRY SERIES except that, in order to save time, the papers are not typeset but are reproduced as they are submitted by the authors in camera-ready form. Papers are reviewed under the supervision of the Editors with the assistance of the Series Advisory Board and are selected to maintain the integrity of the symposia; however, verbatim reproductions of previously published papers are not accepted. Both reviews and reports of research are acceptable, because symposia may embrace both types of presentation.

Contents

Preface vii

1. **Modeling and Applications of Downstream Processing:
   A Survey of Innovative Strategies**... 1
   Jean-François P. Hamel and Jean B. Hunter

   EXTRACTION AND MEMBRANE PROCESSES

2. **Statistical Thermodynamics of Aqueous Two-Phase Systems** 38
   Heriberto Cabezas, Jr., Janis D. Evans, and David C. Szlag

3. **Theoretical Treatment of Aqueous Two-Phase Extraction
   by Using Virial Expansions: A Preliminary Report** 53
   Daniel Forciniti and Carol K. Hall

4. **A Low-Cost Aqueous Two-Phase System for Affinity
   Extraction** 71
   David C. Szlag, Kenneth A. Giuliano, and Steven M. Snyder

5. **Separation of Proteins by Using Reversed Micelles** 87
   Claude Jolivalt, Michel Minier, and Henri Renon

6. **Enzymes in Liquid Membranes: Reaction and Bioseparation** 108
   Donald K. Simmons, Sheldon W. May, and Pradeep K. Agrawal

7. **Pilot-Scale Membrane Filtration Process for the Recovery
   of an Extracellular Bacterial Protease** 130
   John J. Sheehan, Bruce K. Hamilton, and Peter F. Levy

PROCESSES USING BIOSPECIFIC INTERACTION WITH PROTEINS

8. Complexation Between Poly(dimethyldiallylammonium chloride) and Globular Proteins...158
   Mark A. Strege, Paul L. Dubin, Jeffrey S. West, and
   C. Daniel Flinta

9. Protein Fractionation by Precipitation with Carboxymethyl Cellulose...170
   Kathleen M. Clark and Charles E. Glatz

10. Protein Separation via Affinity-Mediated Membrane Transport....188
    Liese Dall-Bauman and Cornelius F. Ivory

11. Affinity Precipitation of Avidin by Using Ligand-Modified Surfactants 212
    Roberto Z. Guzman, Peter K. Kilpatrick, and Ruben G. Carbonell

NOVEL ISOLATION AND PURIFICATION PROCESSES

12. Ultracentrifugation as a Means for the Separation and Identification of Lipopolysaccharides...238
    Marshall Phillips and Kim A. Brogden

13. Removal and Inactivation of Viruses by a Surface-Bonded Quaternary Ammonium Chloride...250
    I-Fu Tsao and Henry Y. Wang

14. Mathematical Model of a Rotating Annular Continuous Size Exclusion Chromatograph..268
    Sandeep K. Dalvie, Ketan S. Gajiwala, and Ruth E. Baltus

15. The Continuous Rotating Annular Electrophoresis Column: A Novel Approach to Large-Scale Electrophoresis............................285
    Randall A. Yoshisato, Ravindra Datta, Janusz P. Gorowicz, Robert A. Beardsley, and Gregory R. Carmichael

Author Index 304

Affiliation Index 304

Subject Index...305

Preface

THE RECENT ADVANCES IN GENETIC ENGINEERING AND CELL CULTURE that have spawned the new biotechnology industry have also stimulated new thinking and research in downstream processing. This new research and development, which focuses on separation and purification of biological materials, is welcome and much needed, in view of the central role of bioseparation engineering in the process economics of biotechnology.

Downstream processing is commonly classified into four distinct steps: broth conditioning and removal of insolubles; isolation of the desired product (including clarification and extraction); purification with high-resolution techniques; and polishing.

Of these steps, isolation and purification currently enjoy the most attention from researchers. The authors of this book have made further progress in their respective research programs since the symposium on which this book is based. These revisions and new data are included in this book. Most chapters include data that have not been published before. Moreover, each chapter has received two reviews by relevant experts.

The aim of this book is not to provide an exhaustive treatise on all areas of isolation and purification of biotechnology products, but to present the spectrum of current thinking and activities on bioseparations, specifically of large molecules such as proteins and polysaccharides. The chapters are divided into three categories: extraction and membrane processes, processes using biospecific interaction with proteins, and novel isolation and purification processes.

An overview chapter by Hamel and Hunter presents the state of the art of research on bioseparations. Extraction processes using biphasic aqueous systems, liquid membranes, reversed-micellar systems, and membrane processes are all being actively studied. Significant advances in these topics, including predictive mathematical models, are presented in the first section. The second section includes several papers on affinity and other interaction techniques that are finding uses in protein purification. In the last section, we offer several reports that delineate advances in isolation and purification processes such as electrophoresis and chromatography.

We gratefully acknowledge the assistance of our reviewers, whose insight and guidance have enlightened the editors and authors alike. We thank the authors for their special assistance generously extended. Finally, we are indebted to Cheryl Shanks of the ACS Books Department for her patience and many helpful hints during the preparation of this book.

JEAN-FRANÇOIS P. HAMEL
Massachusetts Institute of Technology
Cambridge, MA 02139

JEAN B. HUNTER
Cornell University
Ithaca, NY 14853

SUBHAS K. SIKDAR
National Institute of Standards and Technology
Boulder, CO 80303

November 6, 1989

Chapter 1

Modeling and Applications of Downstream Processing

A Survey of Innovative Strategies

Jean-François P. Hamel[1] and Jean B. Hunter[2]

[1]Department of Chemical Engineering, Massachusetts Institute of Technology, Cambridge, MA 02139
[2]Department of Agricultural and Biological Engineering, Cornell University, Ithaca, NY 14853

Downstream processing is playing an increasingly important role in the biochemical industry, especially since the advent of recombinant DNA technology. The use of recombinant DNA technology not only enables improvements in the production efficiency of therapeutic and industrial proteins, but it also permits the modification and improvement of protein structure and thus function. However, the commercial application of such technology was initially accompanied by concerns over product safety.

Quality criteria have been made especially stringent for products derived from genetically-modified microorganisms. The establishment of strict quality guidelines was the result of early concern about the oncogenic potential related to products contaminated by DNA sequences of the host mammalian cells (1). The quest for high quality has created a growing need for high-resolution techniques at the process scale as well as for novel strategies for the isolation and purification of bioproducts. Since the typical environment for producing biologicals is a complex one and quality criteria need to be strict, primary recovery techniques are typically implemented in a purification scheme prior to (or in conjunction with) high-resolution techniques. The most sophisticated and useful schemes take advantage of both the different physical and chemical properties of the components in complex mixtures and of the interactive nature of the downstream processing techniques (see Figure 1).

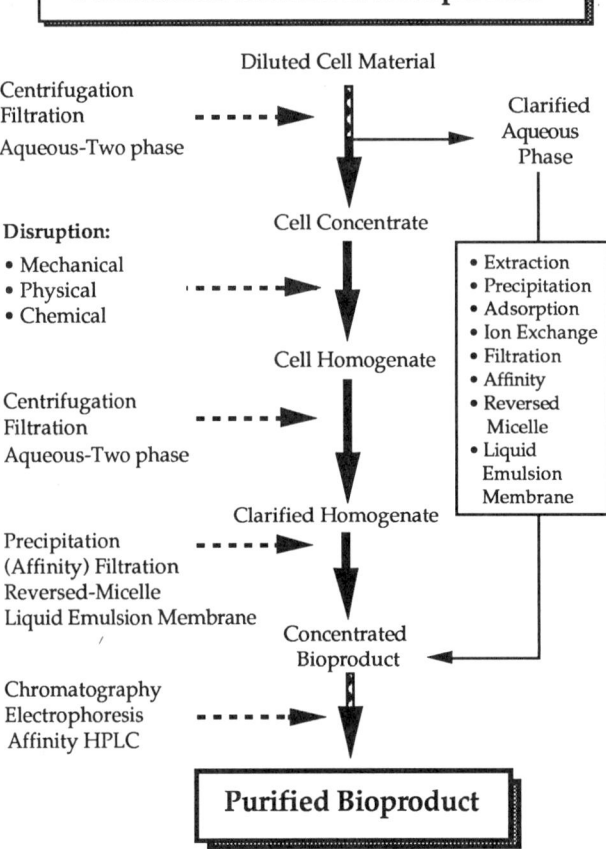

Figure 1. Bioproduct Purification Chart
Intracellular Bioproduct Route
Extracellular Bioproduct Route

Since proteins are polymers of amino acids, the chemical nature of the amino acid side chains and the order of the amino acids play an important role in establishing the biological properties of the active protein. Proteins may differ from each other according to size, charge density, shape and biological activity. Similarly, protein purification schemes require a similar diverse combination of separation techniques based on the various physicochemical properties of proteins.

Typically, protein is lost at every purification step and one normally wishes to reduce the number of steps. An added advantage of fewer steps for some unstable proteins is faster processing time and thus, improved quality of the desired protein when time is critical to maintain stability. However, the number of physicochemical properties are limited; so are the number of purification techniques developed from them.

In non-genetically engineered microorganisms or cells, the protein of interest often represents a small fraction of total cellular or extracellular protein. Several strategies have been developed, using the techniques of molecular biology (e.g. gene dosage, leader sequence), which permit the design of efficient and simple purification schemes. For example, the overexpression of cloned genes in Escherichia coli or animal cells is an increasingly used strategy to produce eukaryotic proteins. Overexpression in bacteria often results in the formation of insoluble protein aggregates which are usually not in an active form. In some cases, the desired protein is already relatively pure and may represent up to 25% of total cell protein. An initial isolation step involving a combination of a disruption and chemical/physical separation could therefore produce a relatively pure product. By comparison, if that same protein were produced as a soluble protein, its initial purity would likely be significantly lower. Thus, an integrated view of each process is of critical importance. Whether the protein produced is soluble or insoluble, the isolation of intracellular proteins typically requires the use of disruption techniques.

High-pressure homogenization is an effective technique to free intracellular products. The detailed mechanisms by which the cells are disrupted are not known, and the parameters for determining the degree of disruption can only be determined empirically (2). Then, such knowledge would be likely to impact the design of equipment. In the last ten years, for example, major efforts have been devoted to homogenizer valve design and to configurations permitting higher pressure (>600 bar) operation, with the rationale that such conditions produce more efficient disruption - in terms of amount of product released per pass. Since the relationships between pressure and particle size distribution are poorly understood, there is a possibility that increasing the homogenizer operating pressure produces decreasing particle size. Smaller particles, in turn, may have a negative impact further downstream, in that their removal during clarification operations may be more difficult.

Often, high-resolution techniques like electrophoresis or affinity chromatography cannot be used readily on a complex mixture. However, in most isolation/purification processes of proteins, chromatography will appear in one form or another. Affinity chromatography has received considerable attention in the last ten years since it is one of the most powerful tools for separating biological products. This technique has been largely researched at the small-scale and only recently have large-scale studies been detailed in the literature. For example, the use of a monoclonal antibody column was recently reported to have provided major purification in a single step of interferon α-2a from extracts of recombinant Escherichia coli cells (3). As a result, a process using affinity chromatography may permit the reduction of the number of steps compared to processes based on other techniques like centrifugation and filtration.

Overall however, there are few published reports of large-scale processes based on affinity separations, and in this context aqueous two-phase systems and membrane technology have imposed themselves (4). This book includes detailed studies by **Cabezas et al.** (5), **Forciniti and Hall** (6), **Szlag et al.** (7), **Dall-Bauman and Ivory** (8), **Guzman et al.** (9) and **Sheehan et al.** (10) based on such technologies as well as many others still confined to the laboratory scale. The contributions are varied in that: 1) some are theoretical, some experimental and some are both, 2) the authors represent both the academic and the private sectors, 3) there are several attempts to describe large-scale processes.

The remainder of this introductory chapter focuses on downstream processing and bioseparation relevant to the chapters presented in this book. Thus, the following topics are covered: multi-phase systems, membrane separation, centrifugation and adsorption techniques, electrophoresis, chromatography, and affinity separations.

MULTI-PHASE SYSTEMS FOR THE RECOVERY OF PROTEINS

Aqueous Biphasic System

More than 70 years elapsed between the first report of aqueous two-phase systems (11) and their subsequent applications to biochemical systems (12). In the last ten years in particular, there have been several innovative applications of aqueous two-phase systems (13). Aqueous two-phase systems consist of two immiscible fluids in a bulk water solvent. In such systems, the percentage of water in both phases is high, i.e. between 75 and 95%. As a consequence, the surface tension between the two immiscible phases may be as low as 0.1 dyne/cm so that a gentle mixing is sufficient to produce and maintain an emulsion (14).

One of the best characterized systems involves mixtures of dextran and polyethyleneglycol (PEG). In such a system, biological substances ranging from soluble proteins to particulate materials (cells or organelles) will partition preferentially in one of the phases. In order to characterize the separation of a substance of interest in an aqueous two-phase system, it is convenient to define a partition coefficient as the ratio of this substance's relevant property in the top and bottom phases. For example, for a protein with biological activity:

$$K_{act} = ACT_{top}/ACT_{bottom}$$

where: K_{act} is the partition coefficient of the protein, ACT_{top} is the activity in the top phase, and ACT_{bottom} is the activity in the bottom phase.

For cells or organelles, the partition coefficient is defined in terms of concentrations. The ability of a given substance to partition in an aqueous two-phase system is the result of several types of interactions (i.e. hydrophobic, electrostatic, and conformational) between this substance and the polymers. Thus, the behavior of homogenized cell material with a wide size distribution is a complicated system to characterize mechanistically. Proteins provide somewhat simpler systems, in that their partitioning can be understood by changing the nature and the concentration of the ions present in the system (15, 16). Several applications of different nature are worth mentioning.

As indicated in the flowchart presented earlier in this chapter (Figure 1), aqueous two-phase systems are especially useful during early primary recovery steps. One of the attractions of this system for extraction and purification of intracellular proteins is its ability to remove cell debris. Since the first report in 1976 describing the use of aqueous two-phase systems for cell debris removal (17), several investigators have demonstrated the generality of the technique. Using dextran/PEG, extractive cell debris removal experiments were carried out with Bacillus sphaericus for the extraction of leucine dehydrogenase (18), with Candida boidinii for the extraction of catalase, formaldehyde dehydrogenase and formate dehydrogenase (19), and with Klebsiella pneumoniae for the extraction of pullulanase (20) to cite just a few examples of enzyme extractions; in these cases yields were above 90% and partition coefficients between 3 and 10. Other original processes exploited the biocompatibility of dextran/PEG systems. In a process of extractive bioconversion, where bioconversion of a substrate is combined with removal of an inhibitory product, Clostridium tetani cells partitioned preferentially in a dextran-rich bottom phase, while the proteolytic toxin they produced remained more evenly distributed between the dextran and the PEG phases. As a result, the degradation of the cell walls of the bacteria was significantly less than compared to a simple aqueous phase system (21). Extractive bioconversion has been successfully demonstrated more recently for glucose fermentation and in the bioconversion of cellulose to ethanol (22). Besides being biocompatible, the dextran/PEG system is flexible in that coupling of this technique with other purification procedures is feasible; for example, it has been successfully integrated in a process using a separator, a settling tank and concentration and ultrafiltration equipment for the purification of leucine dehydrogenase (18).

Most of the research conducted with aqueous two-phase systems has been experimental and empirical; few studies of the fundamental thermodynamic mechanisms of phase separation and partitioning have been conducted (5, 23, 24). Furthermore, the systems which have been described use highly purified, expensive polymers, for model laboratory-scale applications. Novel bioseparation research based on aqueous two-phase systems needs to focus more on fundamental aspects needed to design phase diagrams and calculate partition coefficients. This knowledge will, in turn, provide the basis for the design of industrial processes. The high cost of the dextran/PEG creates opportunities to design less expensive polymer systems (25). Such an approach has already proved to be fruitful and hydroxypropylstarch was used in combination with polyethyleneglycol for the partitioning of catalase and ß-galactosidase (26). In this book, several chapters by **Cabezas et al.** (5), **Forticini and Hall** (6), **Szlag et al.** (7), are devoted to both fundamental and practical aspects of research based on aqueous two-phase systems.

Reversed - Micellar Systems

Reversed micelles result from the formation of aggregates of surfactants that form in an organic/aqueous environment. The surfactants used in such systems have an hydrophilic headgroup and an hydrophobic tail. When placed in an organic/aqueous environment, the hydrophilic headgroups of the surfactant form a polar core containing water, while the hydrophobic tails remain in contact with the bulk organic phase (See Figure 2).

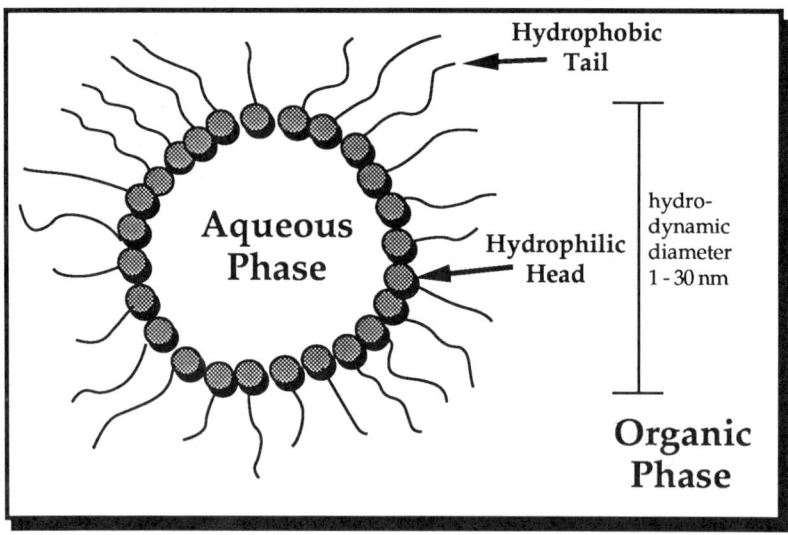

Figure 2. Diagram of a Reversed-Micellar System

Such a system was used successfully to solubilize enzymes within the inner core of reversed micelles without significant loss of activity (27). Besides its use to study enzymatic reactions in organic solvents with poorly water-soluble substrates (27), reversed-micellar systems have also been developed for the isolation and recovery of solubilized proteins

(28), and recently for the refolding of denatured proteins (29). For example, extraction experiments at the small-scale have been reported where α-amylase was extracted from a water phase into an oil phase (trioctylmethylammoniumchloride in isooctane) with reversed micelles, followed by the extraction of α-amylase from the oil phase to another water phase (30). By careful manipulation of pH and salt concentration, significant α-amylase activity could thus be recovered (30).

Novel aspects of protein extraction with reversed-micelles include both fundamental studies and process design studies/approaches. Fundamental studies are essential in order to design a reversed-micelles based extraction process in a rational manner. Such theoretical programs have been initiated and are providing a better understanding of the partitioning and transport phenomena in such systems (31). In this book, Jolivalt et al. (32) review the modeling aspects and the applications of reversed micelles for protein separations.

Furthermore, the results obtained with several experimental models are encouraging and suggest that the recovery of a single protein from a complex mixture, like a cell culture supernatant or a fermentation homogenate, may be feasible.

Liquid Membranes

Liquid membranes consist of an emulsion of two immiscible phases dispersed in an external, continuous phase (33) (Figure 3).

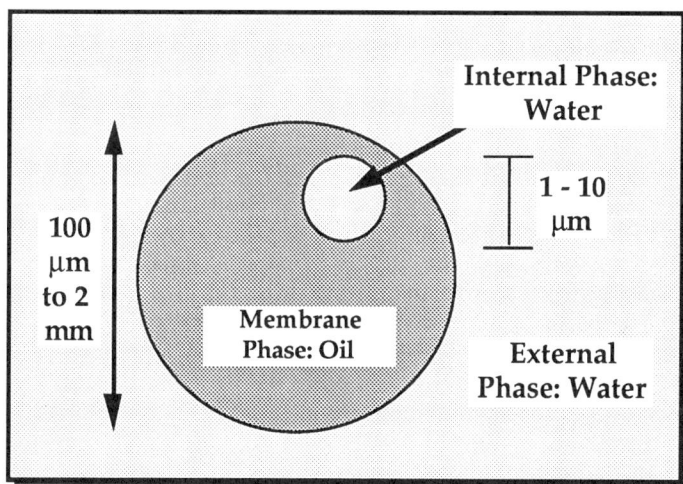

Figure 3. Diagram of a Liquid Membrane System

In such a system the internal and external phases are separated by an oil phase often called the membrane phase. As a result, the internal and external phases cannot come into direct contact.

Liquid membrane systems were first introduced in 1968 (34), and since then they have been evaluated for various chemical and biochemical applications (35). Some of the applications include: the selective extraction of hydrocarbons (36), the recovery of rare earths from process streams (37), the extraction of organic contaminants like phenol from water streams (38), and amino acid recovery (39).

These applications demonstrate the versatility of the liquid membranes, which can be adapted to obtain desired properties, such as stability and selectivity. Liquid membranes offer several advantages, including: 1) the ease to maintain them in suspension by agitation, 2) their relative large surface area per unit volume, facilitating mass transfer between the external and the internal phases, 3) their simple recovery, since upon interruption of agitation, the droplets coalesce to form an emulsion layer which can be separated from the external phase by gravity and, 4) the possibility to achieve recovery and concentration in a single step.

Pilot plant feasibility studies have been encouraging, and economic evaluations have indicated that liquid membranes can compete with other conventional ion exchange or solvent extraction techniques (40, 41). While the initial work with liquid emulsion membranes involved chemical systems, the first biomedical application was demonstrated with the use of liquid emulsion membranes for drug delivery and drug overdose prevention systems (42). In the biochemical field, an early study describes a liquid emulsion membrane-encapsulated bacterial cell-free homogenate able to carry out the reduction of nitrate and nitrite (43). Since this early study in 1974, many other biochemical applications have been reported which describe more complex enzyme/liquid emulsion membrane systems. They detail the critical role of membrane formulation in minimizing membrane breakage and protein inactivation (44). Membrane breakage can be affected by emulsion composition or by hydrodynamic shear, and translates into the leakage of the internal phase through the emulsion (45, 46, 47, 48).

In the future, novel developments of liquid membranes for biochemical processes should arise. There are several opportunities in the area of fermentation or cell culture, for the in situ recovery of inhibitory products, for example. Another exciting research direction is the use of liquid membrane for enzyme encapsulation so that enzymatic reaction and separation can be combined in a single step. Chapter 6 by **Simmons et al.** (49) is devoted to this technique. The elucidation of fundamental mechanisms behind the liquid membrane stability is essential, and models should be developed for the leakage rate in various flow conditions. Such models will be useful to address the effect of parameters such as flow regime, agitation rate, and microdroplet volume on leakage of the internal phase into the external phase. Furthermore, such knowledge would form a basis for the design of recovery processes.

MEMBRANE SEPARATION

In the last few years, there has been an increasing interest in the use of membranes for the pharmaceutical and the food industries due to the limitations and drawbacks of competing technologies. Membranes are effectively used for air and aqueous feed stream sterilization (50), for recovery and purification of bioproducts or treatment of wastes (51) and for extractive fermentation processes (52), as support to immobilize biocatalysts (50), or in an affinity cross-flow filtration design (53). A large choice of membranes is available depending on their hydrophobic character, on their chemical structure (ceramic or polymeric), on the geometry of their pores, on their performance and on their cost. They offer ease of operation and great flexibility, and do not require addition of chemical agents. Thus, they are found in a multitude of process configurations, including cross-flow (also called tangential-flow) filtration, reverse osmosis, electrodialysis, affinity filtration, pervaporation and membrane distillation. Furthermore, for a given process the membranes can be packed in several configurations. For example, reverse osmosis membranes may be in one of the following classes: tubular, spiral wrap, fiber, flat plate (54).

Membranes are particularly suited for bioprocesses involving the cultivation of microorganisms or cells as biocatalysts, in which the product of interest is produced extracellularly. Such processes are becoming increasingly attractive when compared to those in which the products accumulate intracellularly. Some of the reasons for this include the use of novel expression systems which favor higher product concentrations, and the ease of purification as compared to an intracellular bioproduct route. One of the drawbacks remains that extracellular protein products are produced in dilute concentration. Extracellular-product based-processes require cell separation, product recovery and concentration. The use of ultrafiltration and microfiltration membranes has become a method of choice in such process schemes.

Microfiltration applies when particulate materials above 50 nm diameter are to be separated from an aqueous phase or from macromolecules. Thus, microfiltration can be used for cell concentration. On the other hand, the same unit operation can be viewed as a fractionation procedure in processes where products are produced extracellularly. Although the first research on microfiltration was carried out in Germany in the early 1900s (55, 56), the technology didn't find major application until after World War II when it was used for the analysis of waste aqueous streams (57). In the late 1970s, applications for cell separation appeared as a substitute for centrifugation in the separation of plasma from whole blood (58). Thus, a significant data base has been produced over the years. Ample literature exists on both the development of flux models (59) and on hemolysis (60) and such studies should now be useful for biotechnology applications involving non-rigid cells. While microfiltration is a very common technique for sterilization (of air and wastewater streams), it is less used in separation schemes (57).

Ultrafiltration employs membranes of smaller pore size, able to retain proteins and other macromolecules (M.W. of 10^3 to 10^6). Ultrafiltration can be used strategically for separation of macromolecules and microorganisms from water and low molecular weight solutes. Unlike microfiltration, separation by ultrafiltration occurs at the molecular level, and thus is mostly suited for soluble substances. Shortly after the initial demonstrations of ultrafiltration applied to bioprocesses in the early 1960s, the first laboratory-scale ultrafiltration membranes became available (61). Since the first report, in 1965, on protein concentration by ultrafiltration (62), this technology has been tested in various configurations on a multitude of biological models, including the ultrafiltration of a cell suspension with proteins in solution (63), the concentration of human albumin using hollow-fiber ultrafiltration (64), the ultrafiltration of skim milk in a rotating module (65), the concentration of S49 lymphoma cells by cross-flow ultrafiltration (66), the concentration and purification of antibiotics and enzymes (67), the production of soybean and peanut protein isolates in a hollow-fiber membrane system operated in an ultrafiltration or a diafiltration mode (68), the recovery by ultrafiltration and diafiltration of high molecular weight products (e.g. polypeptides or enzymes) obtained in dilute aqueous solutions in bioreactors (69), the concentration of soya protein precipitate (70), the recovery of steroids from biotransformation medium by tangential-flow filtration used in combination with microsized polymeric particles (71). From a practical standpoint, cross-flow ultrafiltration and cross-flow microfiltration have a lot in common. However, in cross-flow microfiltration, parameters like "deformability" (for cells), adsorption (for colloids) and transmembrane flux are critical (57).

Over the last 15 years, there has been an increasing interest in the use of cross-flow filtration for processing cell suspensions. In spite of this, little engineering performance data useful in design or in elucidating fundamental mechanisms is available in the literature. There are few reports of industrial-scale experiments. One of the earlier reports on industrial-scale cross-flow filtration, describes cell harvest data for eight different organisms in high-velocity filters (72). The chapter by **Sheehan et al.** (10) extends our knowledge of cross-flow filtration systems applied to cell separation and product recovery, in their comparative evaluation of the performance of centrifugation and filtration operations at the pilot-scale. Part of the experimental work was carried out at the pilot-scale level, and the study reports a comparative evaluation on the performance of centrifugation and filtration unit operations.

There are several research opportunities in membrane filtration including: cross-flow filtration for processing shear-sensitive animal cell suspensions, pilot-scale cross-flow filtration for cell separation and macromolecule concentration, correlations between microfiltration flux data and theoretical models, predictive models for ultrafiltration performance in multicomponent systems, mechanisms of flux reduction in multicomponent protein solutions, and effects of concentration polarization on experimental rejection coefficients.

ANALYTICAL and ISOLATION TECHNIQUES

Ultracentrifugation

About 50 years ago, the advent of the analytical ultracentrifuge offered to researchers an alternative tool to fractionate and characterize proteins (73, 74). It thus permitted 1) to push further the detection limits of the previous techniques, mostly based on the solubilities of proteins, and 2) to characterize individual proteins in complex solutions. This old and respected technique has nearly been displaced by electrophoresis and HPLC but it deserves another look. Ultracentrifugation is a powerful tool to determine size, composition and concentration of a macromolecule; however, the equipment involved is expensive which explains, in part, why it remains essentially a small-scale laboratory technique. In this book, **Phillips and Brogden** (75) revisit CsCl gradient ultracentrifugation as a tool for the isolation of lipopolysaccharides (LPS) from gram-negative microorganisms. Its potential use for the isolation of LPS produced by recombinant organisms is also discussed in that chapter.

Isoelectric Precipitation

Proteins have historically been recovered by isoelectric precipitation and by salting-out with inorganic salts, usually ammonium sulfate. Polyelectrolytes such as carboxymethylcellulose (CMC) and polyacrylic acid (PAA) are also effective precipitants for proteins, and offer an operationally simple method for protein recovery which is easily scalable, produces a high purity and concentrated product stream, and does not denature the target protein (76). Unlike salt precipitation, only small quantities of precipitant are used, from 5 to 25% of the protein by weight. In this book, a chapter by **Clark and Glatz** (77) demonstrates the power of this method in recovery of lysozyme from a 1:1 mixture with ovalbumin. For example, at a dosage of 0.1 g/g protein, over 70% of the lysozyme was recovered essentially free of albumin. Precipitation occurs when polymer chains and proteins combine by electrostatic interactions to produce "primary particles" which aggregate into flocs upon aging (78). Key parameters are the pH and ionic strength, which govern the protein/polyelectrolyte interactions, and fluid turbulence, which disperses the polymer feed but may shear the flocs apart.

HIGH-RESOLUTION PURIFICATION TECHNIQUES

Electrophoresis

Electrophoresis, the migration of charged molecules under the influence of an electrical field, is an efficient and inherently mild technique which has found widespread use in both analytical and small-scale preparative purification of proteins and nucleic acids. Of four basic techniques - zone electrophoresis, moving boundary electrophoresis, isotachophoresis and isoelectric focusing (79) - only zone electrophoresis and isoelectric focusing are widely applied. Zone electrophoresis (ZE) resolves the components of a sample on the basis of their relative electrophoretic mobilities. The mobility is a function of charge and molecular weight for soluble species and of zeta potential for colloids and particles. Isoelectric focusing (IEF) separates proteins on the basis of their isoelectric point. A sample is placed into a support medium, usually a gel, containing a stable pH gradient decreasing from the cathode to the anode. When an electrical field is applied to the system, each protein migrates towards the position corresponding to its isoelectric point. When the protein reaches this position, its net charge falls to zero and its motion stops because the electrical field no longer exerts a force on it. Zone electrophoresis is a dynamic separation, as it is based on relative rates of movement, while IEF is an equilibrium separation which reaches a steady state.

Recent developments in electrophoresis have focused on two areas:

- extension of the scale of electrophoretic methods from the conventional sample size range of 10^{-3} to 10^{-6} g protein to extremely small (10^{-12} g) and large (1 to 10^3 g per hour) scale operation.
- development of hybrid methods which combine electrophoresis with other separation techniques.

• Nanoscale separation

Capillary electrophoretic methods including open-column zone electrophoresis, disc electrophoresis in gels, isotachophoresis and isoelectric focusing have received considerable attention from the analytical community over the last three or four years (80, 81, 82). In capillary zone electrophoresis (CZE), nanogram quantities of sample are placed in a silica capillary, 50 to 300 microns in diameter and 50 to 100 cm long. Since the small dimensions of the capillary allow for efficient removal of Joule heat, electrical fields up to 350 V/cm can be applied. Under the influence of the field, sample components separate by zone electrophoresis while they are carried downstream by electro-osmosis.

Efficiencies on the order of 10^6 theoretical plates are achievable. Separated components may be detected by fluorescence, electrochemical detection or by interfacing to a quadrupole mass spectrometer via electrospray ionization (83). Mass spectrometry can provide extremely sensitive detection, in the attomole range. Moreover, the mass/charge spectrum of each product yields a precise measurement of its molecular weight, to the nearest dalton. Peptide analytes in the range of 500 to 2500 daltons have been separated and identified by CZE/MS (84, 85), and the technique can be extended to molecular weights on the order of 100,000 (86). Capillary zone electrophoresis/mass spectrometry may eventually compete with SDS/PAGE for molecular weight determinations. Current CZE research focuses on modeling column/solute interactions and other band-broadening phenomena (87), improvement of sample introduction, and development of more sensitive detectors.

Pulsed and crossed-field electrophoresis have recently become popular for separation of chromosome-sized DNA segments on agarose gels. In these techniques, the electric field in the gel is periodically shifted or reversed. At each shift, the macromolecules' migration is retarded while they change conformation and realign with the field. Relatively smaller molecules relax faster and move farther per cycle, resulting in much improved resolution. The development of pulsed-field electrophoresis has been driven largely by the human genome project and related studies. Though well-accepted for analytical separations, it is difficult to envision any process-scale applications for this technology.

• Process-scale separation

Three devices for free-flow electrophoretic separation are now available commercially. They are described in more detail in Ivory's excellent review (79).

The Biostream rotationally stabilized free-flow electrophoresis device, based on the Philpot-Harwell design (Figure 4), uses an annular geometry stabilized against radial convection by rotation of the outer cylinder. Carrier buffer and feed are injected at the base of a vertical annulus and move axially upward to fraction collectors at the top.

An electrical field of several tens of V/cm is applied radially between the inner cylinder of the annulus (generally the cathode) and the rotating outer cylinder. The device has a capacity of 1 to 2 L/h of feed, or several g/h of protein. Several analyses of hydrodynamic dispersion and Joule heating have been published, e.g. Beckwith and Ivory (88). Though solute dispersion measured in the separator is several times greater than theoretical predictions, the apparatus can still perform well when buffer composition has been optimized to maximize the difference in solute mobilities (79).

Thin-film free-flow electrophoresis devices have been studied since the late 1950s (Figure 5).

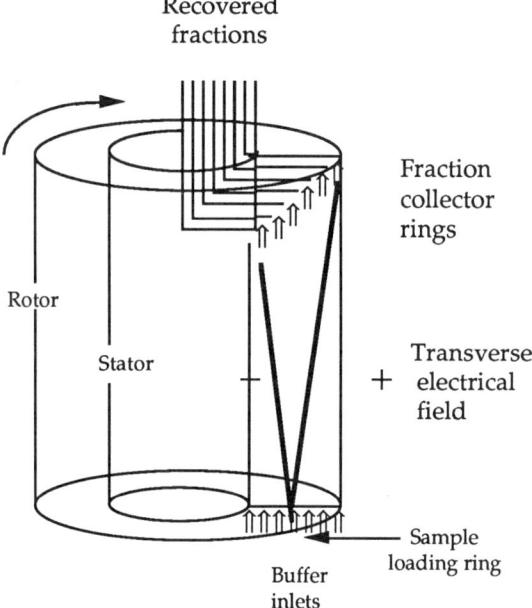

Figure 4. Philpot - Harwell Device

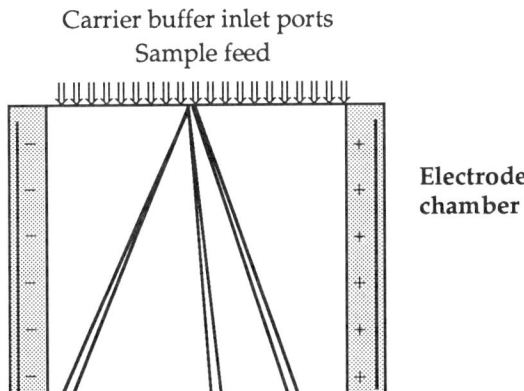

Figure 5. Thin-Film Electrophoresis

These essentially consist of a pair of closely spaced, vertical rectangular plates bounded on the sides by the electrodes. The sample and carrier buffer are fed from the top of the slit and travel down in laminar flow to a battery of fraction collectors at the bottom. Unlike the Philpot-Harwell device, which is essentially adiabatic, the thin-film separator can be cooled at the plates. The commercially available device, the Elphor®, has a throughput of around 0.1 g/h of protein when operated for multi-component separation. It has been used to separate not only proteins, but cells and other particulate materials. Like the Philpot-Harwell apparatus, it uses a relatively large quantity of carrier buffer and the products are substantially diluted during separation.

Much work has been devoted to modeling thin-film separators in the hope of improving their scaling characteristics. Ivory (79) cites three major impediments to expanding their capacity: natural convection due to thermal gradients in the slit; overheating at the column centerline; and the "crescent phenomenon", the hydrodynamic dispersion of solute into a crescent-shaped profile by a combination of horizontal electro-osmotic flow and the vertical parabolic velocity profile of laminar flow in the slit. The first two effects can be overcome by running the system in microgravity. The company McDonnell-Douglas has flown several electrophoresis experiments on the space shuttle, but the work has been impeded by delays in the space program.

The thin-film separator can also be operated in a binary mode called field-step focusing (89) (Figure 6).

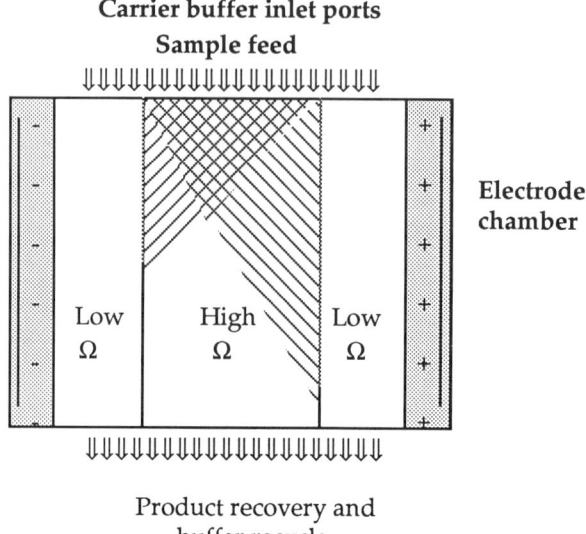

Figure 6. Field-Step Focusing

The sample, dissolved in a low ionic-strength, low-conductivity buffer, is fed to the middle region of the slit, and a high-conductivity buffer is fed to each side, adjacent to the electrodes. The high electrical field in the center causes sample components to migrate rapidly to the left and right, until they are effectively stopped by the low electrical field in the side buffer streams. Two concentrated streams of protein are recovered from the buffer interfaces and can be sent to a second field-step separation after desalting. The authors claim a 10 to 100 fold improvement in throughput over conventional thin-film operation, or 1 to 10 g/h of protein in a binary separation.

• Recycling free-flow methods

The problems of Joule heating and natural convection have been addressed by recycling methods for isoelectric focusing (RIEF) (90), zone electrophoresis (RZE) (91, 92) and most recently, isotachophoresis (RITP) (93). These methods use repeated short electromigrations of solutes to achieve high-resolution batch separations at relatively high throughputs. In both RIEF and RZE, the solution to be fractionated is repeatedly passed through a bank of fractionation channels, bounded by porous membranes to minimize convection, and then through a bank of heat exchangers for cooling. In RITP, no membranes are needed as the thin film configuration limits convection.

Recycling isoelectric focusing operation (90) is started by pre-focusing the ampholyte solution, returning the contents of each compartment back into itself until the entire system achieves a stable pH gradient. Then, sample is added and cycled through the system until each component collects in the channel corresponding to its isoelectric point. Resolution of proteins whose pI's differ by 0.1 pH units is possible in this device; however, the purified fractions must be separated from the ampholytes before further processing. A RIEF device with 60 ml capacity is commercially available (Rotofor®; Bio-rad, Inc.) and is claimed to have a throughput of 0.4 g protein/h over a 4-hour run.

Recycling isoelectric focusing, like its parent method IEF, is an equilibrium process in which each component migrates to a steady-state position and remains there. By contrast, zone electrophoresis is a rate process in which each component moves at a steady-state velocity. In order to convert ZE to a recycle system, it is necessary to provide a counterflow to offset electromigration of the solutes.

The recycle zone electrophoresis (RZE) apparatus of Gobie and Ivory (91, 92) accomplishes this by shifting the reinjection point of each compartment to a port one or more compartments upstream (against the direction of electromigration). The upstream recycle provides an effective counterflow whose magnitude can be adjusted at different positions in the apparatus by changing the distance over which the reinjection point is shifted. The prototype apparatus, with 50 ports, was built with low-, medium- and high-shift regions to produce a binary separation, but n-component purification is theoretically possible in an apparatus with n+1 sections at increasing shift distances. Throughputs of 1.5 g protein/h were reported for the initial apparatus (79).

A new recycle isotachophoretic process (93) uses a thin-film geometry with the electrical field perpendicular to the principal flow direction. Leading buffer, a marker dye, feed and trailing buffer are introduced into one end of the slit. An isotachophoretic stack develops perpendicular to flow as the liquid moves downstream. A fraction collector at the outlet collects the fractions, which are recycled until the stack sharpens. A computerized feedback control system keeps the stack centered in the apparatus. It regulates the withdrawal of trailing buffer and the addition of leading buffer in counterflow to the migration of the stack, based on the position of the marker dye front.

Righetti and coworkers (94) have reported an isoelectric refining method in which a liquid sample is circulated between two gels held at slightly different pH's. The gel segments are prepared with immobilized ampholytes at pH values which bracket the isoelectric point of the target protein. All contaminating species are ionized and eventually migrate into one or the other of the gels, leaving the target species alone in the liquid phase. Although the problem of ampholyte contamination is avoided, isoelectric precipitation of the protein of interest could prove troublesome, as could titration of the gel surfaces by adsorbed or dissolved contaminants. Nevertheless, this technique has potential as a polishing step for therapeutic proteins because of the extremely high resolution it promises.

• **Electrochromatography**

Continuous systems using anticonvective packings have also been proposed. The rotating annular electrochromatograph consists of an annular bed of anticonvective medium which may have specific chromatographic interactions with the solutes to be separated. Carrier buffer is pumped axially through the annulus, and the feed is introduced at a fixed point as the bed slowly rotates past it. The electrical field may be either axial, as in the "CRAE" system (95, 96) or radial (97). The CRAE system (Figure 7), with parallel convective and electrophoretic flows, produces a highly tunable one-dimensional separation; the annular electrochromatograph of the Oak Ridge group has the potential to produce a continuous separation in two dimensions.

Both designs for the annular electrochromatograph appear to be limited by heat transfer (79) and to suffer from mechanical and electro-osmotic dispersion of the solute bands. However, electro-osmosis may actually decrease dispersion under some conditions, according to a model developed by **Yoshisato and co-workers** (98). Precise and comprehensive models of annular electrochromatography, now under development, are necessary to guide the design and operation of the equipment.

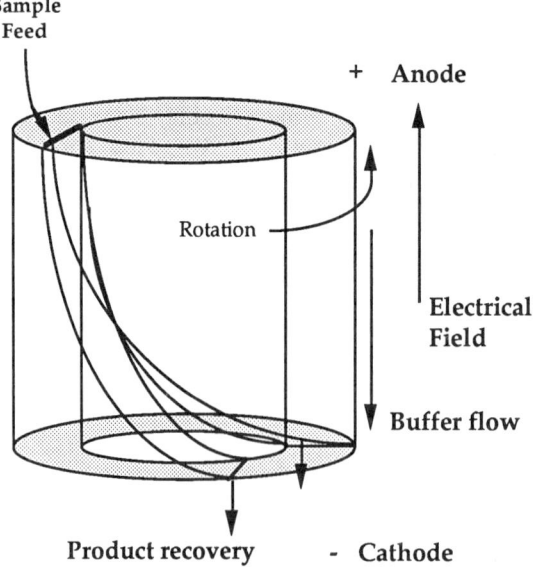

Figure 7. Continuous rotating annular electrophoresis unit (CRAE)

Another electrochromatographic technique, proposed by O'Farrell (99) is counteracting chromatographic electrophoresis (CACE) (Figure 8). In this technique, an axial electrical field is applied antiparallel to convective flow in a cylindrical packed bed of size exclusion gel. The upstream portion of the column is packed with a relatively "excluding" gel and the downstream portion with an "including" gel, so that macromolecules are convected faster on average in the upstream portion than in the downstream portion. By properly tuning the electrical field, a target protein can be made to migrate to the interface and accumulate there while other proteins migrate off the column at either end. Several analyses of CACE have been reported (100, 101, 102). Although this method can be operated continuously, and produces an extremely pure and concentrated product, the throughput is limited by Joule heating and by the pressure-drop limitations of size exclusion gels.

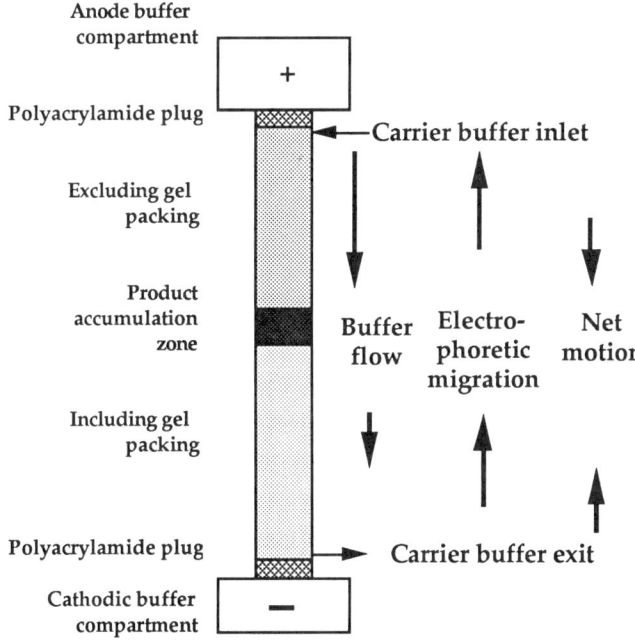

Figure 8. Counteracting Chromatographic Electrophoresis (CACE)

• **Electrically assisted separation**

Electrophoresis has also been proposed as a way of overcoming diffusion limitations in membrane based processes such as cross-flow microfiltration (103). Lee and Hong (104) used electrophoresis to aid recovery of aspartic acid synthesized by an immobilized enzyme coupled to a membrane. In principle, removal of an electrically charged reaction product by electrophoresis can be used to drive a reaction with an unfavorable equilibrium constant. In their model of facilitated transport of proteins through membranes, **Dall-Bauman and Ivory** (8) found electrical fields to enhance transport of the target protein to the upstream side and away from the downstream side of the membrane. By reducing concentration polarization on both sides of the membrane, small imposed electrical fields allowed the carrier system to function much more efficiently, and produced substantial increases in overall transport across the membrane. Yarmush and Olson (105) have used electrophoresis to elute proteins from affinity membranes. After the protein is dissociated from the ligand by a high pressure environment (~1000 psi), it migrates away from the surface of the adsorbent under an electrical field. Electrical fields may alter the microstructure of a membrane as well as the flux of solute inside it (106). The resulting large changes in permeability and selectivity can be controlled by switching the field on and off. In contrast to other electrophoretic processes in which proteins migrate for distances of millimeters or centimeters, boundary layer disruption requires an electromigration distance of only a few tens of microns. Low-intensity or intermittent fields can be used, avoiding the Joule heating problems which plague conventional large-scale electrophoresis.

Novel separation methods in electrophoresis share a common factor: the complexity of systems where mass transfer, heat transfer, electro-osmosis, dispersion, adsorption, and Donnan effects are all relevant and all interact. Precise and comprehensive modeling efforts are - and will continue to be - of paramount importance in evaluating these new ideas. Important advances in modeling have come from the groups at Oak Ridge National Laboratory, University of Arizona, University of Iowa, University of Washington (Pullman), and North Carolina State University. Likely developments in the near future include a better understanding of the role of electro-osmosis in large-scale separations, the development of unusual geometries to facilitate heat removal from large-scale separation devices, expanded interest in electrophoresis to counteract diffusion limitations, and a steady improvement in experimental apparatus.

Chromatography

Chromatography, the workhorse of protein fractionation, may be defined as the percolation of a fluid through a column of a particulate stationary phase which selectively retards certain components of the fluid. Though very broad, this definition identifies chromatography as a muticomponent separation technique based on differential migration due to adsorption or partitioning of solutes. In the limiting case where the solutes do not move at all, chromatography becomes batch sorption, and additional driving forces must be applied to desorb the solutes.

The past ten years have seen a virtual explosion in every aspect of preparative chromatography - the development of rigid, monodisperse packings for HPLC; the proliferation of stationary phase chemistries, now including systems for chiral and affinity separations; advances in on-line detection systems; and the commercial application of process-scale chromatography in the biotechnology industry, to name only a few. Column chemistries can be counted on to improve steadily, permitting ever finer fractionations. However, many problems remain to be solved. Chromatography is still inherently a batch process; by and large it still uses packed beds of media with their problems of pressure drop, dispersion, and intolerance of particulates in the feed; it still requires large quantities of buffers and yields diluted products.

The central issue in process-scale chromatography is the problem of increasing the throughput of product per unit amount of packing, subject to constraints of product quality and column life. These constraints, and the scale of the "preparative" process, vary enormously across the field of bioseparations. Perhaps the largest scale chromatographic bioseparation is the refining of ultra-high fructose syrups from an equilibrium mixture of fructose and glucose on calcium-loaded ion exchange resins. World production is on the order of millions of tons per year, and the product is 90 to 95% pure (107). On the opposite end of the spectrum, enzymes and hormones for drug use must meet the most exacting standards of purity, at an output of only kilograms per year.

Historically, the throughput problem has been addressed by heuristics for scale-up of conventional packed beds for multicomponent separations. The most recent scale-up analyses, focusing on intraparticle mass-transfer resistance as a limiting factor, have led away from the traditional long columns to several alternative geometries (108, 109, 110).

Wankat and Koo (110) have shown that the efficient mass transfer achievable with small (~ 10 micron diameter) monodisperse packing can provide excellent resolution on very short columns, even when adsorption isotherms are nonlinear. For high-throughput processes, the most efficient columns resemble squat disks or pancakes (109). The ultimate "column" geometry may well be a membrane or consolidated packing with mobile phase flow through monodisperse pores.

If a pancake column is rolled into a tube, the result is radial-flow chromatography. This geometry has already been commercialized for ion-exchange separations (Zeta-Prep®, Cuno, Inc.), and the concept is being extended to other chemistries (111). A radial-flow separation module is made by wrapping a sheet of separation medium around a hollow core, then encapsulating the roll in a rigid cartridge. Particulate packings may also be used. Sample and eluant are introduced into the shell side and flow radially to the center outlet. Because of the short bed depth, isocratic resolution is poor, and the column is preferably operated by gradient elution. Throughput is proportional to the cartridge surface area, so scale-up is modular (111, 112).

Chromatography in two different hollow fiber geometries has recently been reported. A hollow fiber can be used as a capillary column analogous to capillary gas chromatography, with the same operating advantages of low pressure drop and rapid mass transfer (108, 113). A bundle of such fibers resembles a consolidated packing or a very thick membrane. Radial hollow fiber chromatography (114) is a miniaturized version of radial flow chromatography and has been demonstrated for affinity purification of fibronectin using gelatin as a ligand. The small volume, low pressure drop and high ligand capacity of the hollow fiber module lead to very short residence times and very efficient use of the ligand. Both of the hollow fiber methods scale up linearly, by using a bigger fiber bundle or multiple modules. Both also suffer from the difficulty of precise flow distribution to a fiber bundle.

• Novel methods in traditional geometries

Binary chromatographic separations are most efficiently run in moving-port and simulated moving-bed processes (115). In these continuous processes, a number of short columns are connected to form a ring. The sample, eluent and withdrawal ports are rotated around the ring to simulate countercurrent movement of the solid phase past a stationary feed port. Weakly bound components move around the column ahead of the feed port and are recovered downstream, while tightly bound components trail behind the feed. Moving-bed and moving-port operation can increase the efficiency of packing use several fold, as there is no waiting for low-mobility samples to clear the column before more feed is injected. The Sorbex process, a simulated moving-bed process, is already standard for process-scale separations of glucose and fructose (107).

Displacement chromatography (Figure 9) is another approach to increased efficiency.

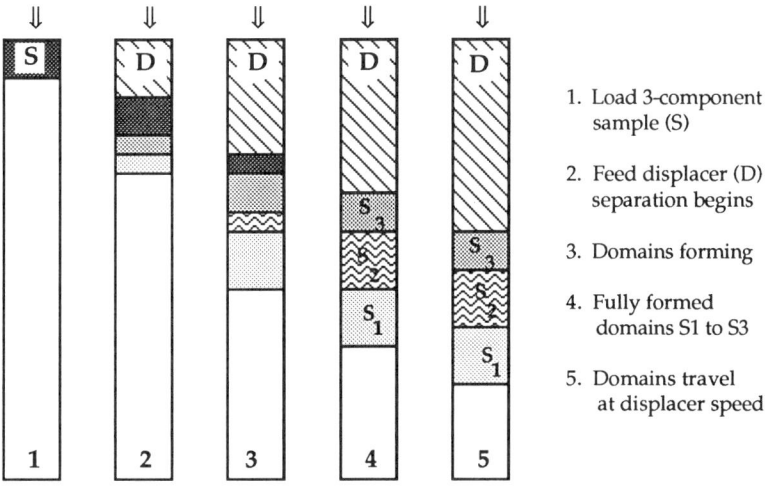

Figure 9. Displacement Chromatography

Ordinary elution chromatography operates in the limit of low solute concentrations, where adsorption isotherms are approximately linear. If the column is overloaded, the usual case in preparative applications, nonlinear isotherms and interactions between solutes cause band broadening and shifts in retention time. At extremely high loading, a new separation mode emerges based on competition between sample components for binding sites on the stationary phase surface. In displacement chromatography, the sample components are displaced from the stationary phase by a front of a highly absorbing solute fed just behind the sample. The displacer drives the sample components downstream, each component displacing the more weakly adsorbed components ahead of it. Ultimately the components form adjacent sharp bands, each traveling at the speed of the displacer front. Displacement chromatography is fast - all sample components are recovered after passage of a single column volume of displacer - and the components are recovered in concentrated form. Separations of antibiotics, amino acids and small peptides have formed the basis for most theoretical work to date (116) and one report of protein separation has appeared (117). Fundamental modeling of nonlinear solute/solute and solute/packing interactions in displacement (118, 119) will provide much needed guidelines for designing recovery strategies and optimizing operating parameters such as feed pulse size and displacer concentration.

Continuous annular chromatography (CAC) has been the subject of several recent experimental studies (120, 121), models (122, **123**) and a brief review (124). The equipment is very similar to the CRAE (Figure 7). Feed, eluent, and regeneration solutions (if necessary) are fed to fixed points or arcs at the top of an annular packed bed which rotates slowly about its axis. As the chromatogram develops, the components separate into helical bands which are collected at fixed points at the base of the column. Reports to date have centered on ion-exchange and size-exclusion separations, but the apparatus should be able to perform continuously any separation currently done by batch chromatography, e.g. displacement and gradient elution separations with continuous column regeneration (124). Linear models of the CAC have developed rapidly, since the CAC becomes analogous to a conventional chromatographic column if the angular position is transformed to time. The chief difference is a term for angular dispersion of solutes.

Continuous annular chromatography is limited by the elution speed of the fastest and slowest migrating components of the sample. Its throughput per unit volume is the same as conventional column chromatography. Nonlinear gradient and displacement chromatography may prove the best applications for CAC, because of their economy in buffer use. However, these modes are likely to be very sensitive to angular dispersion and to concentration-dependent flow disturbances such as channeling around a viscous feed pulse. Nonlinear models should appear over the next few years.

Countercurrent chromatography, also called centrifugal partition chromatography, is analogous to a multistage countercurrent liquid/liquid extraction system (125). Current technologies using organic and aqueous solvents are suitable for purification of antibiotics or amino acids. A recent report of countercurrent chromatography in an aqueous two-phase system (126) indicates its promise as an initial step for isolation of a macromolecule from a crude fermentation broth, or in classifications of living cells.

Affinity Separation

A significant fraction of biomolecules display natural biological affinity for certain other species, e.g. immuno-ligands, enzyme substrates, hormones. These properties can be exploited in an affinity separation process to recover and purify biomolecules in a more effective way (i.e. with higher yield and higher resolution) than can be achieved with more conventional means of purification (e.g. size exclusion chromatography).

Affinity separations are characterized by the formation of a reversible, specific biochemical interaction between the target molecule (the adsorbate) and an immobilized molecule (the ligand). The two molecules may interact as enzyme and substrate, analogue, cofactor or inhibitor; as antibody and antigen; messenger and receptor; or as complementary nucleic acid sequences (127). The triazine dyes interact with nucleotide-binding sites of a wide range of enzymes. Plant lectins and agglutinins bind to specific sugar moieties, hence are useful for purifying glycoproteins such as mammalian cell surface proteins, and for separating subclasses of mammalian cells based on their surface receptors. Recent advances have focused on "generic" ligands which are useful for whole classes of proteins. An example is protein G, which binds to any IgG. "Generic biospecificity" may sound like an oxymoron; but it is the most cost-effective approach to ligand chemistry. Optimized conditions for binding and elution on a generic ligand can produce excellent yield and resolution without the need for unique affinity interactions (128).

The classical affinity separation (enrichment) of a single target product comprises four steps: adsorption, washing to remove non-specifically bound components, elution of the target component, and regeneration of the ligand. The process context is usually column chromatography. A brief review of current practice is given by McCormick (129). Biospecific interactions can also be used to strip specific contaminants such as endotoxins, DNA or T-lymphocytes from a product stream. As expected, the operating requirements for enrichment and depletion modes are quite different (130). Applications of potential importance based on this principle include, i) the purification of viruses for the production of antiviral vaccines (131), and ii) the removal of viruses from blood products and therapeutic recombinant proteins (132). **Tsao and Wang** (132) investigated batch adsorption of viruses from protein solutions onto immobilized quaternary ammonium chlorides (QAC's), a class of antimicrobials which can disrupt cell membranes. The treatment appeared effective against "enveloped" viruses having surface lipids. Adsorption of these viruses via hydrophobic interactions was followed by their inactivation at the solid surface. Results for a non-enveloped virus were less conclusive, and were complicated by competitive binding of soluble proteins. In general, viruses adsorb readily to many different types of solids (133) and future work in this area will require careful analysis of non-specific binding.

Affinity ligands can be covalently immobilized to an immense variety of supports. For chromatographic processes, agarose beads have been a popular support since agar is porous, dimensionally stable over a wide range of pH and ionic strength, and is easily activated for covalent coupling of ligands. However, affinity chromatography by HPLC has grown in popularity and may become the method of choice for large-scale "polishing" affinity purifications (134). Bergold and coworkers (135) have reviewed this technology in detail.

Affinity separations have historically been used in later stages of a purification train, in order to protect the expensive ligands from reactive components of a crude system, and to minimize the extent of nonspecific binding. However, the specificity and the high binding constants of affinity interactions make them especially attractive for isolation of biomolecules from crude medium. Affinity-based product recovery can provide a high yield, some purification and substantial concentration of the product, reducing the processing volume and leaving the initial crude mixture nearly unaffected. Much recent work has appeared on affinity methods for initial product isolations, such as affinity partitioning, adsorption in a fluidized bed reactor (136) and affinity separations using magnetic particles (137, 138). Affinity partitioning is an attractive technique which combine properties of affinity and aqueous two-phase systems. Since the demonstration of affinity partitioning in 1975 (139), several model systems have been described (140, 141). The triazine-dye has been the ligand of choice for the affinity extraction of glycolytic and other enzymes, e.g. phosphofructokinase from baker's yeast (142). Compared to other affinity purification techniques, affinity partitioning has decisive advantages, e.g. higher capacity, which make it attractive for large-scale, continuous operations using complex systems such as crude homogenates (142).

Both affinity ligands and their adsorbates are typically high-priced, labile biomolecules, hence affinity separations may be costly. The ideal affinity isolation would make efficient use of ligand through careful immobilization, retaining its bioactive conformation at an appropriate surface density of binding sites. The support would have a high surface area to promote rapid binding of adsorbate. The loaded affinity support would separate readily from its surroundings, and washing, elution and regeneration procedures would be chemically mild. Continuous processes involving rapid recycle of the ligand/support system could achieve high throughput per unit quantity of ligand.

Recent advances in affinity separation fall into two main categories: affinity isolation and process concepts for desorption strategies.

• Isolation by Affinity Interaction

In an isolation step, where yield and concentration are more important than purity, the adsorption mechanism can be considered an on/off process, and several alternative contacting schemes can be used. Ligands have been bound to magnetized particles (137, 138) for continuous countercurrent adsorption in magnetically stabilized fluidized beds. Ligands attached to liquid perfluorocarbons (143), to dextran and related polymers (144), or incorporated into liposomes (145), or reversed micelles (146) may be used for biospecific liquid-liquid extraction or "affinity partitioning". Ligands have also been attached to surfactants and biopolymers for selective precipitation of dilute protein species (147, 148).

Affinity escort systems consist of a ligand attached to a high molecular-weight polymer (149) or to a small particle (52). The so-called macroligand will bind an adsorbate and increase its size so that it may be separated by ultrafiltration or cross-flow microfiltration. Even larger particle sizes have been used. Nigam and coworkers (150, 151) immobilized ligands to finely divided solid supports which they encapsulated in hydrogel beads of up to 3 mm diameter. The hydrogel prevented fouling of the ligand by high molecular-weight and particulate materials. Higher ligand loadings and a reduction in internal mass transfer resistances were achieved by encapsulating a liquid phase containing a ligand immobilized to a soluble polymer.

Pungor and coworkers (152) described a continuous affinity-recycle extraction system which allows continuous separation of an adsorbate from crude cell lysate without preliminary clarification. In this scheme, a feed stream is added to a slurry of particulate affinity adsorbent (agarose beads) in a continuous well-mixed contactor. A wash buffer dilutes contaminants as it carries the loaded adsorbent to a desorption stage, where the adsorbate is recovered in concentrated form by a slow flow of desorbing buffer. The regenerated adsorbent particles are then recycled to the adsorption stage. The system survived 24 hours of operation with no observable ligand leakage, and recovered 70% of the product (beta-galactosidase) from a slurry of homogenized bacterial cells. In both of these methods, the ligand is protected from fouling, and the rapid recycle of adsorbent makes very efficient use of the ligand. Multiple contacting stages or a wash stage prior to desorption can be used to increase yield or purity.

• Recovery Strategies

Historically, the removal of the adsorbate from its complex with the ligand has been a critical step in affinity separation. Typical methods have included reversible denaturation of the adsorbate and/or ligand with urea, guanidine salts, chaotropic salts, iodide or thiocyanate. Even after prompt removal or dilution of the denaturant, full renaturation is not always achieved. Conducting elutions at extreme pH's, high ionic strength, or by addition of organic solvents results in the disruption of the ligand/adsorbate interaction, and may denature the proteins, especially after repeated exposure to the hostile conditions. An added drawback to chemical elution methods at process scale is the cost of recycle or disposal of eluents. Recently, "switch" monoclonal antibodies have been prepared for which small changes in environment, such as a pH change of 1 to 2 points, produce major changes in binding constant (153). This and future advances in ligand chemistry could ease the elution problems in large-scale affinity separations.

Several mild and effective elution methods have recently been proposed. Olson and coworkers (154) have reported that brief exposure to pressures of 1000 to 2000 atmospheres can dissociate antibody/antigen complexes and other non-covalently bound protein complexes without affecting the activity of monomeric proteins. Repeated cycling to high pressures did not affect the binding capacity of the ligand, nor did it denature the adsorbate. Electrophoretic elution, described above, is a useful adjunct to this method or can be used alone (105). The use of temperature-programmed elution to dissociate tightly-bound complexes in affinity HPLC has been reported by Bergold and coworkers (135).

Electrophoretic elution and "switch" monoclonal antibodies are combined in a new rapid recycle method: an affinity-mediated membrane transport process reported by **Dall-Bauman and Ivory** (8). In this modeling paper, a "switch" monoclonal antibody incorporated into a supported liquid membrane is used to facilitate the transport of human growth hormone from a high-pH to a low-pH environment. Electrochemical effects, including Donnan equilibria between the membrane and external environments, and imposition of external electrical fields, significantly affected the flux of protein across the membrane. Experimental confirmation of the simulation results could introduce affinity-mediated transport as a powerful new biospecific separation method.

CONCLUSION

The contributions in this book illustrate the important role of downstream processing and purification processes in the application of biotechnology. We expect that this trend will continue, especially with the proliferation of recombinant proteins derived from the recombinant DNA technology. All of the techniques presented in this book can play a critical role in the downstream processing and in the effective production of biological products. Most of the early fears related to the safety aspects of recombinant DNA products have been assuaged since studies showed that quantities of DNA (obtained from chinese hamster ovary cells) at the hundred of µg level did not result in the formation of tumors in newborn rats (155).

The science behind the concepts and techniques of bioseparations is exciting: each fundamental mechanism which is uncovered sets the direction of future development work and motivates further advances.

On the development side, it is important to recognize that a successful bioprocess leading to a safe product results from the integration of techniques ingeniously connected with one another. The fermentation engineer will confer with downstream processing colleagues to design the fermentation process, since the mode of operation (e.g. fed-batch or continuous) can have major effects on product stability and response to handling, as well as on the nature of the impurities which may remain with the product. Looking at a process with an integrated vision not only minimizes the likelihood that serious mistakes will occur, but it also favors optimization of each unit operation in the context of the entire process. This plays a significant role in making a process viable and cost-effective.

Literature Cited

1. *Developments in Biological Standardization: Standardization and control of biologicals produced by recombinant DNA technology*; Perkins, F. T.; Hennessen, W., Eds.; Proceedings of a Joint IABS/WHO Symposium, Nov. 29 - Dec. 1, 1983; Geneva, Switzerland; S. Karger, Basel, Switzerland, 1985, Vol. 59, pp 147-188.
2. Schütte, H.; Kroner, K. H.; Hustedt, H.; Kula, M.-R.; "Disintegration of Microorganisms in a 20 l Industrial Bead Mill"; In *Enzyme-Technology*, III. Rotenburger Fermentation Symposium, 1982. Lafferty, R. M. and Maier, E., Eds.; Springer-Verlag, 1983, pp 115-124.
3. Kahn, F. R.; Bailon, P.; Koplove, M.; Rai, V. R.; "Large-Scale Purification of Recombinant Human Leukocyte Interferon (rIFN-α-2a) Using Monoclonal Antibodies"; Paper presented at the ACS meeting, New Orleans, Aug. 30 - Sept. 4, 1987.
4. Mattiasson, B.; In *Immobilized Cells and Organelles*, Vol. 1 & 2, Mattiasson, B., Ed.; CRC Press: Boca Raton, FL.; 1983.
5. Cabezas, Jr., H.; Evans, J. D.; Szlag, D. C.: this book.
6. Forciniti, D.; Hall, C. K.: this book.
7. Szlag, D. C.; Giuliano, K. A.; Snyder, S. M.: this book.
8. Dall-Bauman, L.; Ivory, C. F.: this book.
9. Guzman, R. Z.; Kilpatrick, P. K.; Carbonell, R. G.: this book.
10. Sheehan, J. J.; Hamilton, B. K.; Levy, P. F.: this book.
11. Beijerinck, M. W.; "Ueber Eine Eigentümlichkeit Der Löslichen Starke"; *Centralblatt Bak.* 1896, 2, pp. 697-699.
12. Albertsson, P.-Å.; *Partition of Cell Particles and Macromolecules*, 2nd Ed.; Almquist & Wiksell: Stockholm, Sweden, 1971.
13. Walter, H.; Brooks, D. E.; Fisher, D.; *Partitioning in Aqueous Two-Phase Systems: Theory, Methods, Uses and Applications to Biotechnology*; Academic Press: Orlando, FL. 1985.
14. Mattiasson, B.; "Applications of aqueous two-phase systems in biotechnology"; *Trends in Biotechnology.* 1983, 1, 1, pp 16-20.
15. Albertsson, P.-Å.; *Partition of Cell Particles and Macromolecules*, 3rd Ed.; Wiley & Sons: New York, NY, 1986.
16. Kula, M.-R.; "Extraction and Purification of Enzymes Using Aqueous Two-Phase Systems"; In *Applied Biochemistry and Bioengineering*; Wingard, Jr., L. B.; Katchalski-Katzir, E.; Goldstein, L., Eds.; Academic Press: New York, NY, 1979, 2, pp 71-95.
17. Kula, M.-R.; Kroner, K. H.; Hustedt, H.; Grandja, S.; Stach, W.; German Patent 26.39.129, U.S. Patent 4,144,130 (1979); "Process for the separation of enzymes".
18. Schütte, H.; Kroner, K. H.; Hummel, W.; Kula, M.-R.; "Recent Developments in Separation and Purification of Biomolecules"; In *Biochemical Engineering III*; Venkatasubramanian, K.; Constantinides, A.; Vieth, W. R., Eds.; Ann. N. Y. Acad. Sci.; 1983, 413, pp 270-282.
19. Kula, M.-R.; Kroner, K. H.; Hustedt, H.; "Purification of Enzymes by Liquid-Liquid Extraction"; In *Advances in Biochemical Engineering*; Fiechter, A., Ed.; Springer: Berlin, Germany, 1982, pp 73-118.
20. Hustedt, H.; Kroner, K. H.; Stach, W.; Kula, M.-R.; "Procedure for the Simultaneous Large-Scale Isolation of Pullulanase and 1,4-α-Glucan Phosphorylase from *Klebsiella pneumoniae* Involving Liquid-Liquid Separations"; *Biotech. and Bioeng.* 1978, 20, pp 1989-2005.
21. Puziss, M.; Hedén, C.-G.; "Toxin Production by *Clostridium tetani* in Biphasic Liquid Cultures"; *Biotech. and Bioeng.* 1965, 7, pp 355-366.
22. Hahn-Hägerdal, B.; Mattiasson, B.; Albertsson, P.-Å.; "Extractive Bioconversion in Aqueous Two-Phase Systems. A Model Study on the Conversion of Cellulose to Ethanol"; *Biotech. Letters.* 1981, 3, 2, pp 53-58.

23. Edmond, E.; Ogston, A. G.; "An Approach to the Study of Phase Separation in Ternary Aqueous Systems"; *Biochem. J.* **1968**, 109, pp 569-576.
24. Baskir, J. N.; Hatton, T. A.; Suter, U. W.; "Protein Partitioning in Two-Phase Aqueous Polymer Systems"; *Biotech. and Bioeng.* **1989**, 34, 4, pp 541-558.
25. Szlag, D. C.; Giuliano, K. A.; "A Low-cost Aqueous Two Phase System for Enzyme Extraction"; *Biotechnology Techniques.* **1988**, 2, 4, pp 277-282.
26. Tjerneld, F.; Berner, S.; Cajarville, A.; Johansson, G.; "New Aqueous Two-Phase System based on Hydroxypropyl Starch Useful In Enzyme Purification"; *Enzyme Microb. Technol.* **1986**, 8, 7, pp 417-423.
27. Martinek, K.; Levashov, A. V.; Klyachko, N. L.; Pantin, V. I.; Berezin, I. V.; "The Principles of Enzyme Stabilization. VI. Catalysis by water-soluble enzymes entrapped into reversed micelles of surfactants in organic solvents"; *Biochim. Biophys. Acta.* **1981**, 657, 1, pp 277-294.
28. Wolf, R.; Luisi, P. L.; "Micellar solubilization of enzymes in hydrocarbon solvents. Enzymatic activity and spectroscopic properties of ribonuclease in N-octane"; *Biochem. and Biophys. Res. Comm.* **1979**, 89, 1, pp 209-217.
29. Hagen, A. J.; Hatton, T. A.; Wang, D. I. C.; "A Study of Protein Folding in Reversed Micelles"; Paper presented at the ACS National Meeting, Sept. 26-30, 1988.
30. Van't Riet, K.; Dekker, M.; "Preliminary investigations on an enzyme liquid-liquid extraction process"; Proceedings of the Third European Congress on Biotechnology (Munich, Sept. 10-14, 1984), Verlag Chemie GmbH: Weinheim, Fed. Rep. of Germany, 1984, Vol. III, pp 541-544.
31. Woll, J. M.; Hatton, A. T.; "A simple phenomenological thermodynamic model for protein partitioning in reversed micellar systems"; *Bioprocess Eng.* **1989**, 4, pp 193-199.
32. **Jolivalt, C.; Minier, M.; Renon, H.: this book.**
33. Bargeman, D.; Smolders, C. A.; "Liquid Membranes"; In *Synthetic Membranes: Science, Engineering and Applications*; Bungay, P. M., Lonsdale, H. K., de Pinho, M. N., Eds.; NATO ASI Series, D. Reidel: Dordrecht, The Netherlands; 1986, pp 567-579.
34. Li, N. N.; U.S. Patent 3, 410, 794 (1968); "Separating hydrocarbon with liquid membranes".
35. Maugh II, T. H.; "Liquid Membranes: New Techniques for Separation, Purification"; *Science.* **1976**, 193, 4248, pp 134-137.
36. Li, N. N.; "Permeation Through Liquid Surfactant Membranes"; *A.I.Ch.E.J.* **1971**, 17, 2, pp 459-463.
37. Hayworth, H. C.; Ho, W. S.; Burns, Jr., W. A.; Li, N. N.; "Extraction of Uranium from Wet Process Phosphoric Acid by Liquid Membranes"; *Sep. Sci. Tech.* **1983**, 18, 6, pp 493-521.
38. Kitagawa, T.; Nishikawa, Y.; Frankenfeld, J. W.; Li, N. N.; "Wastewater Treatment by Liquid Membrane Process"; *Environ. Sci. Tech.* **1977**, 11, pp 602-605.
39. Thien, M. P.; Hatton, T. A.; Wang, D. I. C.; "Liquid Emulsion Membranes and Their Applications in Biochemical Separations"; In *Separation, Recovery, and Purification in Biotechnology*; ACS Symposium Series # 314, 1986, pp 67-77.
40. Cahn, R. P.; Frankenfeld, J. W.; Li, N. N.; Naden, D.; Subramanian, K. N.; "Extraction of Copper by Liquid Membranes"; In *Rec. Dev. in Sep. Sci.* **1982**, 6, pp 51-64.
41. Prötsch, M.; Marr, R.; "Development of a Continuous Process for Metal Ion Recovery By Liquid Membrane Permeation"; Proceedings Inter. Solvent Extrac. Conf. 1983, pp 66-67.
42. Chiang, C.-W.; Fuller, G. C.; Frankenfeld, J. W.; Rhodes, C. T.; "Potential of Liquid Membranes for Drug Overdose Treatment: *In vitro* Studies"; *J. Pharm. Sci.* **1978**, 67, 1, pp 63-66.
43. Mohan, R. R.; Li, N. N.; "Reduction and Separation of Nitrate and Nitrite by Liquid Membrane-Encapsulated Enzymes"; *Biotech. and Bioeng.* **1974**, 16, pp 513-523.

44. Makryaleas, K.; Scheper, T.; Schügerln, K.; Kula, M.-R.; "Enzymkatalysierte Darstellung von L-Aminosäure mit kontinuierlicher Coenzym-Regenerierung mittels Flüssigmembran-Emulsionen"; *Chem. Ing. Tech.* **1985,** 54, 4, pp 362-363.
45. Hochhauser, A. M.; Cussler, E. L.; "Concentrating Chromium with Liquid Surfactant Membranes"; In *Adsorption and Ion Exchange*; AIChE. Symp. Ser. # 152, AIChE, New York, NY, 1975; 71, pp 136-142.
46. Martin, T. P.; Davies, G. A.; "The Extraction of Copper from Dilute Aqueous Solutions Using a Liquid Membrane Process"; *Hydrometallurgy.* **1977,** 2, 4, pp 315-334.
47. Takahashi, K.; Ohtsubo, F.; Takeuchi, H.; "A Study of the Stability of (w/o)/w-Type Emulsions Using a Tracer Technique"; *J. of Chem. Eng. of Japan.* **1981,** 14, 5, pp 416-418.
48. Stroeve, P.; Varanasi, P. P.; "An Experimental Study on Double Emulsion Drop Breakup in Uniform Shear Flow"; *J. Colloid Interface Sci.* **1984,** 99, 2, pp 360-373.
49. **Simmons, D. K.; May, S. W.; Sheldon, W. M.; Agrawal, P. K.: this book.**
50. Strathmann, H.; "Membranes and membrane processes in biotechnology"; *Trends in Biotechnology.* **1985,** 3, 5, pp 112-118.
51. Paulson, D. J.; Wilson, R. L.; Spatz, D. D.; "Crossflow Membrane Technology and its Applications"; *Food Technology.* **December 1984,** pp 77-87.
52. Cho, T.; Shuler, M. L.; "Multimembrane Bioreactor for Extractive Fermentation"; *Biotech. Progress.* **1986,** 2, 1, pp 53-60.
53. Herak, D. C.; Merrill, E. W.; "Affinity Cross-Flow Filtration: Experimental and Modeling Work Using the System of HSA and Cibacron Blue-Agarose"; *Biotech. Progress.* **1989,** 5, 1, pp 9-17.
54. Belfort, G.; In *Advanced Biochemical Engineering*, Bungay, H. R.; Belfort, G., Eds.; Wiley & Sons: Pub. New York, NY, 1987, pp 239-297.
55. Bechhold, H.; "Kolloidstudien mit der Filtrationsmethode."; *Zeitschrift für Physikalische Chemie.* **1907,** 60, pp 257-318.
56. Zsigmondy, Von R.; Bachmann, W.; "Über neue Filter"; *Zeitschrift für anorganische und allgemeine Chemie.* **1918,** 103, pp 119-128.
57. Brun, J.-P.; "Ultrafiltration et Microfiltration"; In *Procédés de séparation par membranes: Transport, Techniques membranaires, Applications*; Masson: Paris, France, 1989; pp 137-158.
58. Solomon, B. A.; Colton, C. K.; Friedman, L. I.; Castino, F.; Wiltbank, T. B.; Martin, D. M.; "Microporous Membrane Filtration for Continuous-Flow Plasmapheresis"; In *Ultrafiltration Membranes and Applications;* Vol. 3 of Polymer Science and Technology; Cooper, A. R., Ed.; Plenum Press: New York, N.Y., 1980, pp 489-505.
59. Zydney, A. L.; "Cross-flow membrane plasmapheresis: an analysis of flux and hemolysis"; PhD Thesis, Massachusetts Institute of Technology, 1985.
60. Forstrom, R. J.; Voss, G. O.; Blackshear, Jr., P. L.; "Fluid Dynamics of Particle (Platelet) Deposition for Filtering Walls: Relationship to Atherosclerosis"; *J. of Fluids Eng.* **1974,** 96, pp 168-172.
61. Michaels, A. S.; "Ultrafiltration: an adolescent technology"; *Chemtech.* **1981,** 11, pp 36-43.
62. Blatt, W. F.; Feinberg, M. P.; Hopfenberg, H. B.; Saravis, C. A.; "Protein Solutions: Concentration by a Rapid Method"; *Science.* **1965,** 150, 3693, pp 224-226.
63. Colton, C. K.; Henderson, L. W.; Ford, C. A.; Lysaght, M. J.; "Kinetics of Hemodiafiltration, I. In Vitro Transport Characteristics of a Hollow-Fiber Blood Ultrafilter"; *J. Lab. Clin. Med.* **1975,** 85, pp 355-371.
64. Friedli, H.; Fournier, E.; Volk, T.; Kistler P.; "Studies on New Process Procedures in Plasma Fractionation on an Industrial Scale"; *Vox Sang.* **1976,** 31, pp 283-288.
65. Hallström, B.; Lopez-Leiva M.; "Description of a Rotating Ultrafiltration Module"; *Desalination.* **1978,** 24, pp 273-279.

66. Howlett, A. C.; Sternweis, P. C.; Macik, B. A.; Van Arsdale, P. M.; Gilman, A. G.; "Reconstitution of Catecholamine-Sensitive Adenylate Cyclase"; *J. Biol. Chem.* **1979**, 254, pp 2287-2295.
67. Short, J. L.; Webster, D. W.; "Ultrafiltration, A Valuable Processing Technique for the Pharmaceutical Industry"; *Proc. Biochem.* **1982**, 17, 2, pp 29-32.
68. Lawhon, J. T.; Lusas E. W.; "New Techniques in Membrane Processing of Oilseeds"; *Food Technology.* **December 1984**, pp 97-106.
69. Tietjen, K. G.; Hunkler, D.; Matern, U.; "Differential Response of Cultured Parsley Cells to Elicitors from Two Non-pathogenic Strains of Fungi, 1. Identification of Induced Products as Coumarin Derivatives"; *Eur. J. Biochem.* **1983**, 131, pp 401-407.
70. Devereux, N.; Hoare, M.; Dunnill, P.; "Membrane Separation of Protein Precipitates: Unstirred Batch Studies"; *Biotech. and Bioeng.* **1986**, 28, pp 88-96.
71. Mateus, M.; Cabral, J. M. S.; "Recovery of 6-α-methylprednisolone from biotransformation medium by tangential flow filtration"; In *Bioprocess Engineering.* 1989, Springer Verlag, (in press).
72. Gallantree, I.; Docksey, S.; "Fermentation Cell Separations with High Performance Ultrafiltration"; In *Pro BioTech, Supplement to Process Biochemistry,* **May/June 1983**, 18, pp xv-xvi.
73. Belter, P. A.; Cussler, E. L.; Hu, W.-S.; *Bioseparations: Downstream Processing for Biotechnology*; John Wiley: New York, NY, 1988.
74. Giddings, J. C.; "Generation of Variance, 'Theoretical Plates', Resolution and Peak Capacity in Electrophoresis and Sedimentation"; In *Separation Science.* **1969**, 4, 3, pp 181-189.
75. **Phillips, M.; Brogden, K. A.: this book.**
76. Sternberg, M.; Chiang, J. P.; Eberts, N. J.; "Cheese whey proteins isolated with polyacrylic acid"; *J. Dairy Sci.* **1976**, 59, pp 1042-1050.
77. **Clark, K. M.; Glatz, C. E.: this book.**
78. Fisher, R. R.; Glatz, C. E.; "Polyelectrolyte Precipitation of Proteins: I. The Effects of Reactor Conditions"; *Biotech. and Bioeng.* **1988**, 32, 6, pp 777-785.
79. Ivory, C. N.; "The prospects for large-scale electrophoresis"; *Sep. Sci.* **1988**, Tech. 23, 8/9, pp 875-912.
80. Hjertén, S.; Elenbring, K.; Kilár, F.; Liao, J.-L.; Chen, A. J. C.; Siebert, C. J.; Zhu, M.-D.; "Carrier-free zone electrophoresis, displacement electrophoresis and isoelectric focusing in a high-performance electrophoresis apparatus"; *J. Chromatography.* **1987**, 403, pp 47-61.
81. Gordon, M. J.; Huang, X.; Pentoney, S. L.; Zare, R. N.; "Capillary Electrophoresis"; *Science.* **1988**, 242, pp 224-228.
82. Jorgenson, J. W.; Rose, D. J.; Kennedy, R. T.; "Nanoscale separations and biotechnology"; *American Laboratory.* **April 1988**, 20, pp 32-41.
83. Smith, R. D.; Maringa, C. J.; Udseth, C. R.; "Improved electrospray ionization interface for capillary zone electrophoresis/mass spectrometry"; *Anal. Chem.* **1988**, 60, pp 1948-1952.
84. Lee, E. D.; Mück, W.; Henion, J. D.; Covey, T. R.; "Liquid Junction Coupling for Capillary Zone Electrophoresis/Ion Spray Mass Spectrometry"; *Biomedical and Environmental Mass Spectrometry.* **1989**, in press.
85. Lee, E. D.; Mück, W.; Henion, J. D.; "On-line Capillary Zone Electrophoresis - Ion Spray Tandem Mass Spectrometry for the Determination of Dynorphins", *J. Chromatography.* **1988**, 458, pp 313-321.
86. Henion, J.; personal communication, 1989.
87. Terabe, S.; Otsuka, K.; Ando, T.; "Band Broadening in Electrokinetic Chromatography with Micellar Solutions and Open-Tubular Capillaries"; *Anal. Chem.* **1989**, 61, pp 251-260.
88. Beckwith, J. B.; Ivory, C. F.; "The influence of diffusion on elution profiles in the Philpot-Harwell Electrophoretic Separator"; *Chem. Eng. Commun.* **1987**, 54 (1-6), pp 301-331.

89. Wagner, H.; Kessler, R.; "Free-flow field-step focusing: a new method for preparative protein isolation"; In *Electrophoresis '83*. Stathakos, D., Ed.; W. de Gruyter: Berlin, Germany, 1983, pp. 303-312.
90. Bier, M.; "Scale-up of isoelectric focusing"; In *Separation, Recovery and Purification in Biotechnology*, Asenjo, J. A.; Hong, J., Eds.; ACS Symp. Ser. #314, ACS Press: Washington, D. C., 1986.
91. Gobie, W. A.; Ivory, C. F.; "Recycle Continuous Flow Electrophoresis: Zero Diffusion Theory"; *AIChE J*. **1988**, 34, pp 474-482.
92. Gobie, W. A.; Ivory, C. F.; "Theoretical and Experimental Investigations of CACE"; submitted for publication, 1989.
93. Sloan, J. E.; Thorman, W.; Twitty, G. E.; Bier, M.; "Automated recycling free-flow isotachophoresis: Principle, instrumentation and first results"; *J. Chromatography*. **1988**, 457, pp 137-148.
94. Righetti, P.G.; Barzaghi, B.; Faupel, M.; "Large-scale electrophoresis for protein purification: Exploiting isoelectricity"; *Trends in Biotechnology*. **1988**, 6, pp 121-125.
95. Yoshisato, R. A.; Korndorf, L. M.; Carmichael, G. R.; Datta, R.; "Performance Analysis of a Continuous Rotating Annular Electrophoresis Column"; *Sep. Sci. Tech*. **1986**, 21, pp 727-753.
96. Datta, R.; Yoshisato, R. A.; Carmichael, G. R.; "Development of a Theoretical Model for Continuous Rotating Annular-Bed Electrophoresis column for biochemical separations"; AIChE Symposium Series #250, Vol. 82, 1986, pp 179-192.
97. Scott, C. D.; "Continuous electrochromatography using a rotating annular system"; *Sep. Sci. Tech*. **1986**, 21, pp 905-917.
98. **Yoshisato, R. A.; Datta, R.; Gorowicz, J. P.; Beardsley R. A.; Carmichael, G. R.: this book.**
99. O'Farrell, P. H.; "Separation techniques based on the opposition of two counteracting forces to produce a dynamic equilibrium"; *Science*. **1985**, 227, pp 1586-89.
100. McCoy, B. J.; "Counteracting chromatographic electrophoresis and related imposed-gradient separation processes"; *AIChE J*. **1986**, 32, pp 1570-73.
101. Hunter, J. B.; "An Isotachophoretic Model of Counteracting Chromatographic Electrophoresis (CACE)"; *Sep. Sci. Tech*. **1988**, 23, pp 913-930.
102. Locke, B. R.; Carbonell, R. G.; "A theoretical and experimental study of Counteracting Chromatographic Electrophoresis"; *Sep. Pur. Meth*. **1989**, 18, pp 1-64.
103. Visvanathan, C.; Ben Aim, R.; "Application of an electric field for reduction of particle and colloidal membrane fouling in cross-flow microfiltration"; *Sep. Sci. Tech*. **1989**, 24, pp 383-398.
104. Lee, C. K.; Hong, J.; "Membrane Reactor Coupled with Electrophoresis for Enzymatic Production of Aspartic Acid"; *Biotech. and Bioeng*. **1988**, 32, pp 647-654.
105. Yarmush, M. L.; Olson, W. C.; "Electrophoretic elution from biospecific adsorbents: Principles, methodology and applications"; *Electrophoresis*. **1988**, 9, pp 111-120.
106. Grimshaw, P. E.; Grodzinsky, A. J.; Yarmush, M. L.; Yarmush, D. L.; "Dynamic Membranes for Protein Transport: Chemical and Electrical Control"; *Chem. Eng. Sci*. **1989**, 44, pp 827-840.
107. Barker, P. E.; Ganetsos, G.; "Chemical and biochemical separations using preparative and large-scale batch and continuous chromatography"; *Sep. Pur. Meth*. **1988**, 17, pp 1-65.
108. Gibbs, S. J.; Lightfoot, E. N.; "Scaling Up Gradient Elution Chromatography"; *Ind. Eng. Chem. Fund*. **1986**, 25, pp 490-498.
109. Wankat, P. C.; "Intensification of Sorption Processes"; *Ind. Eng. Chem. Fund*. **1987**, 26, pp 1579-85.
110. Wankat, P. C.; Koo, Y. M.; "Scaling Rules for Isocratic Elution Chromatography"; *AIChE J*. **1988**, 34, pp 1006-1019.
111. Jungbauer, A.; "Scaleup of monoclonal antibody purification using radial streaming chromatography"; *Biotech. and Bioeng*. **1988**, 32, pp 326-333.

112. Saxena, V.; Dunn, M; "Solving Scaleup: Radial-flow Chromatography"; *Bio/Technology.* **1989**, 7, pp 250-255.
113. Ding, H.; Yang, M. C.; Schisla, D.; Cussler, E. L.; "Hollow-fiber liquid chromatography"; *AIChE J.* **1989**, 35, pp 815.
114. Brandt, S.; Goffe, R. A.; Kessler, S. B.; O'Connor, J. L.; Zale, S. E.; "Membrane-based Affinity Technology for Commercial Scale Purifications"; *Biotechnology.* **1988**, 6, 7, pp 779-783.
115. Wankat, P. C.; "Improved preparative chromatography: Moving port chromatography"; *Ind. Eng. Chem. Fundamentals.* **1984**, 23, pp 256.
116. Frenz, J.; Horváth, Cs.; "High Performance Displacement Chromatography"; In *HPLC: Advances and Perspectives, Vol. 5*, Ed.; Horváth, Cs.; Academic Press: New York, N.Y., 1988.
117. Liao, A. W.; El-Rassi, Z.; LeMaster, D. M.; Horváth, Cs.; "High Performance Displacement Chromatography of Proteins: Separations of β-lactoglobulins A and B"; *Chromatographia.* **1987**, 24, pp 881-885.
118. Phillips, M. W.; Subramanian, G.; Cramer, S. M.; "Theoretical optimization of operating parameters in nonideal displacement chromatography"; *J. Chromatography.* **1988**, 454, pp 1-21.
119. Yu, Q.; Wang, N. H. L.; "Multicomponent interference phenomena in ion-exchange columns"; *Sep. Pur. Tech.* **1986**, 15, pp 127-158.
120. Howard, A. J.; Byers, C. H.; Carta, G.; "Separation of sugars by continuous annular chromatography"; *Ind. & Eng. Chem Res.* **1988**, 27, pp 1873-1882.
121. Byers, C. H.; Sisson, W. G.; DeCarli, II, J. P.; Carta, G.; "Pilot scale studies of sugar separations by continuous chromatography"; *Appl. Biochem. Biotech.* **1989**, 20/21; pp 635-654.
122. Begovich, J. M.; Sisson, W. G.; "A rotating annular chromatograph for continuous separations"; *AIChE J.* **1984**, 30, pp 705-709.
123. Dalvie, S. K.; Gajiwala, K. S.; Baltus, R. E.: this book.
124. Sisson, W. G.; Begovich, J. M.; Byers, C. H.; Scott, C. D.; "Continuous Chromatography"; *Chemtech.* **1988**, 18, 8, pp 498-502.
125. Ito, Y.; "High-speed countercurrent chromatography"; *CRC Crit. Rev. Anal. Chem.* **1986**, 17, pp 65-143.
126. Ito, Y.; Zhang, T. Y.; "Multistage mixer-settler planet centrifuge: Preliminary studies on partition of macromolecules with organic-aqueous and aqueous two-phase solvent systems"; *J. Chromatography.* **1988**, 437, pp 121-129.
127. Chase, H. A.; "Scale-up of immunoaffinity separation processes"; *J. Biotechnology.* **1984**, 1, pp 67-80.
128. Vijayalakshmi, M. A.; "Pseudobiospecific ligand affinity chromatography"; *Trends in Biotechnology.* **1989**, 7, 3, pp 71-76.
129. McCormick, D.; "Chromatography 1988"; *Bio/Technology.* **1988**, 6, pp 158-165.
130. Hammer, D. A.; Linderman, J. J.; Graves, D. J.; Lauffenburger, D. A.; "Affinity chromatography for cell separation: Mathematical model and experimental analysis"; *Biotech. Progress.* **1987**, 3, pp 189-204.
131. Bresler, S. E.; Katushkina, N. V.; Kolikov, V. M.; Potokin, J. L.; Vinogradskaya, G. N.; "Adsorption Chromatography of Viruses"; *J. of Chromatography.* **1977**, 130, pp 275-280.
132. Tsao, I.-F.; Wang, H. Y.: this book.
133. Bitton, G.; "Adsorption of viruses to surfaces: technological and ecological implications"; In *Adsorption of microorganisms to surfaces*, Eds.; Bitton, G.; Marshall, K. C.; Wiley: New York, NY 1980; pp 331-374.
134. Ohlson, S.; Hansson, L.; Glad, M.; Mosbach, K.; Larsson, P.-O.; "High Performance Liquid Affinity Chromatography: a new tool in Biotechnology"; *Trends in Biotechnology.* **1989**, 7, 7, pp 179-186.
135. Bergold, A. F.; Muller, A. J.; Hanggi, D. A.; Carr, P. W.; "High Performance Affinity Chromatography"; In HPLC: Advances and Perspectives, Vol. 5., Horváth, Cs., Ed.; Academic Press: New York, N.Y., 1988.

136. Somers, W.; Van't Riet, K.; Rozie, H.; Rombouts, F. M.; Visser, J.; "Isolation and Purification of Endo-polygalacturonase by Affinity Chromatography in a Fluidized Bed Reactor"; *Chem. Eng. Jl./Biochem. Eng. Jl.* **1989**, 40, pp B7-B19.
137. Burns, M. A.; Graves, D. J.; "Continuous Affinity Chromatography Using a Magnetically Stabilized Fludized Bed"; *Biotechnology Progress.* **1985**, 1, pp 95-103.
138. Lochmuller, C. H.; Wigman, L. S.; "Affinity Separations in Magnetically Stabilized Fluidized Beds: Synthesis and Performance of Packing Materials"; *Sep. Sci. Tech.* **1987**, 22, pp 2111-2125.
139. Flanagan, S. D.; Barondes, S. H.; "Affinity Partitioning"; *J. Biol. Chem.* **1975**, 250, pp 1484-1489.
140. Johansson, G.; Andersson, M.; "Liquid-Liquid Extraction of Glycolytic Enzymes From Baker's Yeast Using Triazine Dye Ligands"; *J. Chromatography.* **1984**, 291, pp 175-183.
141. Kopperschläger, G.; Lorenz, G.; Usbeck, E.; "Application of Affinity Partitioning in an Aqueous Two-Phase System to the Investigation of Triazine Dye-Enzyme Interactions"; *J. Chromatography.* **1983**, 259, pp 97-105.
142. Janson, J.-C.; "Large-scale affinity Purification: State of the art and future prospects"; *Trends in Biotechnology.* **1984**, 2, 2, pp 31-38.
143. Kobos, R. K.; Eveleigh, J. W.; Arentzen, R.; "A novel fluorocarbon based immobilization technology"; *Trends in Biotechnology.* **1989**, 7, pp 101-105.
144. Firary, M.; Carlson, A.; "Affinity partitioning of Acid Proteases in the Hydroxypropyldextran-dextran Aqueous Two-Phase System"; presented at the AIChE Summer Meeting, Boston, MA, August 1986.
145. Powers, J. D.; Kilpatrick, P. K.; Carbonell, R. G.; "Protein Purification by Affinity Binding to Unilamellar Vesicles"; *Biotech. and Bioeng.* **1989**, 33, 2, pp 173-182.
146. Woll, J. M.; Hatton, T. A.; Yarmush, M. L.; "Bioaffinity separations using reversed micellar extraction"; *Biotechnology Progress.* **1989**, 5, pp 57-62.
147. Guzman, R.; Torres, J. L.; Carbonell, R. G.; Kilpatrick, P. K.; "Water-soluble nonionic surfactants for affinity bioseparations"; *Biotech. and Bioeng.* **1988**, 33, pp 1267-76.
148. Senstad, C.; Mattiasson, B.; "Affinity-Precipitation Using Chitosan as Ligand Carrier"; *Biotech. and Bioeng.* **1989**, 33, 2, pp 216-220.
149. Luong, J. H. T.; Male, K. B.; Nguyen, A. L.; Mulchandani, A.; "Mathematical Modeling of Affinity Ultrafiltration Process"; *Biotech. and Bioeng.* **1988**, 32, pp 451-459.
150. Nigam, S.; Wang, H. Y.; "Mathematical Modeling of Bioproduct Adsorption using Immobilized Affinity Adsorbents"; In *Separation, Recovery and Purification in Biotechnology*, (ACS Symp. Ser. #314), Asenjo, J. A.; Hong, J., Eds.; ACS Press: Washington, D. C., 1986, pp 153-168.
151. Nigam, S.; Sakoda, A.; Wang, H. Y.; "Bioproduct recovery from unclarified broths and homogenates using immobilized adsorbents"; *Biotech. Prog.* **1988**, 4, pp 166-172.
152. Pungor, E.; Afeyan, N. B.; Gordon, N. F.; Cooney, C. L.; "Continuous Affinity-recycle Extraction: A Novel Protein Separation Technique"; *Bio/Technology.* **1987**, 5, pp 604-608.
153. Hill, C. L.; Bartholomew, R.; Beidler, D.; David, G. S.; " 'Switch' immunoaffinity chromatography with monoclonal antibodies"; *Biotechniques.* **1983**, 1, 14.
154. Olson, W. C.; Leung, S. K.; Yarmush, M. L.; "Recovery of Antigens From Immunoadsorbents using High Pressure"; *Bio/Technology.* **1989**, 7, pp 369-373.
155. Levinson, A. D.; Svedersky, L. P.; Palladino, Jr., M. A.; " Tumorigenic Potential of DNA Derived from Mammalian Cell Lines"; In *Abnormal Cells, New Products and Risk*, Hopps, H. E.; Petricciani, J. C.; Proceedings of a Workshop, July 30-31, 1984; NIH, Bethesda, MD, Tissue Culture Association: Gaithersburg, MD, 1985, Monograph 6, pp 161-165.

RECEIVED November 10, 1989

EXTRACTION AND MEMBRANE PROCESSES

Chapter 2

Statistical Thermodynamics of Aqueous Two-Phase Systems

Heriberto Cabezas, Jr.[1], Janis D. Evans[2,3], and David C. Szlag[2]

[1]Department of Chemical Engineering, University of Arizona, Tucson, AZ 85721
[2]Center for Chemical Engineering, National Institute of Standards and Technology, Boulder, CO 80303

> Hill's theory of solutions was used to model the phase diagrams of polymer-polymer aqueous two-phase systems. This theory expresses chemical potentials in terms of polymer-polymer osmotic virial coefficients. Scaling expressions for predicting these coefficients from the degree of polymerization and parameters were developed from Renormalization Group theory. For a two polymer system the parameters consist of two constants, b_1 and b_2, and two exponents, υ_2 and υ_3. Values for these parameters which are valid for all solutions were obtained from experiment. Phase diagrams at ambient conditions were predicted for three different systems consisting of aqueous mixtures of polyethylene glycol and dextran ranging in molecular weight from 3690 to 167,000. The predicted phase compositions are within 1-3% from experimental values which have uncertainties of about 1%.

In the years since the pioneering studies of Albertsson and co-workers (1), aqueous two-phase extraction has gained wide acceptance as a method for the recovery and purification of proteins, enzymes, and other molecules and particles of biological origin. Interest in these systems has been rekindled in recent years (2) due to the rapid growth of the biotechnology industry in an increasingly competitive environment and to the identification of the cost of separation as the major component in the price of bioproducts (3). This has created a need for a separation process which is gentle to sensitive biomolecules while offering high product recovery, high

[3]Current address: Phillips Petroleum Company, P.O. Box 350, Borger, TX 79008

product purity and also economical operation and ease of scaling up or down. The aqueous two-phase extraction technique meets the above requirements and in addition, allows us to bring to bear on the problem, the existing body of knowledge and experience for the operation and design of industrial liquid-liquid extractions (4, 5).

The design and optimization of any liquid-liquid extraction process, including one involving aqueous two-phase systems, is predicated on the availability of a phase diagram for the system as a first step. Early progress in the prediction of phase diagrams was made by Edmond and Ogston (6), and additional theoretical advances have been accomplished recently by Prausnitz and coworkers (7) and by Sandler and coworkers (8) among others. An overall review of this field has been given by Benge (9). Yet, at least two major problems remain. The simplest and most successful model, that of Edmond and Ogston is formulated for a solution under its own osmotic pressure and is not strictly applicable to the calculation of a phase diagram at constant temperature and pressure. The second problem is that there is no simple, quantitative way of accounting for the changes in the phase diagram with polymer molecular weight.

Our work addresses these two problems by adopting the solution theory of Hill (10, 11) and by adapting ideas from the Group Renormalization (12, 15) theory of polymer solutions to the prediction of the model parameters from the degree of polymerization of the phase forming polymers.

STATISTICAL MECHANICAL BASIS

Our purpose in this section is to derive a set of useful expressions for the chemical potentials starting with the principles of statistical mechanics. The expressions we shall obtain take the form of virial expansions similar to those of the Edmond and Ogston (6) but having a very different theoretical basis. Our model parameters are isobaric-isothermal virial coefficients which are about an order of magnitude smaller than the osmotic virial coefficients in the Edmond and Ogston model. We shall develop the theory neglecting the effect of polydispersity because we empirically did not find this to be very important at the level of accuracy commonly attainable in experimental phase diagrams for these systems.

To outline the fundamental basis of the model, we follow the notation of Hill (10) and extend his derivation to a three component mixture. Component 1 is the solvent which in our case is water, component 2 is a solute or polyethylene glycol, and component 3 is another solute or dextran. We base the theory on an isobaric-isothermal ensemble first introduced by Stockmayer (14). This choice of ensemble is the most appropriate because it yields expressions for the chemical potentials of the components with temperature, pressure, and solute molality or mole fraction as the natural independent variables, and these are the independent variables normally used in calculation, experiment, and industrial practice.

We begin with the canonical partition function for a three component system which is given by Equation 1 and where the independent variables are temperature, volume and mole numbers.

$$Q(N_1, N_2, N_3, V, T) = \sum_i e^{-E_i/kT} \quad (1)$$

The summation in Equation 1 is taken over all of the energy states of the ensemble.

From a series of transformations of Equation 1 we obtain a new partition function (Γ) whose independent variables are temperature, pressure, solvent mole number, and the chemical potentials of the solutes (components 2 and 3). These transformations consist of first creating a partition function with pressure rather than volume as an independent variable, and then using this result to create yet another partition function in which we have switched independent variables from solute mole numbers to solute chemical potentials. These operations are analogous to the Legendre transforms commonly employed in thermodynamics.

$$\Gamma(N_1, P, T, \mu_2, \mu_3) = e^{N_1\mu_1/kT} = \sum_{i=2,3} \sum_{N_i \geq 0} e^{N_2\mu_2/kT} e^{N_3\mu_3/kT} \Delta_{N_i} \quad (2)$$

where:
$$\Delta_{N_i} = \sum_V e^{-PV/kT} Q(N_1, N_2, N_3, V, T)$$

Q = canonical partition function

The right hand side of Equation 2 includes a power series in the solute activities, a_2 and a_3, of components 2 and 3, respectively. Expanding the series and taking the logarithm of both sides of the equation yields Equation 3.

$$\ln \Gamma = \ln \Delta_o \left[1 + \sum_{i=2,3} \sum_{N_i \geq 0} X_{N_i} a_i^{N_i} \right] \quad (3)$$

where:
$$X_{N_i} = \Delta_{N_i} \Delta_o^{N_i - 1} N_1^{N_i} / \Delta_1^{N_i}$$

$$a_i = \Delta_1 e^{\mu_i/kT} / N_1 \Delta_o$$

$$\Delta_o = \Delta_{N_i = 0}$$

$$\Delta_0 = \Delta N_i = 1$$

Expansion of the logarithm about unit activity and collection of terms of like power in solute activity yields Equation 4 where the index j refers to the series expansion of the logarithm. Since solute activity approaches unity only as the solute molalities approach zero, this expansion is strictly valid only for dilute solutions.

$$\frac{N_1 \mu_1}{kT} = \ln \Gamma = \ln \Delta_0 + N_1 \sum_{i=2,3} \sum_{j \geq 1} \theta_{ji}(T, P)\, a_i^j \tag{4}$$

From the Gibbs-Duhem equation and Equation 4 we obtain Equations 5 which express molality (m_2, m_3) as a power series in activity.

$$\frac{1}{kT} \frac{\partial \mu_1}{\partial \ln a_2} \bigg|_{T, P, a_3} = m_2 = \sum_{j \geq 1} j\, \theta_{j2}(T, P)\, a_2^j \tag{5a}$$

$$\frac{1}{kT} \frac{\partial \mu_1}{\partial \ln a_3} \bigg|_{T, P, a_2} = m_3 = \sum_{j \geq 1} j\, \theta_{j3}(T, P)\, a_3^j \tag{5b}$$

where: $m_2 = N_2/N_1$

$m_3 = N_3/N_1$

The inverse of Equations 5 give activity as a power series in molality. Taking this inverse and collecting terms in like powers of the molalities up to first order we obtain Equations 6 which give the solute chemical potential as a power series in solute molality with osmotic virial coefficients that are functions of temperature and pressure.

$$\mu_2(T, P, m_2, m_3) - \mu_2^0(T, P, 0, 0) = kT \ln a_2 = -kT \cdot [\ln m_2 + 2C_{22} m_2 + 2C_{23} m_3 + \ldots] \tag{6a}$$

$$\mu_3(T, P, m_2, m_3) - \mu_3^0(T, P, 0, 0) = kT \ln a_3 = -kT \cdot [\ln m_3 + 2C_{23} m_2 + 2C_{33} m_3 + \ldots] \tag{6b}$$

where: $C_{ij} = C_{ij}(T, P)$

$$\mu_i^0 = \lim_{\substack{m_2 \to 0 \\ m_3 \to 0}} \mu_i(T, P, m_2, m_3)$$

Finally from the Gibbs-Duhem equation and Equations 6 we obtain Equation 7 which gives the solvent (water) chemical potential in terms of solute molalities and the aforementioned coefficients.

$$\mu_1(T, P, m_2, m_3) - \mu_1^0(T, P, O, O) = -kT(m_2 + m_3 + C_{22} m_2^2 + 2C_{23} m_2 m_3 + C_{33} m_3^2 + ...) \quad (7)$$

Equations 6 and 7 are the fundamental expressions giving the chemical potentials as functions of solute molalities and Hill osmotic virial coefficients.

We can obtain an expression for the osmotic pressure by considering the pure solvent in equilibrium with a solution under its own osmotic pressure. This is expressed by Equation 8.

$$\mu_1(T, P, m_2, m_3) = \mu_1(T, P - \pi, O, O) \quad (8)$$

Next we integrate the pressure derivative of the pure solvent chemical potential (at zero solute molality).

$$\mu_1(T, P, O, O) - \mu_1(T, P - \pi, O, O) = \int_{P-\pi}^{P} V_1(T, P) \, dP = V_1 \pi \quad (9)$$

Inserting Equation 7 in the left hand side of Equation 8 and then in Equation 9 and assuming the solvent to be incompressible, we obtain Equation 10 for the osmotic pressure (π).

$$\frac{\pi V_1}{kT} = m_2 + m_3 + C_{22} m_2^2 + 2C_{23} m_2 m_3 + C_{33} m_3^2 + ... \quad (10)$$

SCALING LAW EXPRESSIONS

In this section we "semi-empirically" adapt some scaling ideas from the Group Renormalization theory (12, 15) of polymer solutions to obtain expressions for the osmotic virial coefficients of Equations 6 and 7 in terms of the degree of polymerization. In the following discussion we will occasionally omit the indices on the osmotic virial coefficients for the sake of simplicity.

It is well known that the osmotic pressure of a solution of one polymer can be scaled with a single dimensionless variable "S" which is proportional to polymer concentration at least for the case of mixtures with good solvents in the dilute to semidilute regime (12, 17). This implies that the osmotic compressibility factor (π/cRT) can be expressed as some function of "S" only as shown in Equation 11.

$$\frac{\pi}{ckT} = 1 + F(S) \tag{11}$$

From the Group Renormalization theory of polymer solutions (12) we know that "S" is proportional to "$\underline{b}$" which depends on the nature of the polymer, the polymer concentration "c", and the degree of polymerization "N" raised to the power of an exponent "3υ" as shown in Equation 12.

$$S = \underline{b} \, c \, N^{3\upsilon} \tag{12}$$

Using Equation 12 we expand F(S) in a Taylor series about infinite dilution (c = 0) and insert the result, truncated to first order in polymer concentration (c), into Equation 11 to obtain Equation 13.

$$\frac{\pi}{ckT} = 1 + Bc + ... \tag{13}$$

$$\text{where:} \quad B = \underline{b} \, N^{3\upsilon}$$

At this point we note that Equation 13 is the McMillan-Mayer (16) expansion for the osmotic compressibility factor which is fundamentally different from the analogous expansion that was obtained from the formalism of Hill (Equation 10). We also identify B as a McMillan-Mayer osmotic virial coefficient.

We have shown that there is a scaling relation for B of the form given in Equation 13. However, we have not shown that an analogous relation exists for the Hill osmotic virial coefficients (C). We start the proof with the exact relation between B and C shown in Equation 14.

$$C = \frac{1}{V_1} (B - \overline{V}^\circ + \frac{1}{2} K_1 RT) \tag{14}$$

where: $\overline{V}^o$ = the polymer partial molar volume at infinite dilution

K_1 = the isothermal compressibility of pure water

To make further progress we need the well known empirical fact that the polymer partial molar volume scales linearly with the degree of polymerization (N) as given by Equation 15.

$$\overline{V}^o = \overline{V}^o_m N \qquad (15)$$

where: $\overline{V}^o_m$ = effective monomer partial molar volume

Next we propose that there is a relation for C of the form given by Equation 16.

$$C = b N^{3\upsilon} \qquad (16)$$

Finally, we insert Equations 15 and 16 into Equation 14, substitute for B in terms of the scaling expression of Equation 13, and solve for b to obtain Equation 17.

$$b = \frac{b}{V_1} - \frac{\overline{V}^o_m}{V_1 N^{3\upsilon-1}} + \frac{K_1 RT}{2 N^{3\upsilon}} \qquad (17)$$

Since the Renormalization Group approach is strictly applicable only for very long polymers, we take the limit of Equation 17 as N becomes very large to obtain Equation 18.

$$b = \frac{b}{V_1} \qquad N \to \infty \qquad (18)$$

This shows that at least for the case of large polymers, there exists a scaling expression for C of the form of Equation 16 where the proportionality constant b is given by Equation 18. One should be aware of the fact that in obtaining this result we have tacitly assumed that the exponent υ in Equation 13 which is defined in a McMillan-Mayer ensemble is the same as that in Equation 16 which is defined in an isobaric-isothermal ensemble. We justify this assumption on the fact that the two exponents are numerically indistinguishable when evaluated from experimental osmotic virial coefficients. We also feel this to be appropriate for the level of approximation in our approach.

The scaling expression of Equation 16 is applicable to a solution of one polymer in water. It therefore gives us relationships for C_{22} and for C_{33} which represent the interaction of either polymer 2 or 3 with itself but not C_{23} which represents the interactions of polymers 2 and 3. Furthermore, the results from Group Renormalization theory do not give as straightforward guidance on the functional form of C_{23} as they do for the other two coefficients (13). We have, therefore, developed on empirical grounds a relationship for C_{23} which still embodies the fundamental scaling concepts from theory (see Equation 19c). The expressions for the three virial coefficients are given by Equations 19.

$$C_{22} = b_1 N_2^{3v_2} \tag{19a}$$

$$C_{33} = b_1 N_3^{3v_3} \tag{19b}$$

$$C_{23} = b_2 [N_2^{3v_2} N_3^{3v_3}]^{\frac{1}{2}} \tag{19c}$$

The scaling expressions for C_{22} and C_{33} are given in terms of a proportionality constant b_1 which depends on temperature and pressure, the degrees of polymerization N_2 and N_3, and the scaling exponents v_2 and v_3 for polymers 2 and 3 respectively. Theory (12) indicates that for model linear polymers the value of the exponent should be universal while the value of the proportionality constant should vary with the chemical nature of the polymer. In fact, the value (17) of the universal exponent is set at 0.59. We empirically found that a value of 0.60 which is very close to the theoretical value was adequate to correlate the molecular weight or N dependence of the virial coefficients of the polyethylene glycols since these are close to being linear polymers. However, this same value proved completely inadequate for the dextrans which are branched polymers. For the dextrans we set the value of the exponent at 0.6948 which is still not an unreasonable number. As implied by Equations 19, we also found that a single proportionality constant with a value of 0.9012 was adequate to correlate the self interaction virial coefficients of both polymers. Since the proportionality constant represents the interaction of two monomers of the same polymer in the presence of water, it would seem that monomers of polyethylene glycols and monomers of dextran have similar self interactions possibly indicating that both interactions are dominated by the hydroxyl group.

The scaling expression for C_{23} is an empirical geometric mean type rule involving the same exponent values used in the relations for C_{22} and C_{33} but with a different value for the proportionality constant. We found that a value of 1.1841 for b_2 was adequate to reproduce phase diagrams. We were not able to find any relation between b_1 and b_2 possibly indicating that the interactions between a monomer of polyethylene glycol and a monomer of dextran are very different from the self interactions mentioned before.

ESTIMATION OF MODEL PARAMETERS

There are four final model parameters after the scaling laws are inserted in Equations 6 and 7. These are v_2, v_3, b_1, and b_2. There are many methods that one could use to obtain values for these parameters. However, we shall be concerned with only three of them.

First, if experimental data are available on the osmotic pressure of the two-polymer system versus the molality of the polymers, one could simply insert Equations 19 into Equation 10 and numerically fit the four parameters. Alternatively, if there are data on the osmotic pressure of only one of the polymers in water one could obtain values for C_{22} or C_{33} but not for C_{23}. Although this method is the least ambiguous, we did not follow it because the necessary data were not available.

Second, if McMillan-Mayer (16) osmotic virial coefficients (B_{ij}) are available from light scattering or some other experiment for several molecular weights of the same polymer, one could use Equation 20,

$$V_1 C_{22} = B_{22} - \overline{V}_2^o + \frac{1}{2} K_1 RT \qquad (20)$$

which relates the virial coefficients from the Hill theory to the McMillan-Mayer osmotic virial coefficients to calculate C_{22}, C_{33}, or C_{23} for several molecular weights. One could then fit the four parameters previously mentioned above to the C_{ij}'s using Equations 19. Using this method with the osmotic virial coefficients of Prausnitz and co-workers (7) we calculated values of C_{22} for polyethylene glycols 8000 and 3350 and then fitted b_1 and v_2 using Equation 19a. We tried the same procedure for the dextrans but found that the value of v_2 so obtained could not represent the phase diagrams accurately.

Third, if experimental phase diagrams are available one could use tie lines and the phase equilibrium relations of Equations 21 to solve for three of the four parameters.

$$\mu_1^T(T, P, m_2^T, m_3^T) = \mu_1^B(T, P, m_2^B, m_3^B) \qquad (21a)$$

$$\mu_2^T(T, P, m_2^T, m_3^T) = \mu_2^B(T, P, m_2^B, m_3^B) \tag{21b}$$

$$\mu_3^T(T, P, m_2^T, m_3^T) = \mu_3^B(T, P, m_2^B, m_3^B) \tag{21c}$$

A modification of this procedure was used by us to estimate values for b_2 and v_3. We simply adopted the previously calculated values for b_1 and v_2, used Equations 21 with Equations 6, 7 and 19 to obtain Equations 22 which can be solved for b_2 and v_3, and applied Equations 22 to one tie line from each of the experimental phase diagrams of Figures 1, 2, and 3 (King, R. S.; Blanch, H. W.; Prausnitz, J. M., University of California at Berkeley, unpublished data presented at the ACS Meeting in Anaheim, CA, 1986). The numerical solution of equations 22 for b_2 and v_3 by applying Equations 19b and c gave us several slightly different values for the two parameters and thus we adopted the arithmetic average.

$$O = (m_2^T - m_2^B) + (m_3^T - m_3^B) + C_{22}\,[(m_2^T)^2 - (m_2^B)^2] + \\ + 2C_{23}\,[m_2^T\,m_3^T - m_2^B\,m_3^B] + C_{33}\,[(m_3^T)^2 - (m_3^B)^2] \tag{22a}$$

$$O = \ln \frac{m_2^T}{m_2^B} + 2C_{22}\,(m_2^T - m_2^B) + 2C_{23}\,(m_3^T - m_3^B) \tag{22b}$$

$$O = \ln \frac{m_3^T}{m_3^B} + 2C_{33}\,(m_3^T - m_3^B) + 2C_{23}\,(m_2^T - m_2^B) \tag{22c}$$

where C_{22}, C_{33}, and C_{23} are given by Equations 19.

PHASE DIAGRAM CALCULATIONS

We have calculated phase diagrams at ambient conditions for three different polyethylene glycol-dextran systems using our model and have compared the results to the experimental phase diagrams of King et al. (King, R. S.; Blanch, H. W.; Prausnitz, J. M., University of California at Berkeley, unpublished data presented at the ACS Meeting in Anaheim, CA, 1986). These calculations are illustrated in Figures 1 to 3. We covered a wide range of polymer molecular weights in order to observe the polymer molecular weight dependence of the virial coefficients. Thus, the calculated system represented in Figure 1 consists of polyethylene glycol 3350 (MW = 3690) and dextran T-500 (MW = 167,000); that in Figure 2 consists of polyethylene glycol 3350 (MW = 3690) and dextran T-70 (MW = 37,000); and

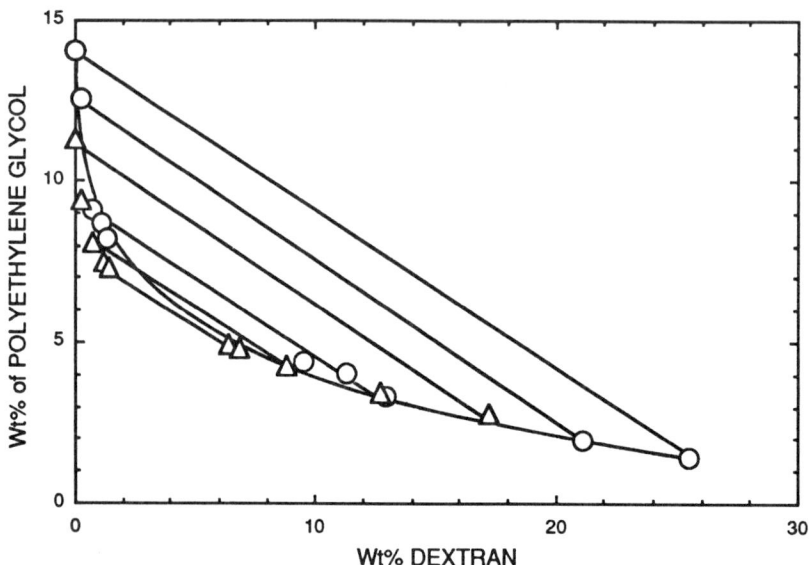

Figure 1. Phase diagram for an aqueous mixture of Polyethylene Glycol 3350 (MW = 3690) and Dextran T-500 (MW= 167,000) at 25°C.
○ Experiment.   △ Model.

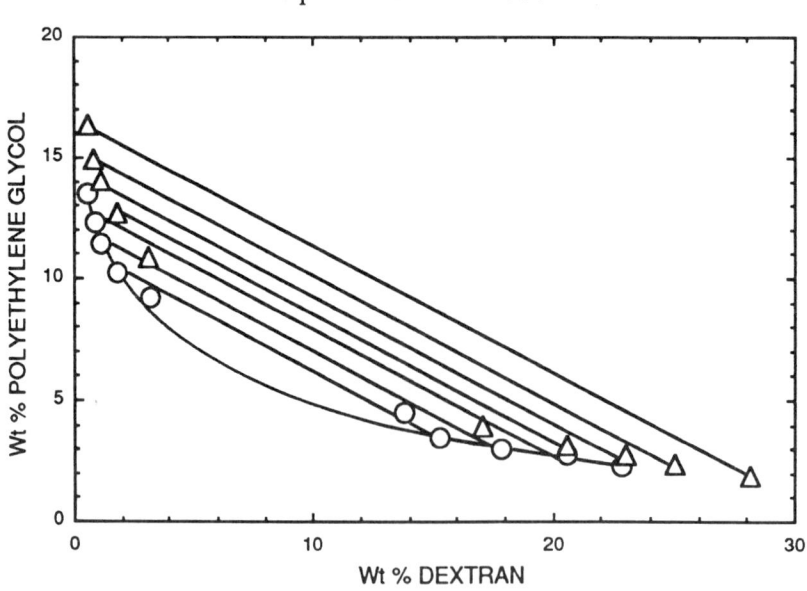

Figure 2. Phase diagram for an aqueous mixture of Polyethylene Glycol 3350 (MW = 3690) and Dextran T-70 (MW = 37,000) at 25°C.
○ Experiment.   △ Model.

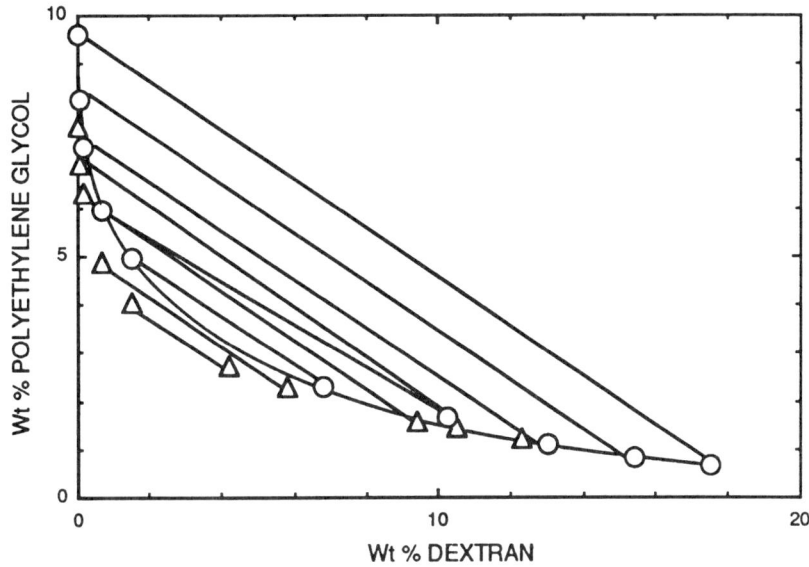

Figure 3. Phase diagram for an aqueous mixture of Polyethylene Glycol 8000 (MW = 8920) and Dextran T-500 (MW = 167,000) at 25°C.
○ Experiment. △ Model.

that in Figure 3 consists of polyethylene glycol 8000 (MW = 8920) and dextran T-500 (MW = 167,000).

The tie lines in Figures 1, 2, and 3 were calculated from the phase equilibrium relations represented by Equations 21 which upon substitution of the chemical potential models become Equations 22. There are three equations and four unknown molalities. For each tie line we therefore set a value for one of the molalities, which in our case was that of dextran in the bottom phase, and simultaneously solved Equations 22 for the remaining three molalities. The numerical algorithm used was the same one that Edmond and Ogston (6, 9) used for their model. The virial coefficients used in Equations 22 for all the calculations were predicted from the scaling expressions of Equations 19.

Comparing the predicted to the experimental tie lines in Figures 1 to 3 indicates that the agreement (1-3 wt%) between the two is, overall, satisfactory. This is particularly true if one realizes that the experimental uncertainty in the determination of the phase compositions is of the order of 1.0 wt.%, especially near the critical point. An additional source of discrepancy between theory and experiment may be a small difference between the average molecular weights of the polymers actually used in the experiments and the molecular weights that we assumed in our model calculations.

CONCLUSIONS

The present work applies the formalism of Hill and the modern theory of polymer solutions to the development of a new and very simple model for the prediction of phase diagrams in aqueous polymer-polymer systems. The fundamental model expressions are rigorous for constant temperature and pressure calculations. The scaling expressions allow us to predict the osmotic virial coefficients in the Hill equations from the number average molecular weight of the polymers and model parameters. The model involves four parameters: b_1, b_2, v_2, and v_3 which appear in the scaling expressions. Calculated phase diagrams are most sensitive to b_2 because it determines the value of C_{23} which represents the interactions between unlike polymers which dominate phase separation. To establish the validity of our ideas, we have applied our model to the calculation of phase diagrams in aqueous polyethylene glycol-dextran systems and shown that it can predict phase behavior accurately for a wide range of polymer molecular weights and composition.

We hope to extend this approach to other polymer systems and to include the effect of polydispersity in a forthcoming paper (Cabezas, H., Jr.; Evans, J.D.; Szlag, D.C. Fluid Phase Equilibria, in press).

ACKNOWLEDGMENTS

The authors are grateful to the National Institute of Standards and Technology of the United States of America for financial and material support.

LEGEND OF SYMBOLS

a_i	-	Activity of component i, dimensionless.
B, B_{ij}	-	McMillan-Mayer osmotic virial coefficient for components i and j, L/Mole.
b, b_1, b_2	-	scaling law proportionality constants, functions of temperature and pressure, dimensionless.
c	-	molar concentration of solute or polymer i, Mol/L.
C, C_{ij}	-	Hill osmotic virial coefficient for components i and j, dimensionless.
E_i	-	energy of system in quantum level i, J.
k	-	Boltzmann's constant, J/molecule -°K
K_1	-	isothermal compressibility of pure solvent or water, bar $^{-1}$.
m, m_i	-	N_i/N_1, dimensionless molality of polymer i.
N	-	degree of polymerization
N_i	-	moles of component i.
P	-	system pressure, bar.
Q	-	canonical partition function.
R	-	gas constant.
S	-	scaled dimensionless polymer concentration.
T	-	system temperature, °K
V	-	system volume, L.
V_1	-	molar specific volume of pure solvent or water, L/Mol.
$\overline{V}_i^o$	-	partial molar volume of solute i at infinite dilution, L/Mol.
X_{N_i}	-	proportionality factor in Equation 3, dimensionless.
Γ	-	partition function for Stockmayers's isobaric-isothermal ensemble.
Δ_{N_i}	-	isobaric-isothermal partition function for Ni Moles.
θ_{ji}	-	proportionality factor in Equation 4, dimensionless.
μ_{ji}	-	chemical potential of component i, J/Molecule.
υ, υ_i	-	scaling exponents of component i, dimensionless.
π	-	osmotic pressure, bar.

Superscripts

T	-	top phase.
B	-	bottom phase.
o	-	infinite dilution.

Subscripts

i, j   - components.
1   - solvent or water.
2   - component 2 or polyethylene glycol.
3   - component 3 or dextran.

LITERATURE CITED

1. Albertsson, P. A. Partition of Cell Particles and Macromolecules; 3rd Ed., John Wiley & Sons: New York, NY, 1986.
2. Walter, H.; Brooks, D. E.; Fisher, D. Partitioning in Aqueous Two-Phase Systems: Theory, Methods, Uses, and Applications to Biotechnology; Academic Press: New York, NY, 1985.
3. Godfrey, P. B.; Kohll, E. A.; Gelbein, A. P. Proc. Econ. Intl. 1985, 5, 12.
4. Treybal, R. E. Liquid Extraction; 2nd Ed., McGraw-Hill: New York, NY, 1963.
5. Kula, M. -R.; Kroner, K. H.; Hustedt, H. In Purification of Enzymes by Liquid-Liquid Extraction; Fiechter, A., Ed.; Advances in Biochemical Engineering No. 24; Springer-Verlag: New York, NY, 1982; p 73.
6. Edmond, E.; Ogston, A. G. Biochem. J. 1968, 109, 569.
7. King, R. S.; Blanch, H. W.; Prausnitz, J. M. AIChE J. 1988, 34, 1585.
8. Kang, C. H.; Sandler, S. I. Fluid Phase Equilibria 1987, 38, 245.
9. Benge, G. G.; Master Thesis, Virginia Polytechnic Institute and State University, Blacksburg, 1986.
10. Hill, T. L. J. Am. Chem. Soc. 1957, 79, 4885.
11. Hill, T. L. J. Chem. Phys. 1959, 30, 93.
12. Schafer, L. Macromolecules 1982, 15, 652.
13. Schafer, L.; Kappeler, C. J. de Phys. 1985, 46, 1853.
14. Stockmayer, W. H. J. Chem. Phys. 1950, 18, 58.
15. Oono, Y. In Statistical Physics of Polymer Solutions; Prigogine, I.; Rice, S. A., Eds.; Advances in Chemical Physics No. 61; John Wiley & Sons: New York, NY, 1985; 301.
16. McMillan, W. G.; Mayer, J. E. J. Chem. Phys. 1945, 13, 276.
17. Des Cloizeaux, J.; Jannink, G. Les Polymères en Solution: leur Modélisation et leur Structure; Les Editions de Physique: France, 1987.
18. Le Gillou, J. C.; Zinn-Justin, J. Phys. Rev. Lett. 1977, 39, 95.

RECEIVED November 16, 1989

Chapter 3

Theoretical Treatment of Aqueous Two-Phase Extraction by Using Virial Expansions

A Preliminary Report

Daniel Forciniti and Carol K. Hall

Department of Chemical Engineering, North Carolina State University, Raleigh, NC 27695-7905

> A theoretical treatment of aqueous two-phase extraction at the isoelectric point is presented. We extend the constant pressure solution theory of Hill to the prediction of the chemical potential of a species in a system containing solvent, two polymers and protein. The theory leads to an osmotic virial-type expansion and gives a fundamental interpretation of the osmotic virial coefficients in terms of forces between species. The expansion is identical to the Edmunds-Ogston-type expression only when certain assumptions are made -- one of which is that the solvent is non-interacting. The coefficients are calculated using simple excluded volume models for polymer-protein interactions and are then inserted into the expansion to predict isoelectric partition coefficients. The results are compared with trends observed experimentally for protein partition coefficients as functions of protein and polymer molecular weights.

When two aqueous solutions of incompatible polymers such as polyethylene glycol (PEG), and dextran (Dx) are mixed above critical concentrations, a liquid-liquid phase separation occurs (1). Proteins or enzymes added to the resulting two-phase mixture will tend to partition unequally between the phases thus allowing for the extraction of a 1particular protein. Separations techniques based on this partitioning have come to be known as aqueous two-phase extraction (2). This technique holds great promise for the isolation of proteins because it is gentle enough for the fragile products of genetic engineering and yet robust enough to be easily adapted to large scale production. Despite the projected importance of aqueous two-phase extractive techniques for future separations technology, very little is known about the molecular basis for protein partitioning.

In this paper we report preliminary work aimed at developing a comprehensive theory of protein partitioning. We focus attention on isoelectric partitioning and use statistical mechanics to examine the fundamental basis of the so called Edmonds-Ogston expression (3) and its extension to four component systems by King et al. (4). This expression, which is an osmotic virial equation truncated at the second term, is well suited to the description of the properties of protein-polymer-solvent systems.

This paper focuses on the relationship between the intermolecular forces and the trends observed experimentally at the isoelectric point for protein partition coefficients as functions of protein and polymer molecular weights. These trends are that at fixed polymer concentrations (on a weight/weight basis): (a) increasing the molecular weight of the protein decreases the partition coefficient and (b) decreasing the molecular weight of one polymer increases the affinity of the protein for the phase rich in that polymer (5). A number of other authors have developed theories which attempt to explain these trends. Brooks et al. (6) and Albertsson et al. (7) have used a Flory Huggins type theory to develop partition coefficient correlations which turn out to be similar to the Edmonds and Ogston and King et al. correlation, Equation 31. The Flory Huggins approach has the advantage that it is analytic and the disadvantage that proteins, although rigid, are treated as flexible polymers. Although these authors claim that their models predict qualitatively the trends described above, they do not consider the effect of the molecular weight dependence of the PEG and Dx concentration differences between the top and bottom phases (hereafter called ΔPEG and ΔDx). Although the molecular weight dependence of ΔPEG and ΔDx is small far from the critical point, it is non negligible in many cases (1, 8). Furthermore, as we will later point out, the molecular weight dependence of ΔPEG and ΔDx can act to oppose the trends observed experimentally. Baskir et al. (9) have developed a lattice approach to treat the conformations of a polymer molecule in the vicinity of a rigid protein molecule, which they model as a hard sphere. They find that they must include attractive protein-polymer interactions in order to predict the trends observed experimentally.

We begin our investigation by extending the constant pressure solution theory of Hill (10, 11) (derived by him for a two component system) to the prediction of the chemical potential of species in a system containing solvent, two polymers and protein. The advantage of using the constant pressure solution theory rather than the constant volume solution theory of McMillan and Meyer (12) is that extraction experiments take place at constant pressure and are therefore more conveniently related to a theory in which pressure is an independent variable. Furthermore, since extraction experiments are conducted by adding solute to a fixed volume of solute, it is easiest to relate to constant pressure solution theory in which molality (grams solute/kilograms solvent) is the natural composition variable.

The theory leads to an osmotic virial type expansion and gives a fundamental interpretation of the coefficients appearing in this expansion in terms of forces between the species. The expansion reduces to the Edmonds-Ogston expression only when certain assumptions are made - namely that the fluids are incompressible and that the solvent is non-interacting. While the first assumption is reasonable, the second assumption is clearly subject to criticism. Nevertheless given these assumptions, the coefficients are calculated using simple excluded volume models for the polymer-protein interactions (the effect of attractions will be considered in a later paper). Three models of molecular shape are considered: polymers are treated as impenetrable spheres, as impenetrable cylinders (particularly applicable to PEG) and as flexible coils; proteins are always modeled as impenetrable spheres. The osmotic virial coefficients associated with these three models are inserted into the expansion to predict isoelectric protein partition coefficients.

The results obtained for these three models are compared with the experimental trends described previously for protein partition coefficients as functions of protein and polymer molecular weight. The most successful of the three models is the model in which protein and dextran are treated as impenetrable spheres and PEG is treated as an impenetrable cylinder. This model predicts the observed experimental trends with protein molecular weight. None of the models can totally explain the dependence of partitioning on polymer molecular weight. The regime of validity for the models depends on the relative size of the protein and the polymer; the smaller the protein, the better the correlation with dextran molecular weight, the larger the protein, the better the correlation with PEG molecular weight.

This investigation has enhanced our understanding of the factors which contribute to the molecular weight dependence of protein partitioning. The molecular weight dependence of the protein partition coefficient results from a competition between two terms in the partition coefficient expansion, namely the crossed second virial coefficient and the differences between the polymer concentrations in the top and bottom phases. While the trend in binodal concentrations tends (in part) to favor the trends observed experimentally, the trend in the second virial coefficient tends to oppose the experimental trends.

THEORY

In 1957 Hill introduced a binary solution theory based on an analysis of the semigrand partition function in which the pressure P, temperature T, and number of solvent particles, N_1, are held fixed (10, 11). In this section, we extend his derivation to a four-component system containing solvent (component 1), two polymers (components 2 and 3) and protein (component 4). The objective of the calculation is to derive expressions for the chemical potentials of all components. Later, by equating the chemical potentials of each component in each phase, we will determine the composition of each phase and hence the protein partition coefficient which is defined to be the ratio of protein compositions in the top and bottom phases.

The theory begins with a derivation of the semigrand partition function, Γ, which is defined for a system at constant P and T that is open with respect to components 2, 3 and 4 but not with respect to 1. The semigrand partition function is given in terms of the canonical partition function, Q, by

$$\Gamma(T,P,N_1,\mu_2,\mu_3,\mu_4) = e^{-N_1\mu_1/kT}$$

$$= \sum_{N_2,N_3,N_4 \geq 0} e^{N_2\mu_2/kT} e^{N_3\mu_3/kT} e^{N_4\mu_4/kT} \Delta_{N_2 N_3 N_4} \quad (1)$$

where $\Delta_{N_2 N_3 N_4}$ is the isothermal-isobaric partition function,

$$\Delta_{N_2 N_3 N_4} = \sum_V e^{-PV/kT} Q(N_1,N_2,N_3,N_4,V,T) \quad (2)$$

Here N_i and μ_i are respectively the number of molecules and chemical potential of component i, and V is the volume per molecule. For convenience, we define new activities

$$a_2 = \frac{\Delta_{100}\lambda_2}{N_1 \Delta_{000}} \quad ; \quad a_3 = \frac{\Delta_{010}\lambda_3}{N_1 \Delta_{000}} \quad ; \quad a_4 = \frac{\Delta_{001}\lambda_4}{N_1 \Delta_{000}} \quad (3)$$

in terms of the absolute activities $\lambda_i \equiv e^{\mu_i/kT}$. The subscripts 000, 100 etc. on the Δ's indicate the number of molecules of species 2, 3 and 4 respectively, e.g. Δ_{010} is the isothermal isobaric partition function for a system containing one molecule of species 3. Expressing Γ in terms of the new activities we obtain

$$\frac{\Gamma}{\Delta_{000}} = 1 + \sum_{N_2,N_3,N_4 \geq 0}' \frac{a_2^{N_2} a_3^{N_3} a_4^{N_4}}{N_2! N_3! N_4!} X_{N_2 N_3 N_4} \quad (4)$$

where the prime on the sum indicates the restriction that $N_2 + N_3 + N_4 \neq 0$ and

$$X_{N_2 N_3 N_4} = \frac{N_2! N_3! N_4! \Delta_{000}^{N_2+N_3+N_4-1} N_1^{N_2+N_3+N_4} \Delta_{N_2 N_3 N_4}}{\Delta_{100}^{N_2} \Delta_{010}^{N_3} \Delta_{001}^{N_4}} \quad (5)$$

Expanding the natural logarithm of Γ/Δ_{000} (Equation 4) and using the relationship between μ_1 and Γ given in Equation 1 we obtain

$$\frac{\mu_1'}{kT}(P,T,a_2,a_3,a_4) \equiv \frac{\mu_1(P,T,N_2,N_3,N_4) - \mu_1(P,T,0,0,0)}{kT} \quad (6)$$

$$= -\sum_{N_1,N_2,N_3,\geq 0}' \theta_{ijk} a_2^{N_2} a_3^{N_3} a_4^{N_4}$$

where the prime on the sum indicates the restriction $N_1+N_2+N_3 \neq 0$, and $\mu_1(P, T, 0, 0, 0) = -kT \ln(\Delta_{000})/N_1$. Evaluating terms out to second order in the activity, we obtain

$$-N_1\theta_{100} = X_{100} = N_1 \Rightarrow \theta_{100} = 1 \quad (7)$$
$$-N_1\theta_{010} = X_{010} = N_1 \Rightarrow \theta_{010} = 1$$
$$-N_1\theta_{001} = X_{001} = N_1 \Rightarrow \theta_{001} = 1$$
$$-2!\,N_1\theta_{200} = X_{200} - X_{100}^2$$
$$-2!\,N_1\theta_{020} = X_{020} - X_{010}^2$$
$$-2!\,N_1\theta_{002} = X_{002} - X_{001}^2$$
$$-N_1\theta_{110} = X_{110} - X_{100}X_{010}$$
$$-N_1\theta_{101} = X_{101} - X_{100}X_{001}$$
$$-N_1\theta_{011} = X_{011} - X_{010}X_{001}$$

The molarity of components 2, 3 and 4, m_i', defined by Hill to be N_i/N_1, can be obtained by applying the Gibbs-Duhem equation

$$m_i' = a_i \left(\frac{\partial(-\mu_1'/kT)}{\partial a_i} \right)_{P,T,a_j,a_k} \quad j,k \neq i \quad (8)$$

to Equation 6, thereby obtaining an expansion of the molalities of components 2 through 4 in terms of the activities of components 2 through 4. For example

$$m_2'(P,T,a_2,a_3,a_4) = \sum_{i,j,k}' i\,\theta_{ijk}(P,T)\, a_2^i a_3^j a_4^k \quad (9)$$

with similar expressions for m_3' and m_4'. Using standard techniques, these series can be inverted to yield expansions for the activities a_2, a_3 and a_4. These may, in turn, be inserted into Equation 6 to yield an expansion for μ'_1, which to second order in the molality is

$$-\frac{\mu_1'}{kT} = m_2' + m_3' + m_4' - \theta_{110} m_2' m_3' - \theta_{101} m_2' m_4' - \theta_{011} m_3' m_4'$$

$$- \theta_{200} m_2'^2 - \theta_{020} m_3'^2 - \theta_{002} m_4'^2 , \qquad (10)$$

or they can be used directly to obtain expressions for μ_4

$$\mu_4\left(P,T,m_2',m_3',m_4'\right) = \mu_4^{\text{ref}} + RT\, (\ln m_4' - 2\theta_{002} m_4' - \theta_{101} m_2' - \theta_{011} m_3'$$

$$- 2\theta_{002}^2 m_4'^2 - \frac{1}{2}\theta_{101}^2 m_2'^2 - \frac{1}{2}\theta_{011}^2 m_3'^2$$

$$- 2\theta_{002}\theta_{101} m_2' m_4' + 2\theta_{002}\theta_{011} m_3' m_4'$$

$$+ \theta_{101}\theta_{011} m_2' m_3') \qquad (11)$$

where

$$\mu_4^{\text{ref}} = RT \ln\left(\frac{\Delta_{000}\, N_1}{\Delta_{001}}\right) \qquad (12)$$

Changing from the simple molality of Hill, m_i', to the more conventional definition of molality, $m_i \equiv N_i M_i / N_1 (1000)$, where M_i is the molecular weight of species i, we find

$$\mu_1(P,T,m_2,m_3,m_4) = \mu_1(P,T,0,0,0) - \frac{RT\, M_1}{1000}\, (m_2 + m_3 + m_4$$

$$+ \frac{c}{2} m_2^2 + \frac{d}{2} m_3^2 + \frac{g}{2} m_4^2 + a\, m_2 m_3 + e\, m_2 m_4 + f\, m_3 m_4) \qquad (13)$$

and

$$\mu_4(P,T,m_2,m_3,m_4) = \mu_4^{\text{ref}} + RT\, \left(\ln m_4 + g\, m_4 + e\, m_2 + f\, m_3 + O(m^2)\right) \qquad (14)$$

where

$$c/2 = -\theta_{200} M_1 / 1000\, N_a \qquad d/2 = -\theta_{020} M_1 / 1000\, N_a$$

$$g/2 = -\theta_{002} M_1 / 1000 \, N_a \qquad a = -\theta_{110} M_1 / 1000 \, N_a$$

$$e = -\theta_{101} M_1 / 1000 \, N_a \qquad f = -\theta_{011} M_1 / 1000 \, N_a \tag{15}$$

Equations 13 and 14 have the same functional form as that postulated by Edmonds and Ogston (3) and later generalized by King et al. (4). The significance of the work presented here is that it enables us to give a fundamental interpretation of the coefficients and the reference potential in terms of forces between the species. It also allows us to relate these coefficients to the virial coefficients which appear in the McMillan-Meyer virial expansion (12) of the osmotic pressure. In the equations of Ogston and of King et al. the coefficients are set equal to the virial coefficients of the McMillan-Meyer virial expansion, but, as we shall see, these coefficients are equivalent only when certain assumptions are made.

In order to relate the coefficients a - f to the forces between molecules it is necessary to evaluate the partition function $\Delta_{N_1 N_2 N_3}$ since the coefficients a - f are, via Equations 7 and 15, functions of the Δ's. Equation 5 may be rewritten in terms of the configurational partition function $Z_{N_1 N_2 N_3 N_4}$ as

$$\Delta_{N_2 N_3 N_4} = \sum_V e^{-PV/kT} \frac{Z_{N_1 N_2 N_3 N_4}}{N_1! N_2! N_3! N_4! \Lambda_1^{3N_1} \Lambda_2^{3N_2} \Lambda_3^{3N_3} \Lambda_4^{3N_4}} \tag{16}$$

where Λ_i is the thermal wavelength of species i. If it is assumed that the solvent molecules do not interact with each other, i.e.

$$Z_{N_1 N_2 N_3 N_4} = V^{N_1} Z_{N_2 N_3 N_4} \tag{17}$$

and that the solvent and solution are incompressible, i.e.

$$V = V_0 + N_2 v_2 + N_3 v_3 + N_4 v_4 \tag{18}$$

where V_0 is the total volume of the solvent and v_i is the molar volume of the solute i, then the summation in Equation 16 contains only one term and Equation 16 becomes,

$$\Delta_{N_2 N_3 N_4} = \frac{e^{-P(V_0 + N_2 v_2 + N_3 v_3 + N_4 v_4)/kT} Q_0 Z_{N_2 N_3 N_4}}{N_2! N_3! N_4! \Lambda_2^{3N_2} \Lambda_3^{3N_3} \Lambda_4^{3N_4}} \tag{19}$$

where

$$Q_0 = \frac{V_0^{N_1}}{\Lambda_1^{3N_1} N_1!} \tag{20}$$

Here we have approximated V by V_0 since the solution is assumed to be dilute. The Z's can be evaluated by relating them to the virial coefficients in the McMillan and Meyer theory expansion for the osmotic pressure, Π, in the density of each species, $\rho_i = N_i / V$. For the case of three solutes in a solvent, its expansion is

$$\frac{\Pi}{kT} = \sum_i \rho_i + B_{200}\, \rho_2^2 + B_{020}\, \rho_3^2 + B_{002}\, \rho_4^2 + 2B_{110}\, \rho_2\rho_3 + 2B_{101}\, \rho_2\rho_4 + 2B_{011}\, \rho_3\rho_4 + \ldots \tag{21}$$

where

$$B_{200} = -\frac{1}{2} \int_0^\infty \left[\exp\left(-w_{22}(r, \mu_1, T)/kT\right) - 1\right] 4\pi r^2 \, dr \tag{22}$$

$$B_{110} = -\frac{1}{2} \int_0^\infty \left[\exp\left(-w_{23}(r, \mu_1, T)/kT\right) - 1\right] 4\pi r^2 \, dr \tag{23}$$

with similar expressions for the other coefficients. The w_{ij} are the potentials of mean force (or effective potential energy) between isolated molecules of type i and j in a sea of solvent at chemical potential, μ_1. Using the well known relationships between the Z's and the virial coefficients (11-13),

$$\theta_{002} = \frac{v_4}{v_1} - \frac{B_{002}}{v_1} \tag{24}$$

with similar relationships for θ_{200} and θ_{020}, and

$$\theta_{110} = \frac{v_2}{v_1} + \frac{v_3}{v_1} - \frac{B_{110}}{v_1} \tag{25}$$

with similar relationships for θ_{011} and θ_{101}. The second virial coefficient B_{ij} may be related to the second virial coefficients A_{ij} which are given in terms of weight per volume units as

$$A_{ij} = \frac{N_a}{M_i M_j} B_{ij} \qquad (26)$$

where N_a is the Avagodro's number. The coefficients are then given by

$$g/2 = \frac{M_1}{1000 N_a}\left[\frac{M_4^2}{N_a}\frac{A_{44}}{v_1} - \frac{v_4}{v_1}\right] \qquad d/2 = \frac{M_1}{1000 N_a}\left[\frac{M_3^2}{N_a}\frac{A_{33}}{v_1} - \frac{v_3}{v_1}\right]$$

$$c/2 = \frac{M_1}{1000 N_a}\left[\frac{M_2^2}{N_a}\frac{A_{22}}{v_1} - \frac{v_2}{v_1}\right] \qquad a = \frac{M_1}{1000 N_a}\left[\frac{2 M_2 M_3 A_{23}}{N_a} - \frac{v_2}{v_1} - \frac{v_3}{v_1}\right]$$

$$e = \frac{M_1}{1000 N_a}\left[\frac{2 M_2 M_4 A_{24}}{N_a} - \frac{v_2}{v_1} - \frac{v_4}{v_1}\right] \qquad f = \frac{M_1}{1000 N_a}\left[\frac{2 M_3 M_4 A_{34}}{N_a} - \frac{v_3}{v_1} - \frac{v_4}{v_1}\right] \qquad (27)$$

Comparison of Equations 12, 13, and 27 with the Edmonds-Ogston equation and its extension by King et al. indicates that Edmonds-Ogston and King et al. equations are valid only when the solvent and solute are incompressible, the solvent is non-interacting and

$$\frac{A_{ii} M_i^2}{N_a} \gg v_i \qquad (28)$$

and

$$2 A_{ij}\frac{M_i M_j}{N_a} \gg v_i + v_j \qquad (29)$$

Clearly the Edmonds-Ogston and King et al. equations are not valid if the proteins are treated as hard spheres since in that case $A_{44} M_4^2 / N_a = 4 v_4$. If the proteins are treated as flexible coils (without excluded volume) then $A_{44} M_4^2 / N_a \gg 4 v_4$. The validity of Equation 29 depends on the relative size of the protein and polymer. Calculations not shown here indicate that for the cases considered here, Equation 29 is generally valid.

If we assume that Equations 28 and 29 are valid, that the solvent and solution are incompressible, and that the solvent is non-interacting, we can use Equation 14 to investigate the dependence of the partition coefficient on the nature of the interactions between the species. To obtain the protein partition coefficient, the protein chemical potentials given by Equation 14 are equated in both phases. Since the reference chemical potentials are the same in both phases, we obtain:

$$\ln K_4 = \ln \frac{m_i^T}{m_i^B} = g\left(m_4^B - m_4^T\right) + e\left(m_2^B - m_2^T\right) + f\left(m_3^B - m_3^T\right) \tag{30}$$

Converting to weight fraction, $w_i = \frac{m_i w_1 M_i}{1000}$ where w_1 is the weight fraction of solvent, and dropping the first term on the right hand side of Equation 30 (since it is negligible compared to the others), we finally obtain

$$\ln K_p \equiv \ln \frac{w_4^T}{w_4^B} = M_4\left[\frac{1}{M_4}\ln \frac{w_1^B}{w_1^T} - 2A_{24}\Delta PEG + 2A_{34}\Delta Dx\right] \tag{31}$$

where Δ_{PEG} and Δ_{DX}, the driving forces for separation, are given by

$$\Delta_{PEG} = \left(\frac{W_2}{W_1}\right)^T - \left(\frac{W_2}{W_1}\right)^B > 0$$

$$\Delta_{D_x} = \left(\frac{W_3}{W_1}\right)^B - \left(\frac{W_3}{W_1}\right)^T > 0 \tag{32}$$

Equation 31 will be used to investigate the dependence of the protein partition coefficient on polymer and protein molecular weights. Aside from the trivial dependence on M_4 in Equation 31, the dependence of K_p on molecular weight comes from its dependence on the second virial coefficients and on the two polymer driving forces, Δ_{PEG} and Δ_{DX}. We will find theoretical and experimental grounds to believe that A_{24} and A_{34} decrease as the PEG or Dx molecular weight increases. As indicated by Equation 31, this molecular weight dependence opposes the trends observed experimentally that were described in the introduction. On the other hand we have data from the literature and from our own experiments (8) which show that ΔDx increases with PEG and Dx molecular weights and that ΔPEG increases with PEG and Dx molecular weights. The increase in ΔPEG with PEG molecular weight and the increase in ΔDx with Dx molecular weight favor the trend, but the increase in ΔPEG with Dx molecular weight and the increase in ΔDx with PEG molecular weight oppose the trend. These dependences will be examined in the next section.

THE DEPENDENCE OF THE SECOND VIRIAL COEFFICIENTS ON MOLECULAR WEIGHT: THREE MODELS

In order to determine the dependence of the virial coefficients on polymer and protein molecular weight it is necessary to specify how the potential of mean force appearing in Equations 22 and 23 depends on protein and polymer molecular weights. Although the forces between molecules in aqueous two phase systems have many origins, in this paper we consider only excluded volume forces, that is, forces which arise because two molecules cannot occupy the same space at the same time. Electrostatic interactions are neglected since we focus on the isoelectric point. Attractions are also neglected. This approach has also been used by Edmonds and Ogston (3) to model phase formation and by Atha and Ingram (14) to model protein precipitation by PEG. While the first agrees qualitatively with experiment, the second fails to predict trends with molecular weight. Recently, however, Mahadevan and Hall (15) have been able to reproduce qualitatively the trends observed experimentally for protein precipitation with protein and PEG molecular weight by considering only excluded volume forces. This suggests that it is worthwhile to see if excluded volume forces alone are responsible for the trends observed with protein and polymer molecular weight. The excluded volume forces may be calculated on the basis of very simple models of molecular shape. By modeling proteins as rigid spheres and polymers as either rigid particles or flexible coils we hope to learn what role excluded volume effects have in partitioning.

While the modeling of polymers as rigid particles is somewhat questionable, the modeling of proteins which have a compact structure as rigid particles is common (9, 14, 16) especially for globular proteins.

Three models of excluded volume forces are considered: In the first model, called the sphere-sphere model, the proteins and polymers are modeled as rigid spheres of radii, R_4 and R_i respectively. In this case, A_{i4} is given by

$$A_{i4} = \frac{N_a}{2M_i M_4} \left[\frac{4}{3} \pi (R_4 + R_i)^3\right] \tag{33}$$

In the second model, the protein is modeled as a rigid sphere and the polymer is modeled as a long thin cylinder of length L. In this case A_{i4} is given by

$$A_{i4} = \frac{N_a}{2M_i M_4} \left[\pi R_4^2 L + \frac{4}{3} \pi R_4^3\right] \tag{34}$$

In the third model the polymer is modeled as a flexible coil while the protein is modeled as a rigid sphere. The second virial coefficients for such a sphere-coil model have been calculated analytically by Hermans (17) who assumed that the segments of the flexible particle do not interact with each other and are Gaussian distributed. The resulting cross second virial coefficient for both long and short polymer (as measured by the ratio of the polymer root mean squared end to end distance, H_0, and the protein radius, R_4) are

$$A_{i4} = \frac{N_a}{2M_4 M_i} \left[\frac{4}{3}\pi R_4^3 + \left(\frac{6}{\pi}\right)^{\frac{1}{2}} H_0 \frac{4}{3}\pi R_4^2 \right] \quad \text{for } \frac{R_4}{H_0} > 0.5$$

$$A_{i4} = \frac{N_a}{2M_4 M_i} \left\{ \frac{\frac{4}{3}\pi R_4 H_0^2}{\frac{24}{5}\left[1 + \frac{18}{7}\left(\frac{R_4}{H_0}\right)^2\right] + 9\left(\frac{6}{\pi}\right)^{\frac{1}{2}} \sum_{k=0}^{\infty} \frac{(-1)^k 6^k}{k!\, q(k)} \left(\frac{R_4}{H_0}\right)^{2k+1}} \right\} \quad \text{for } \frac{R_4}{H_0} < 0.5$$

(35)

where $q(k) = (2k-1)(2k+1)(2k+3)(2k+4)(2k+6)$.

The parameters R_4, R_i, L, and H_0 used in all three of the models above can be related to the molecular weight of the various species using expressions obtained from the literature. For dextran, which is a branched, flexible polymer the radius was taken to be

$$R_3 = 6.6 \times 10^{-9} M_3^{0.43} \qquad (36)$$

where M_3 is the number average molecular weight of dextran and R_3 is in units of centimeters (18). For PEG, which is a linear polymer with a helical configuration (19) the length of the fiber (in centimeters) was approximated (14) by

$$L = 9.37 \times 10^{-10} M_2 \qquad (37)$$

while the end-to-end distance (in centimeters) was approximated (16) by

$$H_0 = 6.527 \times 10^{-9} M_2^{0.526} \qquad (38)$$

For cases in which PEG is modeled as a sphere, we have taken R_2 to be the radius of an equivalent sphere, $R_2 = .38\, H_0$ as prescribed by Flory (20). The values for the protein radius, R_4, were taken from measurements of the hydrodynamic radii of the individual proteins. Thus for lysozyme, $R_4 = 2.06 \times 10^{-7}$cm. ($M_4 = 14,100$); for chymotrypsin, $R_4 = 2.25 \times 10^{-7}$cm. ($M_4 = 23,200$), for albumin; $R_4 = 3.61 \times 10^{-7}$cm. ($M_4 = 65,000$); and for catalase, $R_4 = 5.22 \times 10^{-7}$ cm.($M_4 = 250,000$) (16).

Substituting the values of R_2, R_3, R_4, L and H_0 into the expressions for the cross second virial coefficient allows us to determine how the crossed virial coefficient changes as a function of molecular weight for the three models.

The three models predict a nonsimple relationship between the second virial coefficient and the molecular weight of the various species which depends on the relative size of protein and polymer. The general trend is the decrease in the second crossed virial coefficient as the size of the particles increases. This is in agreement with scaling theories (21-23) which predict that the crossed second virial coefficient scales the same as the pure second virial coefficient which itself decreases with increasing molecular weight. It is in disagreement, however with the recent experimental data of King, et al. (4) who found that the crossed second virial coefficient decreases with increasing polymer molecular weight but increases with increasing protein molecular weight. The discrepancy has several possible explanations: the problem may be that our approximations for A_{i4} are too simple or that attraction cannot be neglected, or the problem may be that the King et al. measurements were conducted away from the isoelectric point possibly allowing electrostatic effects to enter the second virial coefficient measurement.

THE DEPENDENCE OF THE POLYMER DRIVING FORCE ON POLYMER MOLECULAR WEIGHT

The dependence of the separation driving forces, Δ_{PEG} and Δ_{DX}, on polymer molecular weight is well documented in the literature (1). For example, Figure 1 shows schematically how changes in the molecular weight of one of the polymers lead to shifts in the binodal. In the figure, the lower curve is for a higher molecular weight PEG than is the upper curve. The tie lines for the two binodal curves are roughly parallel. It can be seen from the figure that if the polymer concentration on a weight by weight basis is held fixed (say at Point A), then increasing the PEG molecular weight (at fixed Dx molecular weight) will result in an increase in Δ_{PEG} and Δ_{DX}. Similarly, reference to binodals available in the literature (1) shows that increasing the Dx molecular weight (at fixed PEG molecular weight) will also increase the values of Δ_{PEG} and Δ_{DX} but not as much as for the PEG molecular weight increase. Clearly the increases in Δ_{PEG} and Δ_{DX} will be greater the closer the tie line is to the critical point. Equation 31 shows that increases in the PEG or Dx molecular weights act to increase the strength of two competing terms in the protein partition coefficient.

DEPENDENCE OF THE PROTEIN PARTITION COEFFICIENT ON POLYMER AND PROTEIN MOLECULAR WEIGHTS

The partition coefficients predicted by the theory for the four globular protein, lysozyme, chymotrypsin, albumin and catalase were determined by inserting into Equation 31 the second virial coefficients calculated for each of the three models considered, and the experimental values for the driving forces, Δ_{PEG} and Δ_{DX}.

K_p vs PEG molecular weight. The best results are obtained for the sphere-cylinder model in which the protein and dextran are modeled as spheres and the PEG is modeled as a cylinder. The sphere-sphere model yields mixed results while the sphere-coil model yields trends opposite to those observed experimentally. We have considered the case in which both polymers are modeled as flexible coils and found that the experimental trends are not predicted.

Figure 2 shows the partition coefficient versus PEG molecular weight predicted by the sphere-cylinder model for the four proteins when the dextran molecular weight is fixed at 23,000 and the mixture composition is PEG: 6%; Dx: 8%. Thus for low values of the Dx molecular weight, the model predicts the observed experimental trend, i.e. that the partition coefficient decreases with increasing PEG molecular weight.

Figure 3 shows the partition coefficient versus PEG molecular weight predicted by the sphere-cylinder model for the four proteins when the dextran molecular weight is increased to 180,000. In this case the experimental trend is predicted only for PEG molecular weights below 10,000.

K_p versus Dx Molecular weight. Modeling Dx as a sphere and PEG as a sphere or a cylinder gives the right trend as a function of Dx molecular weight for Dx molecular weights greater than 150,000. This trend is that K_p increases as Dx molecular weight increases. At lower values of the Dx molecular weight these models predict the experimental trend, only for the smallest proteins, chymotrypsin and lysozyme. This is illustrated in Figure 4 which shows the partition coefficient versus Dx molecular weight predicted by the sphere cylinder model for the four proteins when the PEG molecular weight is fixed at 7500 and the mixture composition is PEG: 6%; Dx: 8%. Keeping Dx modeled as a sphere but considering PEG to be a flexible coil increases the protein molecular weight below which the experimental trend is predicted. We have also considered the case in which both polymers are taken to be flexible coils and found that the theory fails to explain the experimental trend.

K_p versus Protein Molecular Weight. Modeling Dx as a sphere and PEG as a cylinder gives the right trend with protein molecular weight namely that K_p decreases as protein molecular weight increases for all values of PEG and Dx molecular weights. See Figures 2, 3 and 4. The sphere-sphere and sphere-flexible coil models predict the experimental trend except for low values of the dextran molecular weight.

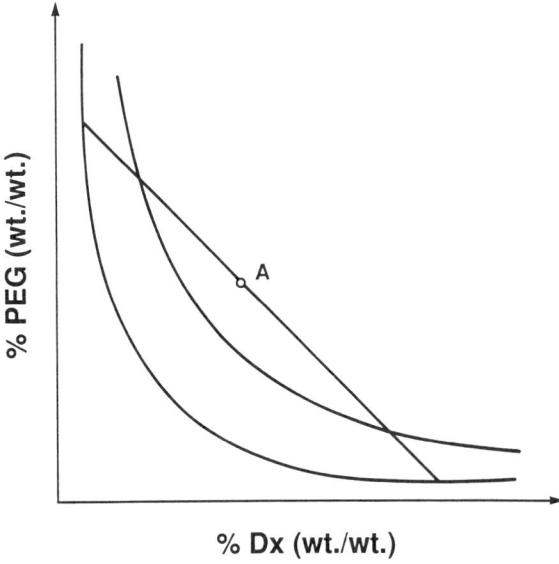

Figure 1. Schematic showing binodal curves for two systems at the same Dx molecular weight but different PEG molecular weights. The lower curve is for the higher PEG molecular weight. A tie line is shown through point A.

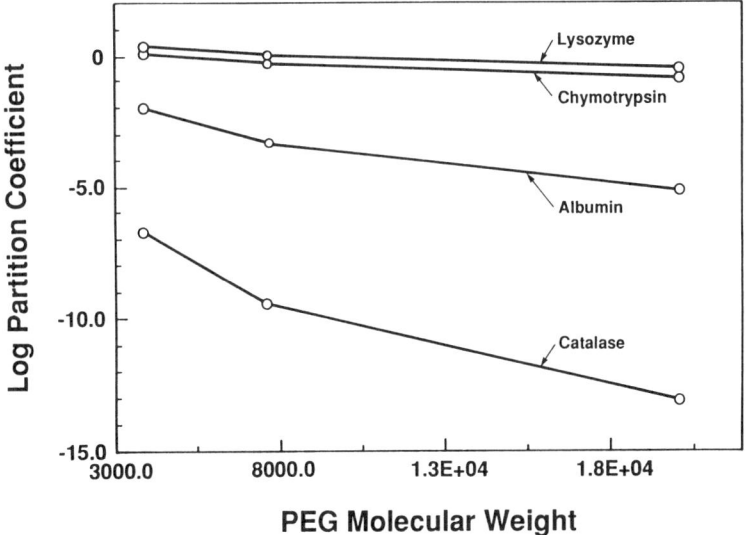

Figure 2. Predicted protein partition coefficient versus PEG molecular weight for lysozyme, chymotrypsin, albumin and catalase. Dextran molecular weight is 23,000 and polymer composition is PEG: 6%; Dx: 8%.

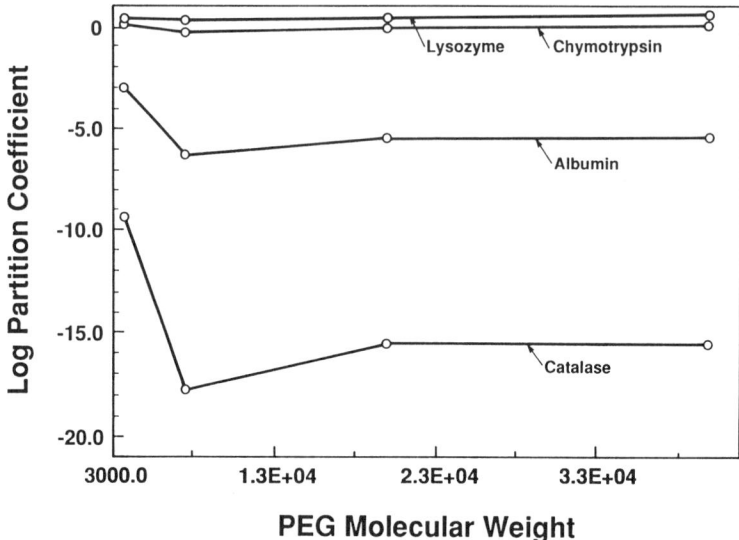

Figure 3. Predicted protein partition coefficient versus PEG molecular weight for lysozyme, chymotrypsin, albumin and catalase. Dextran molecular weight is 180,000 and polymer composition is PEG: 6%; Dx: 8%.

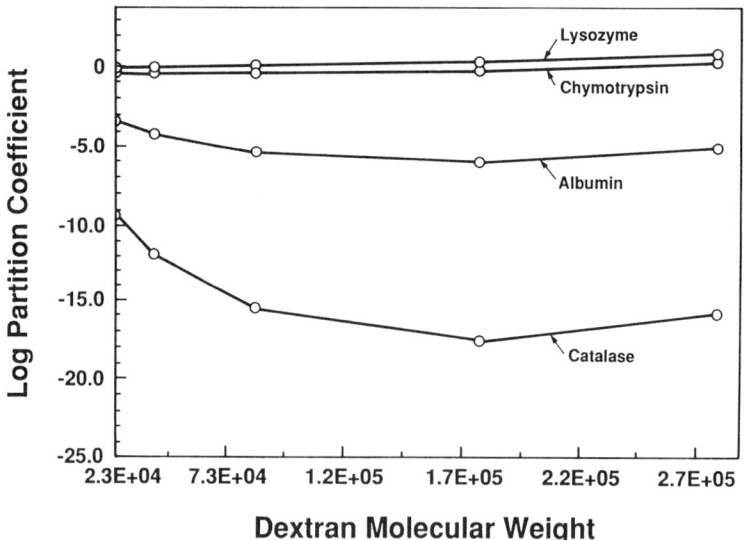

Figure 4. Predicted protein partition coefficient versus Dx molecular weight for lysozyme, chymotrypsin, albumin and catalase. PEG molecular weight is 7,500 and polymer composition is PEG: 6%; Dx: 8%.

CONCLUSION AND DISCUSSION

In the paper we have derived an expression for the protein partition coefficient which can be used to understand the molecular basis of partitioning. By playing with the equation we can learn what effect each type of intermolecular force (and interspecies force) can be expected to have on the partition coefficient. By working backwards from measured values of the partition coefficients we can learn something about the forces between proteins and polymers in solution. As a result of this work we have learned that excluded volume forces alone are not sufficient to predict the trends observed experimentally for protein coefficients as a function of molecular weight. This was a surprise to us since models of PEG-induced protein precipitation based on excluded volume forces only have been quite successful (15). It appears that attractions between species play a strong role in partitioning and should be included in any modeling effort. This is in agreement with the conclusions reached by Baskir, et al. (9) who found it necessary to include an attractive term in their lattice theory of aqueous two-phase extraction in order to obtain reasonable values for the free energies. The inclusion of an attractive term might also improve the agreement between theoretical and experimental predictions of the cross second virial coefficient.

One might also question whether the truncation of the expansion at the second viral coefficient level and the neglect of the protein-protein interaction term in Equation 30 is valid (24). While the inclusion of three-body terms (and of the protein-protein term) should improve the comparison with experiment quantitatively, we suspect that the trends predicted for the protein partition coefficient as a function of molecular weight would be the same.

We have also shown that the Edmonds-Ogston expression and its extension by King, et al. are valid only if one assumes that the fluids are incompressible, that the solvent is non-interacting, and that Equations 28 and 29 are valid. The incompressibility assumption seems reasonable, but the lack of interaction between solvent and solutes seems less reasonable. We are currently investigating the consequences to the theory if this assumption is dropped. The assumption of the validity of Equations 28 and 29 is not a problem since these equations are valid when both species are flexible coils; they are also valid in the case when one species is rigid and the other is a flexible coil.

One of the problems that we have encountered in this work is that experimental data in the literature on Δ_{PEG} and Δ_{DX}, while quite extensive, is not extensive enough (nor accurate enough) for us to thoroughly examine the trends predicted by these models under a variety of conditions. Our current and future work therefore includes some experimental measurements of binodals (and partition coefficients) and some modeling work to obtain equations for estimating Δ_{PEG} and Δ_{DX} (8).

ACKNOWLEDGMENTS

This work was supported by the National Institutes of Health (Grant # 1 R01 GM40023-01), the National Science Foundation (Grant # CBT-8720284) and the North Carolina Biotechnology Center.

LITERATURE CITED

1. Albertsson, P. A., Partition of Cell Particles and Macromolecules; J. Wiley & Sons, New York, 1986.
2. Walter, H.; Brooks, D. E.; Fisher, D., Partition in Aqueous Two-Phase Systems; Academic Press; Florida, 1985.
3. Edmonds, E.; Ogston, A. G., Biochem. J., 1968, 109, 569.
4. King, R. S.; Blanch, H. W.; Prausnitz, J. M., AIChE J., 1988, 34, 1585.
5. Johansson, G. In Partition in Aqueous Two Phase Systems, Walter, H.; Brooks, D. E.; Fisher, D., Eds., Academic Press; Florida, 1985; pp. 161-219.
6. Brooks, D. E.; Sharp, K. A.; Fisher, D., in Ref. 2.
7. Albertsson, P. A.; Cajarville, A.; Brooks, D. E.; Tjerneld, F., Biochem. Biophys. Acta., 1987, 926.
8. Forciniti, D.; Hall, C. K.; Kula, M. R., to be published.
9. Baskir, J. N.; Hatton, T. A.; Suter, U. W., Macromolecules, 1987, 20, 1300.5
10. Hill, T. L., J. Am. Chem. Soc., 1957, 79, 4885.
11. Hill, T. L., J. Chem. Phys., 1959, 30, 93.
12. McMillan, W. G.; Meyer, J. E., J. Chem. Phys., 1945, 13, 276.
13. McQuarrie, D. M., Statistical Mechanics, Harper and Row, New York, 1976.
14. Atha, D. H.; Ingram, K. C., J. Biolog. Chem., 1981, 256, 12108.
15. Mahadevan, H.; Hall, C. K., to be published.
16. Tanford, C., Physical Chemistry of Macromolecules, Wiley and Sons, New York, 1963.
17. Hermans, J., J. Chem. Phys., 1982, 77, 2183.
18. Senti, F. R.; Hellman, N. N.; Ludwing, N. H.; Babcock, G. E.; Tobin, R.; Glass, C. A.; Lamberts, B., J. Poly. Sci., 1955, 27, 527.
19. Koenig, J. L.; Angood, A. C.; J. Poly. Sci., A-2, 1970, 8, 1797.
20. Flory, P. J., Principles of Polymer Chemistry, Cornell University, Ithaca, New York, 1953.
21. Joanny, J. F.; Liebler, L.; Ball, R., J. Chem. Phys., 1984, 81, 4640.
22. Broseta, D.; Liebler, L.; Joanny, J. F., Macromolecules, 1987, 20, 1935.
23. Kosmas, M. K.; Freed, K. F., J. Chem. Phys., 1978, 69, 3647.
24. Haynes, C.; Prausnitz, J. M., (private communication), 1989.

RECEIVED September 28, 1989

Chapter 4

A Low-Cost Aqueous Two-Phase System for Affinity Extraction

David C. Szlag, Kenneth A. Giuliano, and Steven M. Snyder

Center for Chemical Technology, National Institute of Standards and Technology, Boulder, CO 80303

> Low-cost maltodextrins (Mavg = 1200, 1800, 3600) can be combined with polyethylene glycol (PEG) to form aqueous two-phase systems which are useful for protein separations. The physical characteristics of these maltodextrin/PEG systems are similar in many respects to dextran/PEG systems. Maltodextrins are currently available for a hundredth of the cost of fractionated dextran making the large scale application of polymer-polymer aqueous two-phase extractions more likely. The physical characteristics of the maltodextrin/PEG two-phase systems are described in this paper along with their application towards the purification of yeast alcohol dehydrogenase.

Aqueous two-phase extraction has been used to separate and purify a wide variety of biological materials, i.e. cells, organelles, enzymes, etc., on the laboratory scale and to a limited extent, on the commercial scale (1). Several factors have contributed to the underutilization of aqueous two-phase extraction systems (ATPS) in the commercial biotechnology community. Foremost among these is the limited theoretical understanding and hence predictability of phase equilibria and protein partitioning. Useful empirical rules of thumb are certainly available, but in general, predictive models for protein partition coefficients and phase equilibrium data are unavailable. Next in significance are the problems of selectivity and cost. This work focuses on these aspects of aqueous two-phase extraction.

Aqueous two-phase systems can be formed by combining either two "incompatible" polymers or a polymer and a salt in water above a certain critical concentration. Many systems have been tested by Albertsson and their phase diagrams determined (2). Comprehensive reviews have been compiled by Walter (1) and Kula (3). Most current commercial applications of ATPS are based on polymer-salt systems. These systems are attractive because of their low-cost and rapid phase disengagement. Polymer-salt

This chapter not subject to U.S. copyright
Published 1990 American Chemical Society

ATPS are not particularly selective however; they can damage fragile proteins or cells; and the high salt concentrations used, constitute a waste disposal problem. Polymer-polymer ATPS, on the other hand, can be made selective by incorporating the appropriate ions or including an affinity ligand in the system (4). The polymers are known to stabilize macromolecules in many cases, and the polysacharide bottom phase polymers, which often cannot be recycled directly, can be biodegraded. Unfortunately, the most common polymer-polymer ATPS based on dextran and polyethylene glycol (PEG) is too expensive to use for large scale separations. The literature contains several alternatives to dextran; the commercial hydroxypropyl (HP) starch derivatives (5) known as Aquaphase™ or Reppal™ and crude dextran (6). The HP starches mimic dextran and are certainly less expensive, but they are still relatively costly to use on a large scale. The same is true of crude dextran, which has the added disadvantage of forming solutions with high viscosities.

We investigated the possibility of using low-cost starch derivatives as replacements for dextran. Mattiasson was able to form an ATPS by combining PEG and a maltodextrin (degree of polymerization = 10, (DP10)) but abandoned this system for an ATPS based on HP starch, which he considered to be more stable (7). We have tested inexpensive, food-grade maltodextrins (MD) with a range of DP numbers (7-20) and found them to perform well in affinity ATPS extractions. This paper describes the physical properties of these low-cost systems and how they were applied in the purification of yeast alcohol dehydrogenase (YADH).

EXPERIMENTAL CHEMICALS AND PHASE SYSTEMS

Poly(ethyleneglycol), average molecular weight 8×10^3 purchased from Sigma Chemical Company (St. Louis, MO), was used as the top phase-forming polymer in all cases. Low molecular weight maltodextrins (MD), derived from acid hydrolyzed corn starch with theoretical average molecular weights of 1200 (DP 7, M150), 1800 (DP 10, M100) and 3600 (DP 20, M040)(estimated by HPLC analysis) (8) were used as the lower phase polymer and were obtained from Grain Processing Corporation (Muscatine, IA). Fractionated dextran, average molecular weight 4.9×10^5, was purchased from Sigma Chemical Co. and used as a bottom phase forming polymer with which to compare the maltodextrin. The phase systems were prepared from aqueous stock solutions of 33-40 mass percent (% m/m) PEG and 33-40% m/m of maltodextrins M150, M100, or M040 (Mavg = 1200, 1800, and 3600 respectively)[1]. The pH of the maltodextrin stock solutions was not controlled and varied between 4 and 5.

For enzyme partitioning experiments, phase systems were composed of 22.5% m/m M100 and 4.0% m/m PEG, or 6.6% m/m dextran (M = 4.9 x

---

[1] The mention of any trade name is not an endorsement by the National Institute of Standards and Technology.

10^5) and 3.8% PEG with 30 mM buffer unless otherwise specified. In some cases dithiothreitol (1 mM) was added to help maintain enzyme activity. Protein, as yeast enzyme concentrate (YEC), was added to a final concentration of 1 mg per gram of phase system or 5% m/m homogenized baker's yeast unless otherwise specified. The experiments were carried out in phase systems of similar chemical composition over a very broad pH range; a solution capable of buffering the pH over the range 4 to 8 was formulated from acetic acid, MOPS, and MES and used in each partitioning experiment at a final concentration of 10 mM for each species (30 mM total). The pH was adjusted appropriately with 1 M NaOH. Polyethylene glycol - triazine dye was prepared as described by Johansson (9) with the omission of the DEAE adsorption. Five to 10 additional chloroform/water extractions were made, which removed nearly all of the unreacted dye. All other chemicals used in these experiments were reagent grade.

Yeast Extract. Yeast enzyme concentrate (YEC) was purchased from Sigma Chemical Company (St. Louis, MO). Stock solutions were made at 10 mg lyophilized powder per ml. The specific activity of ADH in this mixture was found to be 17 U/mg protein using an assay described by Vallee and Hoch (10).

Yeast Homogenate. Yeast homogenate (YH) was prepared by combining 5 g dried yeast with 25 g dry ice and grinding for 5 minutes with a mortar and pestle. The resulting yeast paste was taken up in 50 ml of water and kept on ice for immediate use or frozen at $-20°C$ for future use. The YADH specific activity in YH varied between 5-10 U/mg protein.

Phase Diagrams. Phase diagrams for PEG-M150, M100, M040 and dextran were determined from 4 or 5, 10 g total weight, two-phase systems for each polymer combination. Systems were mixed for 30 seconds and equilibrated at $25°C$. Phases were completely separated in a temperature controlled centrifuge (25°C) at 4000 x g for 15 minutes. The PEG and maltodextrin concentrations in each phase were determined by a combination of refractometry and polarimetry. The specific rotations of the M040, M100, and M150 polymers were found to be 191, 157, and 163 (deg. g^{-1} dm^{-1}), respectively. The weight of the bottom phase was determined after carefully draining the sample tube after puncture. The upper phase weight was calculated by difference. Knowledge of the phase weights confirmed the phase diagram compositions.

Densities of the samples were measured at 25°C using a frequency type densimeter. Viscosities of the phases were determined using a cone-and-plate viscometer at 25°C.

Extraction of Alcohol Dehydrogenase from a Crude Yeast Lysate. The polymer stock solutions, buffer, water and yeast enzyme concentrate were weighed, combined, and mixed thoroughly by gently vortexing for 10 seconds. The YADH activity in the entire system was then determined. The transient emulsion which formed was centrifuged as described above and both the YADH activity and total protein content were determined in each phase. The partition coefficient, K, is defined as the ratio of enzyme activity or concentration in the top to that in the bottom phase. For systems containing particulates, a partition fraction, K_f, is defined as the ratio of activity or concentration in the top phase to the total activity or concentration in the entire phase system. Activity balances are based on the total activity of YADH added to the two-phase system as YEC or YH.

Protein Assay. Protein was determined by the Coomasie Blue binding assay with alcohol dehydrogenase as the standard (Pierce Chemical Company, Rockford, IL). For systems containing PEG-dye ligands, an identical system containing no protein was used as a control.

RESULTS AND DISCUSSION

Phase Diagrams and Physical Properties. Three low-molecular-weight starch derivatives were tested to see if they would form aqueous two-phase systems with PEG 8000 at 25°C. The binodal curves resulting from the analysis of the three MD/PEG systems are presented in Figures 1a-1c. For comparison, the dextran (Mavg = 4.9×10^5)/PEG binodal is shown in Figure 1d.

The concentration of maltodextrin required to form two phases with PEG is much greater than that of dextran and increases as the molecular weight of the maltodextrin decreases. Thus the binodal curves are shifted toward the right as the bottom phase polymer molecular weight decreases, although the general shape of the binodal remains the same. This behavior is consistent with the observations made by Albertsson (2). Polyethylene glycol is almost totally excluded from the bottom phase of a MD/PEG system, while, the concentration of maltodextrin in the top phase is relatively high when compared to systems formed with dextran.

The viscosities of both top and bottom phases formed in the MD systems were measured and found to be independent of shear rate (data not shown). Density and viscosity data are provided for the MD/PEG and dextran /PEG systems in Table 1.

As expected, the viscosity increased at higher polymer concentrations and was greatest in the bottom phase. The bottom phase viscosities of maltodextrin systems were less than half those formed with dextran, however, and they were comparable to the PEG 8000-hydroxypropyl starch systems (5). The upper phase viscosities are increased slightly from 2 to 5 cps as compared to dextran or hydroxypropyl starch systems. These viscosities are still well below those of the bottom phase and did not appear to hinder the rapid mixing or mass transfer characteristic of these types of systems. This slight increase in upper phase

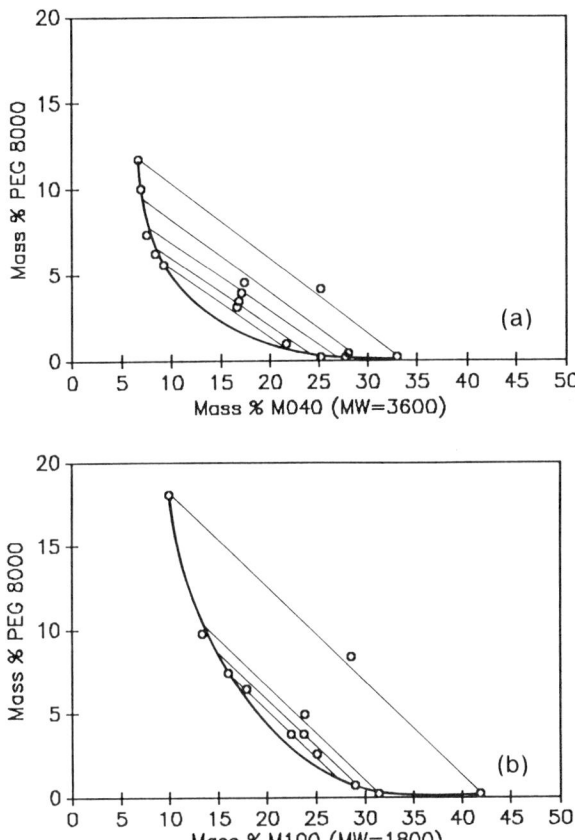

Figure 1. Binodal curves determined at 25 °C for PEG 8000 with 2 different lower phase polymers: (a) M040 (DP 20, Mavg = 3600) and (b) M100 (DP 10, Mavg = 1800). *Continued on next page.*

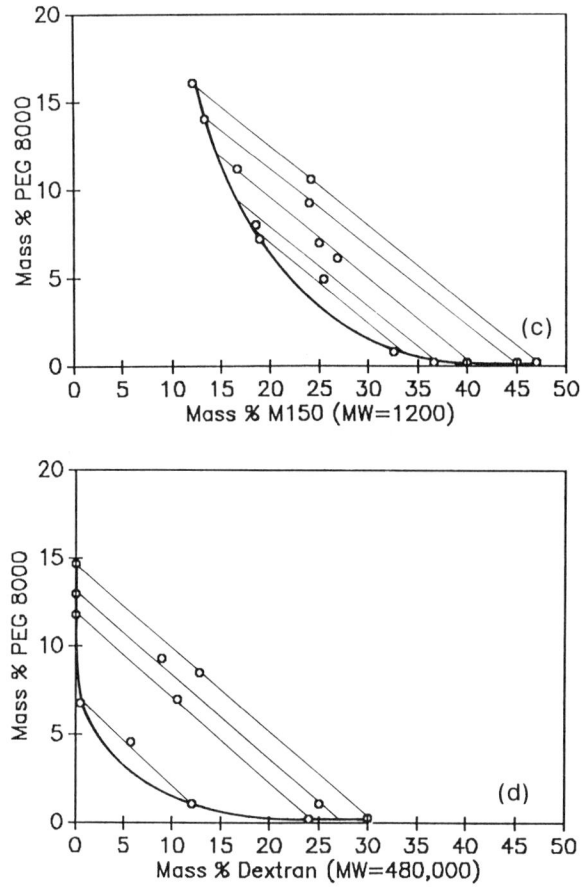

Figure 1. *Continued.* Binodal curves determined at 25 °C for PEG 8000 with different lower phase polymers: (c) M150 (DP 7, Mavg = 1200) and (d) Dextran.

viscosity results from the high concentrations of maltodextrin present in the upper phase (see phase diagrams).

The density difference between the phases increased as the total system composition moved away from the critical point. The density difference between phases is similar to that found in the dextran/PEG system, although the actual phase densities in the maltodextrin system are much higher. A point worth commenting on here is the speed of phase separation. Initially we thought that MD/PEG systems would separate faster than the dextran/PEG ATPS because of the their lower viscosity. In fact the opposite was observed; for the compositions given in Table 1 the MD/PEG systems separated slightly slower than their dextran counterparts. This might be attributable to the very low interfacial tension in MD/PEG ATPS. Accurate measurements were not made which would confirm this however.

A concern that arises when using starches in ATPS is the tendency for gel formation, particularly at low temperatures. Gelation was observed in the bottom phase of the maltodextrin (DP 20)/PEG ATPS at all maltodextrin concentrations in the two phase region. Although this feature could be useful in applications requiring a bottom-phase gel, in all subsequent experiments we used the DP 10 maltodextrin for which no gelation was observed even after several days. At 4°C, systems furthest removed from the critical point became turbid after prolonged standing (24-48 hrs). Since our extractions were conducted in under 2 hrs this was not seen as a significant liability. No tendency for gel formation or decreased solubility has been noted in the DP 7 containing system. For our work we have found the DP 10/PEG systems to offer the reasonable combination of stability and moderate polymer concentrations. Concentrated stock solutions of M100 (DP 10) (30-35% m/m), were prepared by autoclaving in glass distilled and deionized water at 121°C for 15 minutes. These were stored at room temperature and remained clear for up to ten days.

The suitability of the maltodextrin based phase systems could be questioned given the ubiquity of starch degrading enzymes in crude microbial extracts. In the test in which crude yeast enzyme extracts are used, there is little problem since yeast do not possess the necessary enzymes for breaking down starch. Partitioning experiments conducted on the crude extract of <u>Thermoanaerobium brockii</u> showed that no significant change in phase behavior occurs over the course of 2-6 hrs.

Affinity Partitioning.[2] We have found that most proteins have low partition coefficients (K<0.4) in the DP 10/PEG phase system (22.5% m/m) DP 10, 4% PEG 8000, pH = 5.0. In order to raise the partition coefficient of the target enzyme (yeast alcohol dehydrogenase) so that reasonable yields are

---

[2] Cibacron Blue FGF triazine dye was a gift from Ciga-Geigy Corporation (Greensboro, NC). Cibacron blue F3GA was purchased from Sigma Chemical Co., St. Louis, MO. Turquoise H-A, Red HE-3B, Green HE-4BDA were gifts of ICI Americas, Inc.

Table 1. Physical properties of MD/PEG phase systems

System	PEG (Mass %)	Maltodextrin 3600 (Mass %)	Viscosity (cps)	Density (g/cm^3)	Phase Ratio (Mass)
1 Total	3.13	16.61	---	---	1.12
1 Top	5.60	9.20	4.87	1.0450	---
1 Bottom	0.96	21.68	21.3	1.0944	---
2 Total	3.47	16.83	---	---	1.13
2 Top	6.24	8.37	4.77	1.0425	---
2 Bottom	0.16	23.53	26.3	1.1010	---
3 Total	3.95	17.12	---	---	1.18
3 Top	7.06	7.31	5.06	1.0402	---
3 Bottom	~0	25.21	35.3	1.1106	---
4 Total	4.53	17.43	---	---	1.26
4 Top	5.94	6.49	5.56	1.0347	---
4 Bottom	~0	27.74	47.8	1.1200	---
5 Total	4.19	25.19	---	---	0.59
5 Top	11.71	6.72	7.79	1.0453	---
5 Bottom	~0	36.04	"gel"	"gel"	---

System	PEG (Mass %)	Maltodextrin 1200 (Mass %)	Viscosity (cps)	Density (g/cm^3)	Phase Ratio (Mass)
1 Total			---	---	2.93
1 Top	7.23	18.92	8.27	1.0821	---
1 Bottom	0.80	32.55	17.6	1.1263	---
2 Total	10.62	24.17	---	---	2.03
2 Top	16.11	12.07	17.9	1.0743	---
2 Bottom	~0	47.02	39.8	1.1493	---
3 Total	9.26	23.98	---	---	2.11
3 Top	14.04	13.27	14.7	1.0743	---
3 Bottom	~0	44.58	55.6	1.1744	---
4 Total	6.16	26.88	---	---	NA
4 Top	11.21	16.67	12.2	1.0850	---
4 Bottom	~0	41.59	37.3	1.1626	---
5 Total	4.97	25.44	---	---	NA
5 Top	8.00	18.58	9.50	1.0861	---
5 Bottom	~0	36.65	23.7	1.1434	---

Table 1. Continued

System	PEG 8000 (Mass %)	Malto-dextrin 1800 (Mass %)	Viscosity (cps)	Density (g/cm^3)	Phase Ratio (Mass)
1 Total	3.75	22.46	---	---	1.44
1 Top	6.4	17.8	8.06	1.0829	---
1 Bottom	2.9	25.1	15.9	1.1146	---
2 Total	4.93	23.95	---	---	1.24
2 Top	9.7	13.3	8.83	1.7051	---
2 Bottom	~0	31.4	28.5	1.1411	---
3 Total	3.75	23.79	---	---	1.02
3 Top	7.4	16.0	7.66	1.0783	---
3 Bottom	~0	28.57	17.3	1.1208	---
4 Total	8.40	28.57	---	---	1.02
4 Top	18.0	10.0	20.6	1.0757	---
4 Bottom	~0	41.8	66.0	1.1747	---

System	PEG 8000 (Mass %)	Dextran 490,000 (Mass %)	Viscosity (cps)	Density (g/cm^3)	Phase Ratio (Mass)
1 Total	6.99	10.57	---	---	1.44
1 Top	11.8	0.26	5.98	1.0171	---
1 Bottom	0.2	24	295.0	1.0930	---
2 Total	8.50	12.78	---	---	2.11
2 Top	14.7	0.26	8.79	1.0226	---
2 Bottom	0.3	30	565	1.1140	---
3 Total	4.57	5.77	---	---	1.51
3 Top	12.78	0.49	3.21	1.0101	---
3 Bottom	1.1	12	51.5	1.0526	---
4 Total	8.31	8.89	---	---	2.1
4 Top	12.97	0.035	7.27	1.0198	---
4 Bottom	25	1.09	487	1.1033	---

achievable, it is necessary to alter the chemistry of the phase system. A simple way of accomplishing this is to immobilize a ligand which binds the target enzyme onto a carrier, in this case the top-phase-forming polymer, PEG. When a ligand (a triazine dye) is attached to the PEG, YADH selectively partitions into the top phase, leaving most of the undesired proteins in the bottom phase. Inexpensive triazine dyes were chosen as affinity ligands, as they are known to specifically bind dehydrogenases and kinases. Derivatization of the PEG is relatively simple and is illustrated in Figure 2.

In order to maximize our recovery and purification, several PEG-triazine dye derivatives were tested for their binding of yeast alcohol dehydrogenase in the maltodextrin based ATPS. The screening experiment, illustrated in Figure 3, showed that under our conditions the best extraction would be made with PEG-FGF as the affinity ligand. All subsequent affinity extractions use this ligand.

In the absence of protein, the partition of dye ligand in the MD/PEG two-phase system showed minimal pH dependence over the range 4 to 7. The partition coefficient of the PEG-FGF in the system 22.6% w/w M100, 4.0% PEG 8000, and 30 mM buffer is 9.1 ± 1.5. For comparison, the partition coefficient of the PEG-FGF in a system containing 6.6% dextran 500, 4.0% PEG 8000 and 30 mM buffer is very similar, 9.4 ± 0.2.

Previous studies demonstrated that the partition coefficient of a dye-ligand in a dextran/PEG system was a function of total polymer concentration or tie line length. The partition coefficient of PEG-FGF behaved similarly in a DP 10/PEG system, increasing from 9 to 40 as the polymer concentrations were increased from 20% DP 10 and 4% PEG, to 28% DP 10, 8.5% PEG.

To demonstrate the utility of phase systems formed with maltodextrin and to determine the pH optimum for the extraction of YADH with PEG-FGF, we performed an initial series of extraction experiments from YEC preparations. Yeast alcohol dehydrogenase partitioned predominantly into the bottom phase in the absence of an affinity ligand (Figure 4a). When the affinity ligand, PEG-FGF, was incorporated into the phase system at a concentration of 1% of the total PEG concentration, the partition coefficient of YADH became strongly dependent on the system pH (Figure 4a). The partition coefficient of total protein in the system also increased with decreasing pH when the dye ligand was present, but not as rapidly as the YADH (Figure 4b).

The optimum pH for the purification of YADH, with 1% of the PEG substituted with PEG-FGF, was found to be approximately 5.0. Of course this does not take into account any synergistic effects that might exist between the system parameters. The partition coefficient could be increased by lowering the pH further. Little could actually be gained in reality by doing this however, as the overall purification factor decreased due to a rapid increase in the total protein partition coefficient and the fact that the enzyme became increasingly labile below pH 5. Above pH 5, the partition coefficient decreased rapidly even with the ligand present,

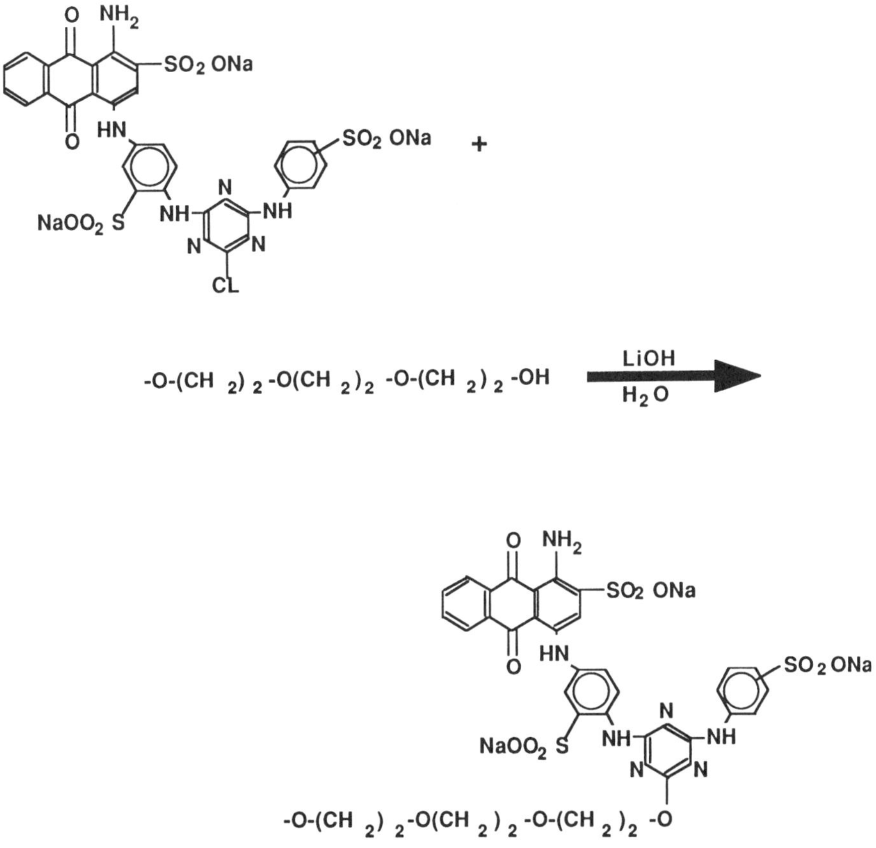

Figure 2. Preparation of PEG-dye ligands. Monochlorotriazine dyes are reacted with PEG under basic conditions. Substitution is on the order of 10%.

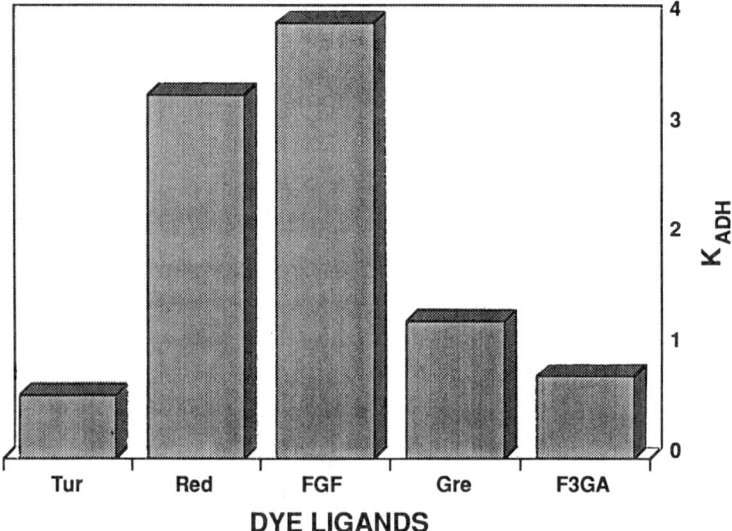

Figure 3. Results of triazine dye ligand screening in MD/PEG ATPS are shown. YADH partition coefficients were measured for five PEG-dye ligand containing systems using 1% m/m of PEG-dye ligand in the system at pH 5. In our studies, PEG-FGF was the most efficient ligand.

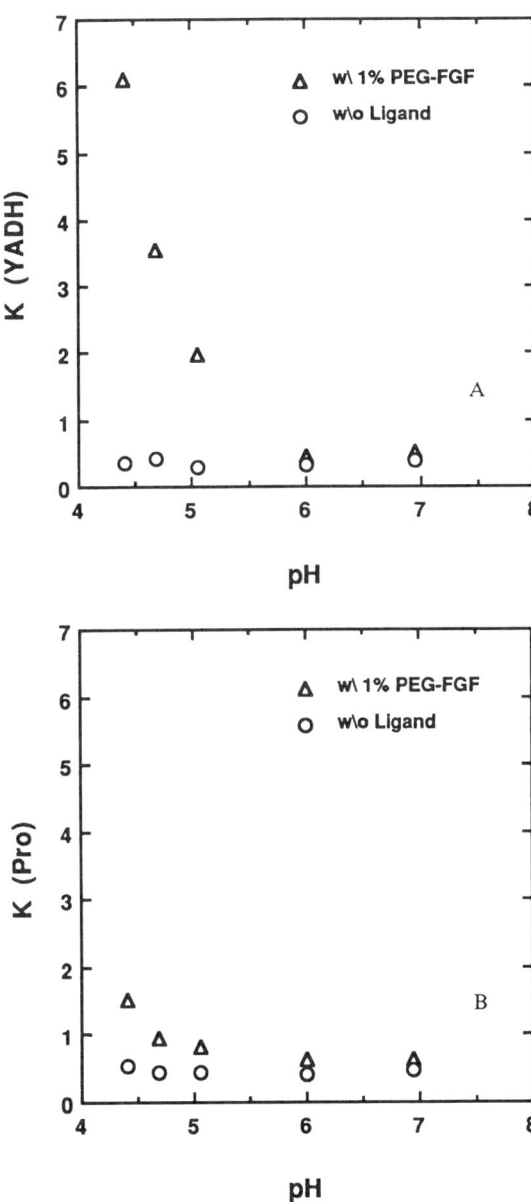

Figure 4. Extraction of YADH (A) and total protein (B) from Yeast Enzyme Concentrate with PEG-FGF (1%) and without at 4°C.

resulting in low recoveries in the top phase unless a very large volume ratio was used (Figure 5).

Comparison of Maltodextrin and Dextran ATPS. Direct comparisons between DP 10/PEG and dextran/PEG systems for enzyme purification were made by extracting YADH from both YEC and a crude slurry of disrupted yeast cells (YH). These data are shown in Table 2. The K_f and K_{ADH} were virtually the same for either dextran/PEG or MD/PEG and YEC. The partition coefficient for total protein, K_{Pro}, in the dextran/PEG system was 2-3 times higher, while the purification factors were 50% greater in the DP 10/PEG system. The partition coefficients for both total protein and YADH drop dramatically, however, while the purification factors in both cases are increased relative to the same parameters for YEC.

The MD/PEG system described above showed partition behavior for proteins and affinity ligands analogous to dextran/PEG systems. Protein partitioning for both two-phase systems with PEG-FGF was greatly influenced by ligand concentration, system pH and the presence of cell debris. In the absence of an affinity ligand, the partition coefficient for YADH was slightly greater in the system containing dextran as compared to that in the system containing maltodextrin. This is consistent with the observation that higher molecular weights of bottom phase forming polymer increase the partition coefficient (2). When the affinity ligand PEG-FGF was incorporated into each system, the partition behavior of the YADH and the total protein was once again very similar in both systems. A low pH, although normally avoided, raised the partition coefficient of YADH sufficiently so that high yields in the top phase were possible.

The purification factor for YADH from the crude yeast cell homogenate, 8.8 times, is over 3 times as great as the reported purification with a salt/PEG system. The amount of activity recovered in the top phase varied between 45 and 55 percent and was dependent on the manner in which the cells were disrupted.

ECONOMIC CONSIDERATIONS

The main drawback to the widespread use of polymer-polymer aqueous two-phase extraction has been the high cost of fractionated dextran. Crude dextran has been used with some success for the purification of enzymes but is much too viscous for many applications. Conversely, polymer-salt systems have relatively low viscosities, separate rapidly, and are inexpensive. Unfortunately, they lack selectivity and cannot be used for affinity partitioning in most cases since the high salt concentrations interfere with the protein-ligand interaction. The starch derivatives are reasonable alternatives for bottom phase polymers but have been hampered by low solubilities and the tendency for gel formation. Tjerneld has reported that chemically modified starches i.e. hydroxypropyl starch

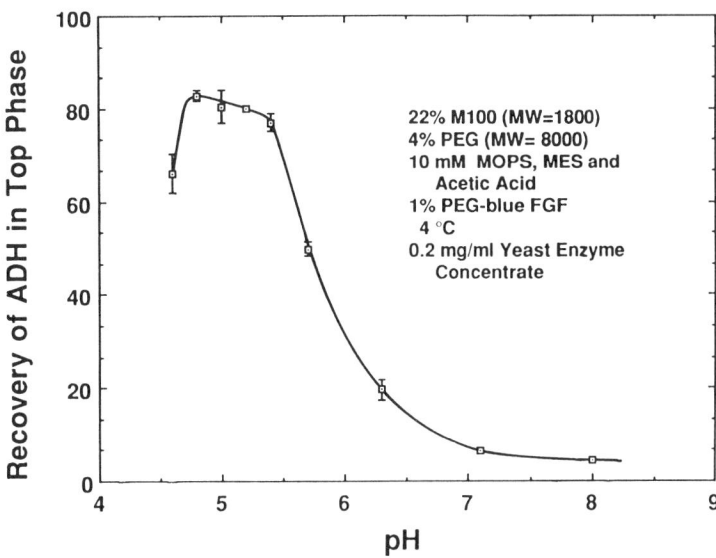

Figure 5. Recovery of YADH activity as a function of pH. Optimum pH for recovery is in the range 4.8-5.2. Phase volume ratio was 1.38. Because of the rapid denaturation of YADH at pH values below 5, all subsequent affinity extractions were carried out at a pH between 5.0 and 5.1.

Table 2. Extraction of YADH from yeast enzyme concentrate and homogenized yeast cells with 2% PEG-FGF, pH 5, 5% of system volume replaced with YH or 1 mg/ml YEC

	PEG - M100 System				
	K_f	K_{ADH}	K_{PRO}	Phase Vol. Ratio	Purification Factor
YEC	0.92	11.4	1.2	2.5	2.9
YH	0.47	0.38	0.03	2.5	8.3
	PEG - dextran System				
YEC	0.92	12.7	2.5	2.6	1.9
YH	0.45	0.36	0.08	2.5	4.2

can be used successfully in two-phase aqueous systems and that the cost of $3-3.5/L of phase system is commercially viable (6).

The maltodextrins are more than an order of magnitude less expensive at $0.59/kg than hydroxypropyl starch ($21/kg) and more than two orders of magnitude less expensive than fractionated dextran ($500/kg). Some of the lower cost is offset by the fact that larger concentrations of maltodextrin are required to form two-phase systems. We have calculated the cost for one kilogram of each of the following phase systems which have similar physical properties: 4% PEG 8000 - 7% Dextran 500, 4% PEG 8000 - 22% M100 (DP 10) and PEG 8000 - 14% hydroxypropyl starch. The respective costs are: $7/kg, $0.2/kg, and $3/kg. It is quite obvious that the maltodextrin systems offer a significant cost advantage.

CONCLUSIONS

The MD/PEG system offers the combined advantages of low-cost, reduced lower phase viscosities and high density differences for inexpensive polymer-polymer affinity partitioning. When coupled with low-cost affinity ligands i.e. triazine dyes, two-phase aqueous affinity partitioning could be used as the first step in a separation train for the recovery of industrially important enzymes. The bottom phase, which is generally considered to be a waste stream and non-recyclable if dextran or salt is used could be used as a substrate for additional fermentations if maltodextrin is used, thereby aiding the overall economics of the process.

Abbreviations Used: YADH, yeast alcohol dehydrogenase; MD, maltodextrins; MES, (2-[N-morpholino]ethanesulfonic acid); MOPS, (3-[N-morpholino]propanesulfonic acid); PEG, polyethylene glycol MW = 8000; YEC, yeast enzyme concentrate; YH, homogenized yeast cells; % m/m, mass percent; DP, degree of polymerization.

LITERATURE CITED

1. Walter, H.; Brooks, D.E.; Fisher, D.; "Partitioning in Aqueous Two-Phase Systems." Academic Press, Inc., 1985.
2. Albertsson, P.-A.; "Partition of Cell Particles and Macromolecules 2nd ed." Wiley, New York. 1986.
3. Kula, M.-R.; Applied Biochemistry and Bioengineering, Vol. 2., Academic Press Inc.; 1979.
4. Johansson, G.; Molecular and Cellular Biochemistry, 4, 169; 1974.
5. Kroner, K. H. ; Hustedt, H.; and Kula, M.-R.; Biotechnology and Bioengineering, 24, 1015; 1982.
6. Tjerneld, F.; Berner, S.; Cajarville, A.; and Johansson, G.; Enzyme Microb. Technol., 8, 417 ; 1986.
7. Mattiasson, B.; and Ling, T.G.I.; J. Chromat. 376, 235; 1986.
8. Grain Processing Corporation. (1986) Bulletin 11005, Muscatine, IA.
9. Johansson, G.; Methods in Enzymology, 10, 356; 1984.
10. Vallee, B.L.; and Hoch, F.L.; Proc. Natl. Acad. Sci. USA, 41, 327; 1955.
11. Kroner, K. H.; Hustedt, H.; and Kula, M.-R.; Process Biochemistry, 10, 170; 1984.

RECEIVED October 13, 1989

Chapter 5

Separation of Proteins by Using Reversed Micelles

Claude Jolivalt, Michel Minier, and Henri Renon

Centre Réacteurs et Processus, Ecole Nationale Supérieure des Mines de Paris, 60 Boulevard Saint-Michel, 75006 Paris, France

Since it appears possible to use active biomolecules outside a natural environment, and even more so since genetic engineering enables the production by recombinant microorganisms or animal cell cultures of more and more various proteins, a great deal of investigation has been carried out in order to separate and purify these compounds from complex medium broths. Due to the continually renewed diversity and complexity of proteins and media, a simple scheme of extraction and purification process does not exist, each example being a specific case for which, usually, a sequence of several techniques must be developed, depending on the properties of the protein, the nature of the impurities and the final purity demanded.

Extraction of proteins using reversed micelles emerges as a new and attractive technique (1-4) in terms of selectivity and concentration. It presents a close similarity with liquid-liquid extraction since both are diphasic processes which consist of partitioning a targeted solute between an aqueous feed phase and an organic phase and then operate the back transfer to a second aqueous stripping phase. The yield and selectivity of the separation is determined by thermodynamic properties at equilibrium. However, classical organic solvents are not suitable to achieve the extraction of most proteins which are hydrophilic molecules and, therefore, insoluble in apolar solvents. Moreover, due to their low polarity, organic solvents modify the interactions which stabilize the native protein conformation, leading to its denaturation. In order to extract and preserve the protein, it is necessary to maintain its aqueous environment by adding a surfactant which aggregates in apolar solvent and forms reversed micelles. The organic phase can thus be described as a thermodynamically stable dispersion of spherical water droplets surrounded by amphiphile molecules which prevent the proteins solubilized in the water core from direct interactions with the apolar solvent (figure 1).

The first interest for reversed micelles in the biological area has been to point out that enzymes actually retained their catalytic activity (5-7) when solubilized in organic solvents via reversed micelles. Catalysis by enzymes in reversed micelles was exploited for conversion purposes, taking advantage of the micro-compartmentation of the organic phase, which permitted contacting hydrophilic and hydrophobic reactants and removing hydrophobic products. Yields and kinetics of enzymatic reactions could thus be improved or changed with regard to those obtained in bulk aqueous phases. Related publications were reviewed in detail recently by Shield et al. (8).

It was noted that activity and conformation change with the amount of water inside micelles, pointing out the importance of the aqueous environment, easily adjustable as a function of the water content, which is impossible when studies are performed in aqueous solutions. Furthermore, because many properties (9) of the water core resemble those of water present at interfaces in biological systems, reversed micelles provide an excellent system for studying the interactions between polypeptides and interfacial water (10-11) or more generally their conformation when solubilized in micelles (12-15), depending on where the biopolymer is located inside the micelle and what its conformation is.

Three techniques (16) have been applied to prepare enzyme-containing micellar solutions. The first consists of stirring solid protein with a hydrocarbon solution of surfactant, the second by injecting a few microliters of a concentrated enzyme solution into it. The solution is then stirred until it becomes transparent, allowing spectroscopic studies. Such methods are suitable for incorporating a given amount of protein into a reversed micellar system and for controlling the water content but, because they involve only monophasic systems, they are of no interest for purification purposes. The third alternative is to transfer the protein from an aqueous phase into the reversed micellar phase. This method has been more or less successfully applied in the above mentioned studies, focused mainly on properties of proteins after solubilization but has established the feasibility of the extraction. Many detailed reviews have been published on these subjects (17-23) which are a good introduction to our present topic, extraction of biopolymers using reversed micelles.

Many surfactants are known to form reversed micelles in apolar media and have already provided a suitable environment for elucidating catalytic activity or conformation properties of some proteins in non aqueous media. But to conduct an extraction two conditions must be fulfilled: reversed micelles must exist in the organic phase in equilibrium with an excess aqueous phase and the performance of the extraction must be significant. The nature of both surfactant and solvent, the composition of the aqueous phase as well as the volume ratio between phases determine both the ability of the system to form such diphasic systems and the extraction and separation of the desired protein from other protein impurities.

Manipulating different parameters in order to achieve an efficient extraction has emphasized that the criteria of yield and selectivity are mostly controlled by two different phenomena: electrostatic interactions due to the charges born both by the protein and the surfactant and the "steric effect" related to the relative sizes of the protein and micelle.

PARAMETERS AFFECTING THE SOLUBILIZATION

In this section, we shall focus on experiments dealing with the extraction of peptides, amino acids or proteins from an aqueous phase into an organic phase, following the typical procedure of liquid-liquid extraction processes.

All the extraction experiments presented below were performed with synthetic aqueous solutions, i.e. purified proteins solubilized in a suitable buffer. All parameters and components are known, which is not the case with fermentation broths or culture media which will be reviewed in a special paragraph.

The general extraction procedure can be described as follows: the aqueous phase containing the protein to be extracted and the organic phase prepared separately are contacted and stirred. After the chemical equilibrium has been reached, the system is centrifuged or allowed to settle until phase separation occurs. Separation and eventually titration of both phases are performed. Back extraction follows a similar process. The laden organic phase is mixed with a suitable aqueous phase where the protein is recovered after settling.

Among all surfactants developed for experiments on solubilization of biomolecules into reversed micelles, only two have, up to now, received the greatest attention to achieve studies on extraction.

Sodium bis (2ethylhexyl) sulfosuccinate (the commercial name of which is AOT) is an anionic surfactant. Its ability to form water in oil microemulsions in ternary oil/surfactant/water systems without adding a cosurfactant reduces the set of parameters, making the system easier to control. This property can be explained by the geometric modelling of the surfactant packing at the interface, developed by Mitchell and Ninham (24). The key element of their theory is the packing ratio defined as the ratio of the cross-sectional area of the hydrocarbon chain v/l_c to that a_o of the polar head of a surfactant molecule at the interface. A necessary geometric condition for the existence of reversed micelles is $v/a_ol_c > 1$. However, AOT is a double chain surfactant with a relatively small head. The packing ratio is thus expected to be higher compared to that of a single chain surfactant for which the incorporation of a cosurfactant in the interface is necessary to increase the mean hydrocarbon volume without affecting either a_o or l_c.

The structure of the AOT micellar system, as well as the state of water entrapped inside swollen micelles, have been characterized using different techniques, such as photon correlation spectroscopy (25), positron annihilation (26), NMR (27, 28), fluorescence (29-32) and more recently small angle neutron scattering (33). The existence of reversed micelles has been demonstrated in the domain of concentrations explored by protein extraction experiments. Their size (proportional to the molar ratio of water to surfactant known as w_0), shape and aggregation number have been determined. Furthermore, the micelle size distribution is believed to be relatively monodisperse.

Quaternary ammonium salts, in most cases trialkylmethyl ammonium salts, have been the second focal point of work in this field. These molecules are cationic amphiphiles and have been revealed to be a complement to studies on the influence of the charge of the surfactant, compared to anionic AOT. In their early studies, Luisi et al. (34, 35) used trimethyloctylammonium chloride (TOMAC) at low surfactant concentrations (about 12 mM) in cyclohexane. But at higher concentrations, this molecule cannot be solubilized in apolar solvents, such as isooctane. It is necessary to add a fourth component, usually an alcohol. Investigations were undertaken to understand reversed micellar structure using quasi-elastic light scattering measurements (36). It was concluded that in a 40 mM solution of Aliquat 336 (a mixture of trialkyl (C_8-C_{10}) methyl ammonium chloride salts) dissolved in isooctane with 2.5 percent isotridecanol, there exist relatively monodisperse particles, the hydrodynamic radius of which is about 22 Å. This size was found to be in fairly good agreement with that calculated taking into account the amount of water in the organic phase and assuming the hypothesis of a molecular model of spherical micelles.

In this section, we shall attempt to review the most relevant observations and, assuming that reversed micelles exist in the organic phase, explain them in terms of both electrostatic interactions and steric effects.

Electrostatic Interactions

Since in all cases considered here both surfactant and proteins are electrically charged molecules, one of the major forces governing their interactions is expected to be of electrical origin.

The most direct feature is that pH determines the rate of dissociation of the charged residues which compose the primary structure of the protein and thus the net charge of the protein. When pH = pI, the protein is globally neutral, when pH < pI (respectively pH > pI), the global charge of the protein is positive (respectively negative). Depending on the pH, the species to be extracted is negatively, neutral, or positively charged, affecting electrostatic interactions with the surfactant.

Since proteins are polymers of amino acids, it is interesting to test the above trend and consider the effects of pH of the aqueous phase on the extraction of amino acids or small peptides. Luisi et al. (35) described the transfer of tryptophan and a dipeptide, L-tryptophylglycine into an organic phase composed of TOMAC in cyclohexane. It was shown that the pH profile of the percent transfer was determined by the ionization of the amino group. The pKa values obtained from these profiles agreed closely with values found in the literature. Thien et al. (37) reported very similar results in AOT microemulsions, leading to a fairly good agreement between the calculated titration curves for arginine and the amount of amino acid extracted as a function of the pH. From the results obtained with surfactants of opposite charge, it can be concluded that for amino acids, a direct relationship exists between the net charge of the molecule to be extracted and its transfer fraction.

It would be expected that the trend described above should also be verified for proteins. Göklen and Hatton confirmed this hypothesis (38, 39), at least for low molecular weight proteins. They studied the solubilization of three proteins, cytochrome C, ribonuclease and lysozyme of similar size but with distinct pI, from a 1 mg/ml aqueous protein solution into a 50 mM AOT/isooctane organic phase (figure 2) and showed that no extraction occurs for pH > pI, i.e. for pH values where the net charge of the protein is negative. As soon as the pH decreases below the pI value, there is an abrupt enhancement of the protein solubilization: the surfactant and the protein bear opposite charges and electrostatic interactions become attractive. The decrease of protein solubilization at low pH values is interpreted by the authors in terms of protein precipitation at the interface due to denaturation.

An increase of the relative amount of protein in the organic phase versus pH would be expected to take place when a cationic surfactant is used. Hatton (2) reported results on solubilization of catalase using a cationic surfactant, dodecyltrimethylammonium bromide (DTAB) in n-octane, with hexanol as cosurfactant. At pH values below pI = 5.3 no solubilization occurred, while there was a significant transfer for pH above pI.

Other results of extraction of proteins using different trialkylammonium salts differ somewhat from this tendency. Jolivalt et al. (36) studied the solubilization of α-chymotrypsin in the Aliquat 336 system using isotridecanol as a cosurfactant. Alpha-chymotrypsin was most significantly extracted for pH above its isoelectric point but as the pH decreased below the pI, the extraction yield decreased very slowly and became negligible at 4 pH units below the pI. This means that α−chymotrypsin can be extracted in significant proportion by a cationic surfactant even when positively charged overall.

A completely different behavior was observed by Van't Riet and Dekker (40) with α-amylase (isoelectric point around 5.5). The surfactant used in this work was TOMAC, 12 mM dissolved with octanol in isooctane.

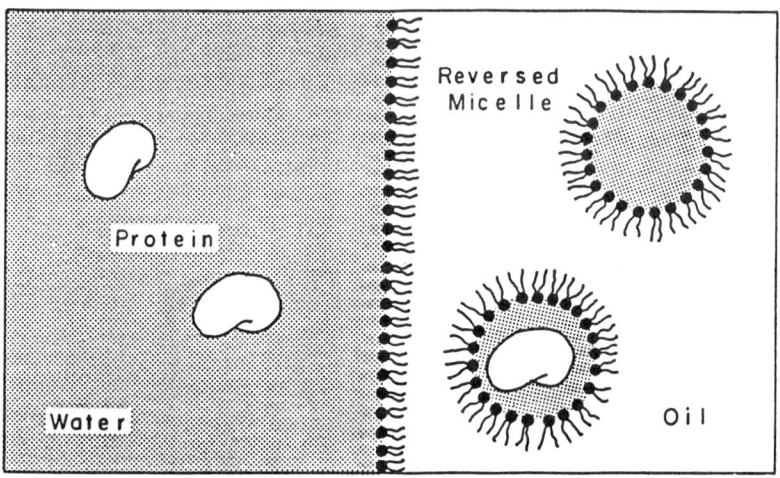

Figure 1. Protein solubilization in organic solvent by reversed micelles. (Reproduced with permission from Ref. 38. Copyright 1987 M. Dekker.)

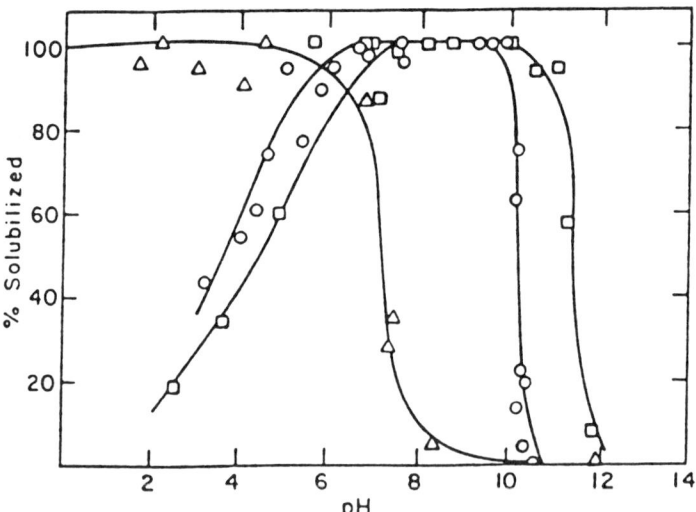

Figure 2. Effect of pH on solubilization of three proteins in AOT system (o) cytochrome c (pI=10.6); (□) lysozyme (pI=11.1); (Δ) ribonuclease (pI=7.8). (Reproduced with permission from Ref. 38. Copyright 1987 M. Dekker.)

Figure 3 shows that solubilization of α-amylase only occurred in a very narrow pH range around pH 10. At this pH, α-amylase bears a negative net charge and electrostatic interactions with surfactant are supposed to favor extraction, but this property cannot provide any satisfactory interpretation for the poor extraction in the range of pH between 5.5 and 10. In an attempt to explain this shift in the solubilization feature, it was noted that at pH = 10, each basic residue of the protein was deprotonated, so that the molecule should not bear any positive charge. Assessing that each negative charge is a complexation site with a surfactant molecule, the resulting complex is a neutral species which can be extracted in the organic phase without the help of a reversed micelle. Extraction would involve an ion-pairing process, as classically reported in the literature (41) on extraction of metal ions. Implicitly, the surfactant molecule is considered as a carrier which enables the transport of the protein into the organic phase.

In reversed micellar systems, the surfactant plays another role: it builds, together with the cosurfactant when necessary, a spherical shell surrounding the micelle. According to the investigations of Wong et al. (28), at least 30 percent of the counterion sodium is dissociated, creating an electrical double layer potential in the vicinity of the charged heads of the surfactant which affects the electrostatic interactions with the protein. This potential field depends strongly on the environment and in particular the forces between the protein surface and the surfactant head group decrease as the inverse of the square of the ionic strength. The most general feature (36, 38, 42) of the influence of the ionic strength is the decrease in the extraction yield when ionic strength increases, as it can be seen in figure 4 for three different proteins. This property is used for back extraction when contacting a laden organic phase with a high concentrated salt solution. In addition, it is worth noting in figure 4 that the "critical" KCl concentration, for which a significant decrease of solubilization occurs, depends on the protein.

However, it can be noted that direct measurements of the concentration of every ionic species present inside the micelle generally were not achieved. Leodidis and Hatton (43) published results on partitioning of cations in AOT reversed micelles. Studies on the effect of the ionic strength have focused on the ionic strength of the aqueous excess phase, supposing that it reflects qualitatively that in the organic phase.

Similar results are reported using cationic surfactants. The partition coefficient of α–chymotrypsin in Aliquat solutions decreased with increasing NaCl concentration (36). However, in order to interpret these results it is necessary to take into account that alkylmethyl ammonium salts are known to be involved in anion exchanges. Assuming an exchange equilibrium between the protein and chloride counterion, an increase of the chloride ion concentration should disfavor the extraction of the protein, following a mechanism of mass action law.

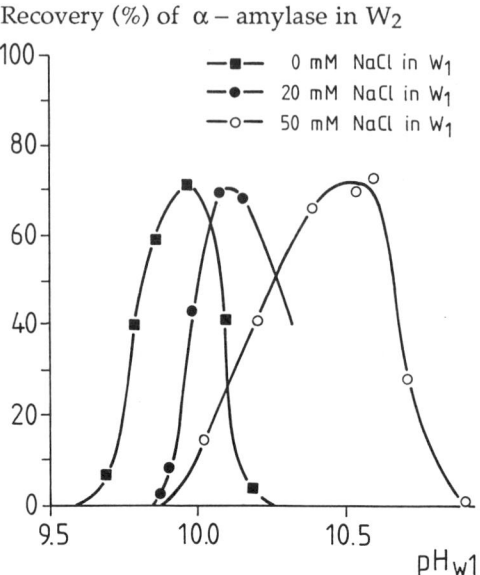

Figure 3. Effect of pH on α–amylase solubilization in TOMAC system for different salt concentrations TOMAC 0.4% (w/v); octanol 0.1% (v/v) in isooctane. (Reproduced with permission from Ref. 44. Copyright 1987 Elsevier Science Publishers.)

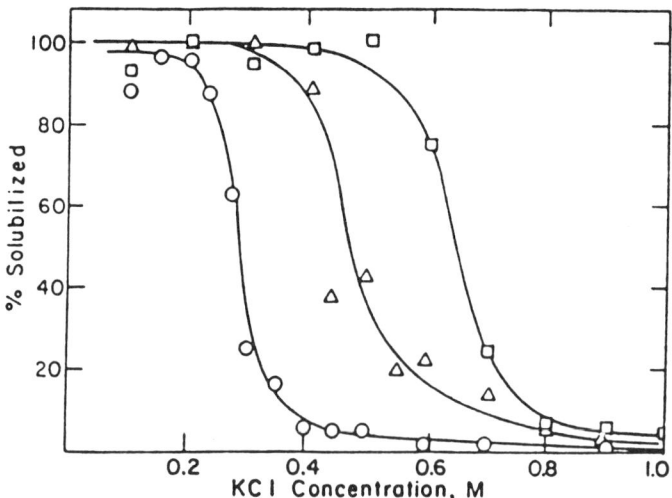

Figure 4. Effect of ionic strength on solubilization of three proteins in AOT system. (o) cytochrome c; (□) lysozyme; (Δ) ribonuclease-A. (Reproduced with permission from Ref. 38. Copyright 1987 M. Dekker.)

A different trend was observed by Dekker et al. (44) for the extraction of α–amylase in trioctylmethyl ammonium chloride. It can be seen in figure 3 that the transfer profile kept its bell shape and a similar maximal transfer value but with a shift towards higher pH when NaCl concentration increased in the aqueous phase up to 50 mM.

A more complete analysis must also include the nature of the ions (45). Thus, as an example, Göklen (46) noted that calcium ions permitted to solubilize cytochrome c beyond its pI although it was not possible with potassium. At a given pH, Leser et al. (47) reported that at 0.1 M salt concentration, solubilization of ribonuclease, lysozyme and trypsin were more efficient with $CaCl_2$ than with $MgCl_2$. Taking these results into account, it is not surprising to note that the choice of the buffer influences solubilization as well. Results on the influence of the pH are thus undissociable from the type of buffer used in the experiments.

Steric Effects

The interpretation of the above results is made more difficult by a supplementary phenomenon due to the aqueous phase ionic strength. Increasing the density of charges in the vicinity of the surfactant headgroups causes a screening effect not only between surfactant and protein but also in the repulsive interaction between surfactant polar heads: they come closer and the micellar size decreases (43). As a consequence, Göklen and Hatton (48) noted a decrease in the amount of water in AOT system when increasing KCl concentration. However, it is difficult from these results to state precisely whether the decrease in protein extraction is due to the decrease of electrostatic interactions or to a size exclusion effect.

By using TOMAC, Jolivalt et al. (36) had a supplementary degree of freedom available by varying the amount of alcohol. As in a microemulsion system, alcohol molecules should form a mixed shell with surfactant surrounding water droplets. Since the repulsive interactions between charged surfactant heads are reduced, the radius of the micelle decreases. This hypothesis was confirmed experimentally by water titration in an Aliquat-isotridecanol-isooctane solution. The amount of water in the organic phase was shown to decrease when the alcohol concentration increased. Moreover, the capacity of extraction of α-chymotrypsin decreased in parallel with the water content for alcohol concentrations larger than 1 percent. Since the alcohol molecule is not charged, it was expected not to modify the electrostatic interactions between surfactant and protein. It was concluded that less of the protein was extracted because the size of the micelle became smaller.

The importance of size exclusion was also shown (46, 49) by increasing the surfactant concentration. When increasing AOT concentration, the pH dependent profile of large proteins became similar to that of the smaller ones. The micellar size, which was shown to increase monotonously with surfactant concentration, should become large enough so that the exclusion effect does not overcome the other driving forces. At the concentration of 50 mM, the solubilization peak of a relatively large protein (Mw = 25 kDa) was narrowed compared to smaller proteins (Mw = 15 kDa).

Dekker et al. (50) reported that the addition of a non-ionic surfactant to a micellar solution of TOMAC enhanced the capacity of solubilization of α–amylase and broadened the pH range of solubilization as well. Although this effect should be studied in more detail, it can be supposed that the non-ionic surfactant was involved in a reorganization of the structure of the organic phase and allowed the formation of larger micelles.

Most of the results presented in this section show a good correlation between possible theoretical interactions and efficiency. But the discrepancies cited should prevent early generalization. In order to complete our understanding of the extraction phenomenon, it is necessary to take into account the protein specificity in terms of conformation, structure and hydrophobicity (as pointed out by Thien et al. (37) and Kirgios et al. (51)) in addition to pI and molecular weight, and to determine more precisely the surfactant-protein interaction sites and the exact transfer mechanism as well.

THERMODYNAMIC MODELS OF SOLUBILIZATION IN REVERSED MICELLES

Parallel to parametric studies of extraction, several papers have addressed the question of the structure of the reversed micelles after uptake of proteins. By and large, the microemulsion phase is described as the dispersion of two populations of spherical droplets surrounded with surfactant, one of which contains one protein solubilized in the middle of the water core, a so-called filled micelle which coexists with another population of monodispersed empty micelles. The main purpose of the experimental work has been to check this representation and to measure the size of both filled and empty micelles. Different techniques have been used which focused on AOT systems: ultracentrifugation (52-54), small angle neutron scattering (SANS) (55), quasi-elastic light scattering (56) or a combination of small angle X-ray scattering and pulse radiolyse to investigate the solubilization of cytochrome c (57). All of these experiments suffer from the weakness of the assumptions necessary for interpreting experimental data which influence the results.

However, these experimental approaches appeared to be relevant enough to form the basis of the first thermodynamic treatment of the solubilization of protein in reversed micelles developed by Caselli et al. (58). The micellar phase before solubilization forms the reference state of thermodynamic calculations. According to the structural model of spherical monodisperse droplets, the knowledge of the micelle radius and density completely characterizes the system. The uptake of protein is made by the injection technique, i.e. a known amount of water and protein is injected into the organic phase. Such a system is monophasic and the aim of the model is to predict its microscopic structure. Caselli et al. assume that there is a reorganization of the initial micelles into empty and filled micelles, the radii and density (resp. r_e, n_e and r_f, n_f) of which are the four unknowns of the system. The mass balance, together with geometrical considerations, yields two equations relating these parameters. The third equation is deduced from the assumption of mono-occupancy of the micelles:

$$N_f = N n_p \qquad (1)$$

n_p is the protein concentration, N_f the density number of the filled micelles and N the Avogadro number.

The fourth equation is obtained by minimizing the free energy change, ΔG, between the reference system and the protein containing system. The quantity ΔG is described as the sum of four contributions, three of which are entropic contributions. The fourth one is an electrostatic term related to the interaction between the charged protein and the internal layer of AOT polar heads, assuming in a first approximation that reversed micelles can be described as a microcapacitor when empty and as two concentric microcapacitors when filled. The calculated values of r_f and r_e were in good agreement with the experimental data obtained by the double dye ultra-centrifugation technique (54) and by SANS (55) for low cytochrome c concentrations.

It is also interesting to note that the radius of the filled micelles depends both on protein concentration and the w_o value: for each situation, the value of the physical parameters changes and determines a new equilibrium.

The simplicity of this model is due for the most part to the choice of the reference system which allows taking into account only the parameters of the micellar phase but does not permit describing any extraction or separation process, since it does not account for the phase transfer of the protein from the aqueous excess phase into the micellar phase.

The phenomenological model developed by Woll and Hatton (59) is an improvement towards this direction since it permits calculating the partition coefficient of proteins between the excess aqueous phase and the micellar phase. The basic concept of the model is the description of the solubilization according to a pseudo-chemical equilibrium. A protein interacts with n empty micelles to form a protein-micelle complex following the equation:

$$n\overline{M} + P \Leftrightarrow \overline{MnP} \qquad \frac{\overline{MnP}}{\overline{M}^n \cdot P} = \frac{K}{M^n} = e^{(\frac{-\Delta G}{RT})} \qquad (2)$$

The distribution constant K is expressed as a function of $\overline{M}$, n, ΔG. These variables are then related to the experimental parameters which have been revealed to influence the protein partition, i.e. the pH and the surfactant concentration S. As a definition, we have:

$$\overline{M} = \overline{S}/Na \qquad \text{Na: aggregation number} \qquad (3)$$

Assuming that Na is independent of S, M depends linearly on S. An expression of ΔG is developed:

$$\Delta G = \Delta G_o + zF\Delta\psi \qquad \text{F: Faraday's constant} \qquad (4)$$

z is the protein net charge which is assumed to depend linearly on the pH according to:

$$z = \alpha\,(pH - pI) \qquad (5)$$

The last assumption concerns n which is directly related to the size of the filled micelle. Assuming that the strength of electrostatic interactions between the protein of charge z and the charged surfactant layer determines the size of the filled micelle, a linear dependency between n and pH is deduced:

$$n = n_o + \varepsilon z \qquad (6)$$

From these simplifying assumptions, the authors deduced the following expression:

$$\ln k = A + BpH + (c + DpH)\ln S \qquad (7)$$

A, B, C, D depend on the extracted protein and are functions of ΔG_o, α, ε, $\Delta\psi$, n_o, pI, Na, z. Their numerical value have been calculated from experimental data on solubilization of ribonuclease and concanavalin A in AOT/isooctane with a good correlation to the model equation. The great interest of this model is that all the assumptions necessary for its elaboration make it very simple, and at the same time, a promising tool of quantification of protein solubilization thermodynamics, even if some further refinements are still needed. It can be noted that there are more parameters than can be adjusted from experimental data. As a consequence, the model can provide no value for n, related to the micelle size, which could have permitted an interesting comparison with that predicted by Caselli et al.'s model.

Leodidis and Hatton (43) chose a more theoretical approach and attempted to model the cation specific distribution between reversed micellar phase and bulk aqueous solution. This effect has been observed experimentally with AOT/isooctane systems (45).

The main purpose of the model was to relate the ion concentration and the local electrostatic parameters. The first step towards this relation was to assume that thermodynamic equilibrium was achieved between the water pools of the reversed micelles and the excess aqueous phase. This hypothesis differs from classical thermodynamics in that the equilibrium does not take place between two macroscopic phases. Biais et al. already suggested in a microemulsion pseudophase model (60) that, due to the low interfacial tension, chemical potentials of any constituents depend on the composition, even in microscopic domains, but not on the geometric parameters of the structure.

Further, the content of reversed micelles was found to be rapidly exchanged between the droplets through collisions on the time scale of milliseconds, at least for AOT microemulsions (25, 61).

The assumption of equality of the electrochemical potentials of any ions in the bulk water phase and in reversed micelles is thus likely to be verified:

$$\mu_i^{bw} = \mu_i^{mwp} \qquad \text{bw: bulk water} \qquad (8)$$
$$\text{mwp: micellar water pool}$$

Assuming that the standard chemical potentials are equal in bulk and inside micelles, it was deduced that:

$$k_b T \ln(C_i^{bw}/C_i^{mwp}(r)) + k_b T \ln(\gamma_i^{bw}/\gamma_i^{mwp}(r)) - z_i e \psi(r) \approx 0 \qquad (9)$$

$\psi(r)$ is the local electrostatic potential inside the micelle, deduced from a model of the electrostatic double layer. This model is based on the assumptions of electroneutrality, monodispersity and spherical symmetry of reversed micelles. The original contribution of Leodidis and Hatton is to assume the fluctuation of the surfactant head position between the radii Ri and Ro as shown in figure 5 and to take into account the cation penetration between surfactant polar heads. From the surfactant head distribution, the authors arrive at an analytical expression of $\psi(r)$.

The last term developed in equation 9 contains the ratio of the ion activity coefficient in each phase. It was expressed using a sum of three contributions:

$$k_b T \ln \frac{\gamma^{bw_i}}{\gamma^{mwp}_i(r)} = \mu_{i,\,entr}^{bw-mwp} + \mu_{i,\,ion-solv}^{bw-mwp} + \mu_{i,ion-ion}^{bw-mwp} \qquad (10)$$

The electrostatic free energy change in this equation differs somewhat from that used by Caselli et al.. Indeed, Leodidis and Hatton introduce the excess free energy of the bulk water whereas Caselli et al. consider only the water pool contribution. Moreover, the main difference is the model for electrostatic interactions. Leodidis and Hatton eliminated the Poisson-

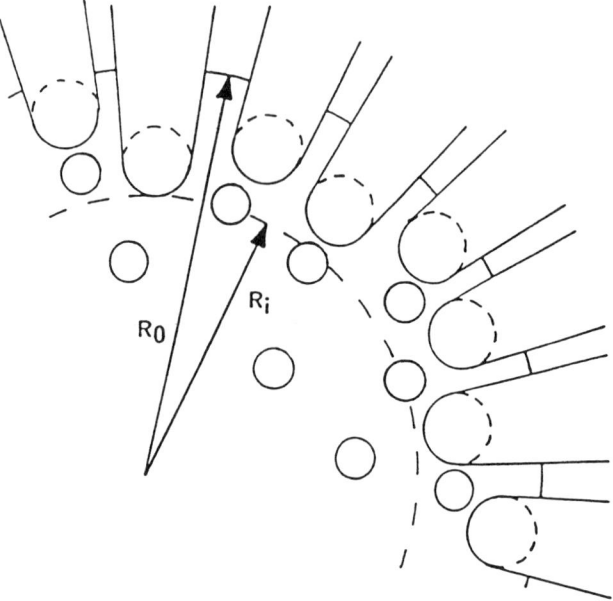

Figure 5. Schematic diagram of the assumed interfacial structure of the reversed micelle. Leodidis and Hatton (43).

Boltzman equation because it does not distinguish between ions of like charge and therefore cannot account for the specificity of the distribution. In order to reflect the experimental reality, the Leodidis and Hatton model takes into account three characteristics of ions: their charge, size and electrostatic free energy of hydration.

In its present form, the model is focused on small cations but could provide a suitable starting point for further approaches including other electrolytes and, in particular, proteins. It must be noted that Leodidis and Hatton's work refers to some theoretical studies on microemulsions. Indeed, as shown in figure 6, reversed micelles are one of the various possible association structures of the microdomains which compose microemulsions. The most fundamental questions that the numerous works published in this field attempt to answer are related to the mechanism of formation of microemulsions and their thermodynamic stability (62, 63). In particular, it was demonstrated that the systems of interest for extraction, i.e. water in oil microemulsions in equilibrium with an excess phase, are governed by the bending stress of the interfacial film (64, 65). In spite of their often being only theoretical, these approaches give a good understanding of the factors characterizing the reversed micellar phase, such as the size of the micelles. They could provide the basic concepts of a general thermodynamics quantification of protein partitioning.

APPLICATIONS

Apart from micellar enzymology (the term Martinek coined to describe enzymatic catalysis in reversed micelles), the main challenge in reversed micelles research is the separation of a targeted protein from a complex medium for further purification and recovery. In this case, diphasic systems are involved in successive steps of extraction and stripping. The relevant parameters of the recovery of proteins from aqueous solutions are mass transfer characteristics, i.e. the efficiency of extraction and stripping steps (including kinetics and yield or concentration factor at equilibrium), the selectivity of the separation between several proteins, and the biochemical constraint of achieving a non-degrading separation. For that purpose, two main areas were investigated: the first, and still the most studied, is the optimization of phase composition - reversed micellar and stripping solutions and to a lesser extent the aqueous feed; the second deals with process development in view of scale-up.

From the observation that the partitioning behavior of each protein was not affected by the presence of the others (38, 49), Göklen and Hatton resolved a mixture of cytochrome-c, ribonuclease-A and lysozyme, three low molecular weight proteins comprised in the range 12.4 - 14.3 kDa. With the same reversed micellar phase, Woll et al. (49) showed that the selectivity of extraction between ribonuclease-A and concanavalin-A could be modulated by varying the surfactant concentration; a 40% enhancement

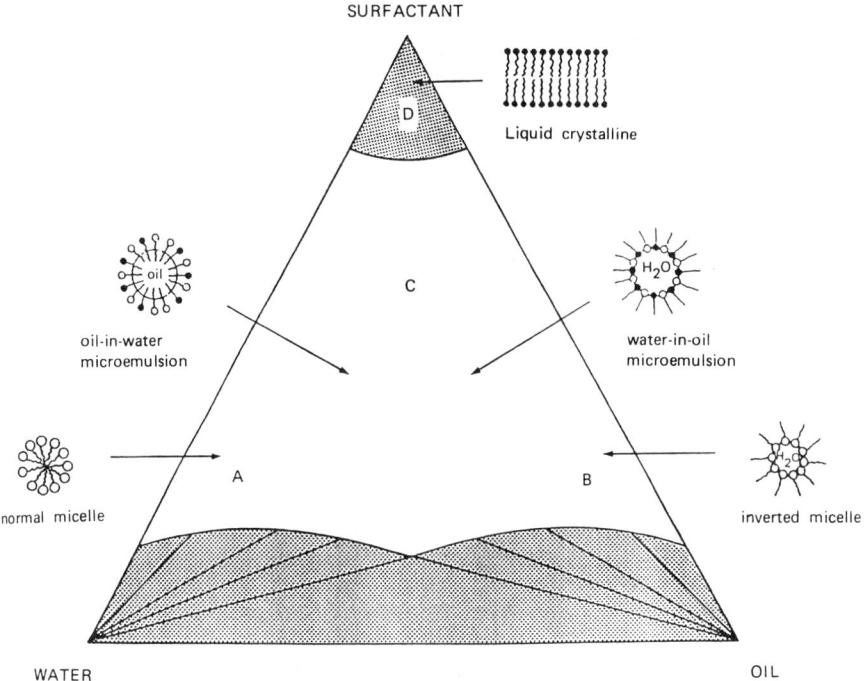

Figure 6. Schematic ternary phase diagram of an oil-water-surfactant microemulsion system. (Reproduced with permission from Ref. 63. Copyright 1988 M. Dekker.)

of the maximal selectivity being obtained by increasing AOT concentration from 50 to 100 mM. For larger proteins, such as concanavalin-A (a dimer of 55 kDa), size exclusion plays a discriminating role in addition to the electrostatic interactions. This was underlined by Armstrong and Li (66) who developed a reversed micellar liquid membrane, using AOT in hexane. Effects of pH and surfactant concentration enabled the separation of a mixture of cytochrome-c, lysozyme, myoglobin and bovine serum albumine (BSA). It is noteworthy that the large BSA (66 kDa) could transfer through the membrane for AOT concentration above 0.5 M.

Further enhancement of the selectivity was obtained by exploiting biospecific interactions between an additional surfactant and the protein. Woll et al. (49, 67) showed that small amounts (2 to 10% of total surfactant concentration) of a biological detergent (octyl ß-D-glucopyranoside) increased, by more than one order of magnitude, the selectivity of extraction of concanavalin-A against ribonuclease-A, in an AOT/isooctane system. The affinity of the carbohydrate-binding concanavalin-A for the ligand was essentially responsible for its increased extraction, while ribonuclease-A was not specifically affected.

In the previous examples, where the principal surfactant of reversed micellar phases was AOT, apparently no problem of protein denaturation was encountered and in favorable cases quantitative back extractions were performed. However, these studies were carried out using synthetic feed solutions. Experiments in more realistic conditions are reported by Rahaman et al. (68), who extracted an extracellular alkaline protease (M = 33 kDa) from a fermentation broth of bacillus strain, with an AOT/isooctane system. In one extraction-stripping stage, an activity yield of 22% and a relative specific activity of 2.2 (compared to feed phase) were achieved; the use of multi-stage cascaded operations and the variation of the volume ratios of the phases permitted improved results (relative specific activity up to 6). These experiments underlined the importance of extraction kinetics, since the protease exhibited marked instability in presence of the organic phase at high pH and since the back transfer was surprisingly slower than in simple synthetic protein solutions; mixing times were optimized at 5 minutes for extraction and 15 minutes for stripping. The influence of broth impurities on separation features was, however, not really evaluated.

Another application of reversed micelles on complete broths was reported by Giovenco and Verheggen (69) for the purification of intracellular dehydrogenases of *Azotobacter vinelandii*. The surfactant cetyltrimethyl ammonium bromide, dissolved in an hexanol-octane solution, served also as a disrupting agent of cell membranes. The activity recovered in the stripping phase was superior to 100% - explained by the supposed removal of inhibitors - with a relative specific activity of 6, for the smaller enzymes ß-hydroxybutyrate and isocitrate dehydrogenases. The largest glucose-6-phosphate dehydrogenase was not extracted in an active form.

Most of the examples of extraction by reversed micelles reported above were performed in a batch mode, with successive steps of extraction and stripping, and little consideration was taken with regard to the process itself.

Using reversed micellar phase as a liquid membrane, Armstrong and Li (66) could perform continuous extraction from and stripping to aqueous solutions in either side of the membrane. However, improvements of kinetic mass transfer parameters, as influenced by membrane thickness and stability in conditions of appropriate hydrodynamics, remain necessary.

Dekker et al. (70, 71) developed a 2-stage mixer-settler device to achieve continuous concentration of α-amylase, the TOMAC/octanol/isooctane reversed micellar phase being recirculated from the stripping settler to the first extracting mixer. On condition that losses of surfactant be compensated by its addition during the course of the experiment, the efficiency of the separation was maintained over six circulations, with a concentration factor of 8 for the active enzyme. Again, this study made clear the importance of achieving rapid mass transfer, since first order kinetics were found for the inactivation of α-amylase during the forward extraction step (the total activity loss was 30% during the procedure). On the other hand, an overly strong mixing would increase the duration of settling and possibly cause similar drawbacks. The use of hollow fibers tries to answer the need of increasing the surface/volume ratio of the extracting phase to the feed volume, by means other than higher agitation and the resulting undesired emulsification. Dekker et al. (71) and Dahuron and Cussler (72) reported experiments using microporous polypropylene membrane to stabilize the interface between the two streams of aqueous feed and reversed micellar phases. Mass transfer coefficients were determined for the extraction of cytochrome-c and α-chymotrypsin by AOT/octane (72) and α-amylase by TOMAC/Rewopal HV5/octanol/isooctane system (71).

An attractive new application of reversed micelles has also been mentioned for the refolding of proteins. Many of the proteins produced by recombinant DNA technology have to be extracted from cells using strong solubilizing denaturants. The step of refolding, by denaturant removal, generally necessitates operating in very dilute solutions, in order to avoid aggregation of the partially refolded intermediates. The method tested by Hagen et al. (73) takes advantage of the isolation of single proteins inside micelles to prevent harmful interactions. They demonstrated that it was possible to extract a denaturated RNAse by an AOT/isooctane phase, and to remove the denaturant guanidine hydrochloride by successive aqueous phase contacting. The reoxidation of disulfide bonds by a glutathione mixture permitted the total reactivation of the enzyme. The last step of stripping led to the recovery of a refolded and fully active RNAse, with an overall yield of 50 percent.

CONCLUSION

A significant amount of work has demonstrated the feasibility and the interest of reversed micelles for the separation of proteins and for the enhancement or inhibition of specific reactions. The number of micellar systems presently available and studied in the presence of proteins is still limited. An effort should be made to increase the number of surfactants used as well as the set of proteins assayed and to characterize the molecular mechanism of solubilization and the microstructure of the laden organic phases in various systems, since they determine the efficiency and selectivity of the separation and are essential to understand the phenomena of bio-activity loss or preservation. As the features of extraction depend on many parameters, particular attention should be paid to controlling all of them in each phase. Simplified thermodynamic models begin to be developed for the representation of partition of simple ions and proteins between aqueous and micellar phases. Relevant experiments and more complete data sets on distribution of salts, cosurfactants, should promote further developments in modelling in relation with current investigations on electrolytes, polymers and proteins. This work could be connected with distribution studies achieved in related areas as microemulsions for oil recovery or supercritical extraction (74). In addition, the contribution of physico-chemical experiments should be taken into account to evaluate the size and structure of the micelles.

LITERATURE CITED

1. Kadam, K. L. Enzyme Microb. Technol. 1986, vol 8, p 266.
2. Hatton, T. A. In ACS Symposium Series; Hinze, W. L., Armstrong, D. N., Eds.; 1987; Chapter 9.
3. Hatton, T. A. In Surfactants and Separation; Scamehorn, J. F. and Harwell J. H., Eds.; New York, 1987.
4. Abbott, N. L.; Hatton, T. A. Chem. Eng. Prog. August 1988, p 31-40.
5. Martinek, K.; Levashov, A. V.; Klyachko, N. L.; Pantin, V. I.; Berezin, I.V. Biochimica and Biophysica Acta 1981, 657, 277-294.
6. Fletcher, P. D. I.; Freedman, R. B.; Mead, J.; Oldfield, C.; Robinson, B. H. Colloids and Surfaces 1984, 10, 193-203.
7. Laane, C.; Verheart, R. Isr. J. of Chemistry 1987-1988, 28, 17-22.
8. Shield, J. W.; Fergusson, H. D.; Bommarius, A. S.; Hatton, T. A. Ind. Eng. Chem.Fundam. 1986, 25, 603-612.
9. de Kruijff, B.; Cullis, P. R. Biochim. Biophys. Acta 1980, 602, 477-490.
10. Gierash, L. M.; Thompson, K. F. Surfactants in Solution; Mittal, K. L. and Lindman, B., Eds., New York, 1984, p 265.
11. Thompson, K. F.; Gierasch, L. M. J. Am. Chem. Soc. 1984, 106, 3648-52.
12. Nicot, C.; Vacher, M.; Vincent, M.; Gallay, J.; Waks, M. Biochemistry 1985, 24, 70-84.
13. Delahodde, A.; Vacher, M.; Nicot, C.; Waks, M. FEBS Lett. 1984, 172(2), 343-47.
14. Steinmann, B., Jäckle, H.; Luisi, P. L. Biopolymers 1986, 25, 1133-56.
15. Marco, A. D.; Zetta, L.; Menegatti, E.; Luisi, P. L. J. Biochem. Biophys. Methods 1986, 12, 335-347.
16. Luisi, P. L. Angewandte Chemie 1985, 24(6), 439-528.

17. O'Connor, C. J.; Lomax, T. D.; Ramage, R. E. Adv. Colloid Int. Sci. 1984, 20, 21-97.
18. Waks, M. Proteins 1986, vol. 1, p 4-15.
19. Luisi, P. L.; Laane, C. Tibtech 1986, p 153-161.
20. Luisi, P. L.; Magid, L. J. CRC Critical Reviews in Biochemistry 1986, vol. 20(4), p 409-474.
21. Luisi, P. L.; Giomini, M.; Pileni, M. P.; Robinson, B. H. Biochemica Biophys. Acta 1988, 947, 209-246.
22. Fendler, J. H. Accounts of Chemical Research 1976, 9, 153-161.
23. Martinek, K.; Levashov, A.V.; Klyachko, N.; Khmelnitski, Y. L.; Berezin, I. V. Eur. J. Biochem. 1986, 155, 453-468.
24. Mitchell, D. J.; Ninham, B. W. J. Chem. Soc. Faraday Trans. 2 1981, 77, 601-629.
25. Zulauf, M.; Eicke, H. F. J. Phys. Chem. 1979, 83(4), p 480.
26. Jean, Y. C.; Ache, H. J. JACS 1978, 100(3), 984.
27. Maitra, A. J. Phys. Chem. 1984, 88, 3122-25.
28. Wong, M.; Thomas, J. K.; Nowak, T. JACS 1977, 99(14), p 4730.
29. Wong, M.; Thomas, J. K.; Grätzel, M. JACS 1976, 98(9), p 2391.
30. Kumar, V. V.; Raghvnathan, P. Lipids 1986, 21(12), p 764.
31. Keh E.; Valeur, B. J. Colloid Int. Sci. 1981, 79(2), p 465.
32. Howe, A. M.; McDonald, J. A.; Robinson, B. H. J. Chem. Soc. Faraday Trans. 1 1987, 83, 1007-1027.
33. Kotlarchyk, M.; Huang J. S. J. Phys. Chem. 1985, 89, 4382-4386.
34. Luisi, P. L.; Henninger, F.; Joppich, M. Biochem. Biophys. Res. Com. 1977, 74(4), p 1384.
35. Luisi, P. L.; Bonner, F. J.; Pellegrini, A.; Wiget, P.; Wolf, R. Helv. Chem. Acta 1979, vol 62(3), n° 77, p 740.
36. Jolivalt, C.; Minier, M.; Renon, H., in press, J. Colloid Int Sci., 1989.
37. Thien, M. P.; " Separation and Concentration of Amino Acids Using Liquid Emulsion Membranes", Doc. of Sci. Thesis, M.I.T., 1988.
38. Göklen, K. E.; Hatton, T. A. Sep. Sci. Techn. 1987, 22(2-3), 831-841.
39. Göklen, K. E.; Hatton, T. A. Proc. ISEC 1986, p 587-595.
40. Van't Riet, K.; Dekker, M. 3rd Eur. Cong. Biochem. 1984, p 540.
41. Paatero, J. Dept. Inst. Ind. Chem. Abo Akademi 1975, n° 106.
42. Göklen, K. E.; Hatton, T. A. Biotechnology Prog. 1985, vol 1(1), p 69.
43. Leodidis, E. B.; Hatton T. A.; Submitted to Langmuir, July 1988.
44. Dekker, M.; Van't Reit, K.; Baltussen, J. W. A.; Bijsterboch, B.H.; Hilhorst, R.; Laane, C. Proc. 4th Eur. Cong. on Biotechnology 1987, p 507.
45. Fletcher, P.; J. Chem. Soc. Faraday Trans. 1 1986, 82, 2651-2664.
46. Göklen, K. E., Ph.D. Thesis, MIT 1986.
47. Leser, M. E.; Wei, G.; Luisi, P. L.; Maestro, M. Biochem. Biophy. Res. Com. 1986, 135(2), p 629.
48. Göklen, K. E.; Hatton, T. A. AIChE Annual Mtg. 1984, n° 5a.
49. Woll, J.; Willon, A. S.; Rahaman, R. S.; Hatton, T. A., Protein Purification - Micro to Macro 1987, p 117-130.
50. Dekker, M.; Baltussen, J. W. A.; Van't Riet, K., Bijsterbosch, B. H.; Laane, C.; Hilhorst, R. Biocatalysis in Organic Media; Laane, C.; Tromper, J.; Lilly, M.D., Eds.; Elsevier 1987; p 285-288.
51. Kirgios, I.; Hansec, R.; Rhein, H. B.; Schugerl, K. Preprint ISEC München 1986, 3,623
52. Levashov, A. V.; Khmelnitsky, Y. L.; Klyachko, N.L.; Chernyak, V.Ya.; Martinek, K. J. Coll. Int. Sci. 1982, 88(2), 444.
53. Bonner, F. J.; Wolf, R.; Luisi, P. L. J., Solid Phase Biochemistry 1980, 5(4), 255.
54. Zampieri, G. G.; Jackle, H.; Luisi, P. L. J. Phys. Chem. 1986, 90, 1849-1853.
55. Sheu, E.; Göklen, K. E.; Hatton, T. A.; Chen, S.H. Biotechnology Progress 1986, 2(4), p 175.
56. Chatenay, D.; Orbach, W.; Cazabat, A. M.; Vacher, M.; Waks, M. Biophys. J. 1985, 48, 893-898.

57. Pileni, M. P.; Zemb, T.; Petit, C. Chem. Phys. Lett. 1985, 118, p 414.
58. Caselli, M.; Luisi, P. L.; Maestro, M.; Roselli, R. J. Phys. Chem. 1988, 92, 3899-3905.
59. Woll, J. M.; Hatton, T. A.; "A Simple Phenomenological Thermodynamic Model for Protein Partitioning in Reversed Micellar Systems"; Bioproc. Eng., in press.
60. Biais, J.; Bothorel, P.; Clin, B.; Lalanne, P. J. Dispersion Sci. and Tl. 1981, 2, 67.
61. Eicke, H. F. Top. Curr. Chem. 1980, 87:86.
62. Bellocq, A. M.; Biais, J.; Bothorel, P.; Clin, P.; Fourche, G.; Lalanne, P.; Lemaire, B.; Lemanceau, B.; Roux, D. Adv. Colloid Int. Sci. 1984, 20, 167
63. Leung, R.; Mean Jeng Hou; Do Shah. Surfactants in Chemical Process Engineering, Surfactant Science Series 1988, vol 28, 315-67
64. Mukhiya, S.; Miller, C. A., Fort, T. J. Colloid Int. Sci. 1983, 81, 223
65. Miller, C. A.; Neogi, P. AIChE J. 1980, 26, 212
66. Armstrong, D. W.; Li, W. Anal. Chem. 1988, 60, 88-90.
67. Woll, J. M.; Hatton, T. A.; Yarmush, M.L.; Biotech. Progr., 1989, 5,2, 57-62.
68. Rahaman, R. S.; Chee, J. Y.; Cabral, J. M. S.; Hatton, T.A.; Biotech. Progr. 1988, 4,4, 218-224.
69. Giovenco, S. and Verheggen, F., Enzyme Microb. Technol. 1987, 9, 470-473.
70. Dekker, M.; Van't Riet, K.; Weijers, S. R. Chem. Eng. J. 1986, 33, B27-B33.
71. Dekker, M.; Van't Riet, K.; Wijnans, J. M. G. M.; Baltussen, J. W. A.; Bijsterbosch, B. H.; Laane, C. Proc. IOCM 1987, p 793.
72. Dahuron, L.; Cussler, E. L. AIChE J. 1988, 34, 130-136.
73. Hagen, A. J.; Hatton, T. A.; Wang, D. I. C. Proc. ACS Mtg. 1988.
74. Gale, R. W.; Fulton, J. L.; Smith, R. D. J. Am. Chem. Soc. 1987, 109, 920-21.

RECEIVED October 13, 1989

Chapter 6

Enzymes in Liquid Membranes

Reaction and Bioseparation

Donald K. Simmons[1,3], Sheldon W. May[2], and Pradeep K. Agrawal[1]

[1]School of Chemical Engineering, Georgia Institute of Technology, Atlanta, GA 30032
[2]School of Chemistry, Georgia Institute of Technology, Atlanta, GA 30032

> The rising need for new separation processes for the biotechnology industry and the increasing attention towards development of new industrial enzyme processes demonstrate a potential for the use of liquid membranes (LMs). This technique is particularly appropriate for multiple enzyme / cofactor systems since any number of enzymes as well as other molecules can be coencapsulated. This paper focuses on the application of LMs for enzyme encapsulation. The formulation and properties of LMs are first introduced for those unfamiliar with the technique. Special attention is paid to carrier-facilitated transport of amino acids in LMs, since this is a central feature involved in the operation of many LM encapsulated enzyme bioreactor systems. Current work in this laboratory with a tyrosinase/ascorbate system for isolation of reactive intermediate oxidation products related to L-DOPA is discussed. A brief review of previous LM enzyme systems and reactor configurations is included for reference.

Liquid membranes are double emulsions formed when a water-in-oil emulsion (w/o) is gently dispersed in a second aqueous phase, the external aqueous phase. The internal (emulsified) and external aqueous phases are kept separate by a layer of hydrocarbon, forming the liquid membrane. Since the two aqueous phases are not in contact, LM systems can be useful for separation processes as well as for enzyme immobilization: separation is accomplished by selective transport of solutes across the hydrocarbon "membrane," and enzyme immobilization is accomplished by encapsulating enzyme(s) via emulsification of an aqueous enzyme solution. It is in fact possible to combine enzymatic reaction(s) with separations in a single LM system. Figure 1 depicts an LM-enzyme system.

[3]Current address: Hoechst Celanese Corporation, Charlotte, NC 28232

LIQUID MEMBRANE FORMULATION

Liquid membranes were developed by Norman Li (1). An aqueous solution which is to become the internal aqueous phase (dispersed phase) is emulsified with a hydrocarbon solution to form a water-in-oil (w/o) emulsion. The hydrocarbon solution consists of an aliphatic solvent of moderate molecular weight and viscosity, a surfactant which selects for a w/o emulsion (typically Span 80), and a "membrane strengthening agent" (e.g. Paranox 100 or Lubrizol #3702). The hydrocarbon solvent, *e.g.* Exxon solvent 100 Neutral, is normally the major constituent of the hydrocarbon membrane, and it must solvate the other additives. A surfactant is essential for stabilizing the emulsion. Span 80 has been recommended (2): this surfactant has a hydrophile-lipophile balance (HLB) value of 4.3. The HLB is calculated based on the relative hydrophilicity and lipophilicity of various groups in the surfactant molecule (3). Hydrophile-lipophile balance values of between about 3 and 6 are necessary for forming w/o emulsions; HLB values between 8 and 18 tend to form oil-in-water emulsions (4). Since the LM must be oil-continuous in order to prevent intermixing of the internal and external aqueous phases, the maintenance of an appropriate HLB range is crucial to avoid breakage of the LM. As will be seen later, when additional compounds are added to the hydrocarbon solution, it is important to consider their effect on the overall HLB. The "membrane strengthening agent" is also a surfactant. However, this agent is typically a block copolymer, *i.e.* a moderately high molecular weight compound with a large hydrophobic "tail" and relatively small polar "head". These compounds strengthen the LM by forming a surfactant film around the dispersed water droplets which keeps the droplets from contacting and coalescing: the long hydrocarbon tails sterically interfere with each other (5).

PROPERTIES

Emulsions are characterized in terms of dispersed / continuous phase, phase volume ratio, droplet size distribution, viscosity, and stability. The dispersed phase is present in the form of microscopic droplets which are surrounded by the continuous phase: both water-in-oil (w/o) and oil-in-water (o/w) emulsions can be formed. The typical size range for dispersed droplets which are classified as emulsions is from 0.25 to 25 μ (6). Particles larger than 25 μ indicate incomplete emulsification and/or impending breakage of the emulsion. Phase volume ratio is the volume fraction of the emulsion occupied by the internal (dispersed) phase, expressed as a percent or decimal number. Emulsion viscosity is determined by the viscosity of the continuous phase (solvent and surfactants), the phase volume ratio, and the particle size (6). Stroeve and Varanasi (7) have shown that emulsion viscosity is a critical factor in LM stability. Stability of the emulsion itself depends on many emulsion parameters, and there are several mechanisms by which emulsions break down (8).

To develop LMs into a viable enzyme reactor systems, it is necessary to overcome some major obstacles: emulsion instability (LM breakage), LM swelling, and transport of charged molecules. An unstable emulsion is highly undesirable for LM enzyme systems, since enzyme(s) would leak out to the external aqueous phase. Liquid membrane swelling, caused by transport of water molecules from the external to the internal aqueous phases and characterized by an increase in the emulsion's volume, dilutes the internal aqueous phase and can lead to emulsion breakage (9). The transport of charged molecules through LMs is a major hurdle and is discussed in detail below.

TRANSPORT

Transport of solutes through the LM occurs by either passive transport or by carrier-facilitated transport. Phenol, for example, is soluble in both phases, and treatment of an aqueous phenol solution with an emulsion results in a lowering of the external concentration of phenol as it passively diffuses through the hydrocarbon (HC) layer and into the internal aqueous phase. Equilibrium is reached when the concentrations of phenol in both aqueous solutions are equal (assuming no other conditions are present which would alter the distribution between the aqueous and HC phases). One way to alter this equilibrium is to trap phenol inside with a sodium hydroxide solution. Phenol ionizes at high pH, and the phenolate ion cannot permeate a HC layer: trace amounts of phenol have been completely removed from wastewaters by this system (10, 11). This exclusion of charged molecules by the aliphatic hydrocarbon LM layer is desirable in some applications, but to employ LM enzyme reactors and/or separation systems with amino acids, it is necessary to incorporate carriers into the HC phase.

CARRIER-FACILITATED TRANSPORT

Carrier-facilitated transport is of vital importance in developing LMs for biotechnology applications involving amino acid transport. A carrier is a molecule, present in the HC phase, which allows solubilization and hence transport of substances which are not normally hydrocarbon soluble. The carrier functions by reversibly binding with a solute at the HC-external aqueous interface. The carrier-solute pair then diffuses through the HC layer by simple diffusion. The solute is finally released to the internal aqueous phase at the HC-internal aqueous interface. The types of bonding between carrier and solute that are suitable for reversible complexation include acid-base interactions, hydrogen bonding, and chelation (12). Noble et al. (13) have reviewed carriers for various industrial separation processes and have discussed the efforts to analytically model the transport of solutes into LMs.

Behr and Lehn (14) first demonstrated carrier-facilitated transport of amino acids through "liquid membranes" composed of a toluene layer floating on top of two isolated aqueous solutions. The carriers used were the quaternary ammonium salt Aliquat 336 (trioctylmethylammonium chloride) and the alkylated arylsulfonic acid dinonylnapthalenesulfonic acid. Amino acids were transported in the form of anions or cations, respectively, with the above carriers. The process is an ion-exchange process, as the carrier must exchange an ionized atom or molecule each time it forms a new ion pair (Figure 2). The net result of this type of carrier-facilitated transport process is that a solute ion is transported into the LM while an equal number of counterions are transported out of the LM.

Amino acids are zwitterionic, or dipolar, at a wide range of pHs. Carrier-facilitated transport with acidic or basic carriers (which contain only one ionized group) requires that the amino acid be present as either a cation or anion, respectively. Figure 3(a) shows the amino acid L-tyrosine (L-tyr) and the dissociation constants associated with its ionizable groups (15). A rule of thumb for interpreting pK' values for singly ionizable groups is that if the pH of the solution is equal to the pK' for a particular ionization, then that particular group will be exactly 50% ionized. This and two accessory rules, a result of the Henderson-Hasselbalch equation (16) are shown below:

$pH = pK'$ equal amounts of both possible forms
$pH \leq (pK' - 2)$ virtually all is present in the protonated form
$pH \geq (pK' + 2)$ virtually all is present in the deprotonated form

The acid group of L-tyr, pK' = 2.20, is thus half ionized at pH 2.2, and ionization is essentially complete above pH 4.2. The amino group of L-tyr, pK' = 9.11, would be at least partially protonated (ionized) until a solution pH of about 11 is maintained. The zwitterionic form of L-tyr is shown in Figure 3(b). To have the anionic form of tyrosine, suitable for transport by the quaternary ammonium salt carrier, the amino group must be deprotonated (Figure 3(c)). A pH ≥ 11 would therefore be desirable to maximize the concentration of anion except for the fact that tyr also has an ionizable hydroxyl group, pK' = 10.07. This group is already beginning to ionize above pH 8, such that two anionic groups are present in greater concentrations as the pH is raised. Transport of the dianion would require that both anions pair with carrier molecules (Figure 2).

Several desirable criteria for facilitative carriers have been described (17): (1) The carrier and its complexed form must be as water-insoluble as possible, (2) the carrier in any form must not separate out from the HC phase, (3) the carrier must ion-pair with the desired solute(s) under the selected external and internal aqueous phase conditions, (4) kinetics of ion-pairing and release of the solute should be fast compared to *e.g.* diffusion

112 DOWNSTREAM PROCESSING AND BIOSEPARATION

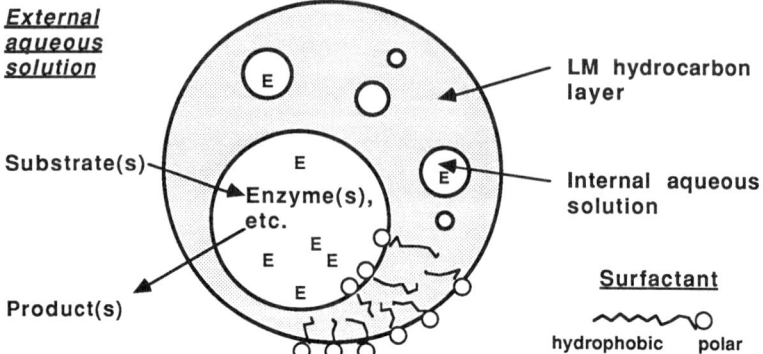

Figure 1. Schematic representation of a Liquid Membrane containing immobilized enzyme(s), showing diffusion / reaction of substrates and products.

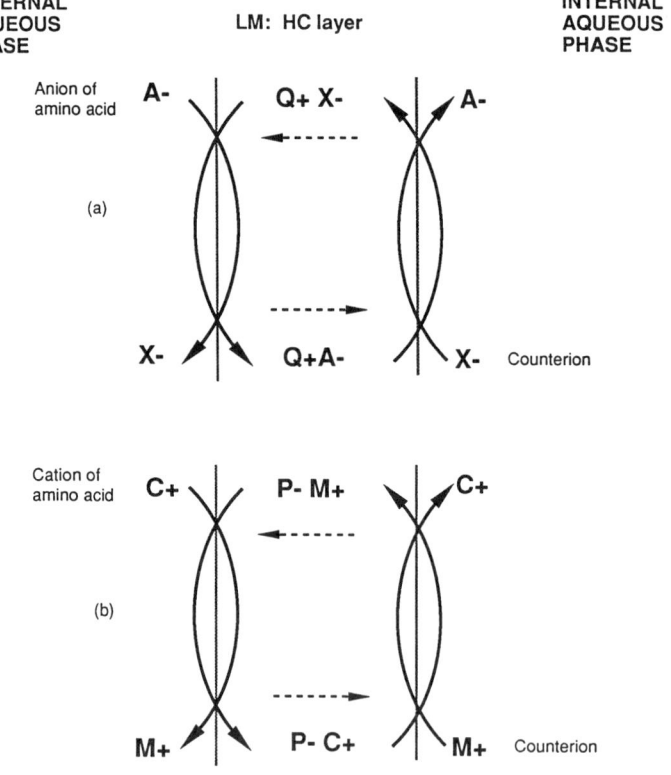

Figure 2. Carrier-facilitated transport of anions and cations in LMs: (a) Anion transport by *e.g.* a quaternary ammonium salt; (b) Cation transport by *e.g.* a phosphate diester.

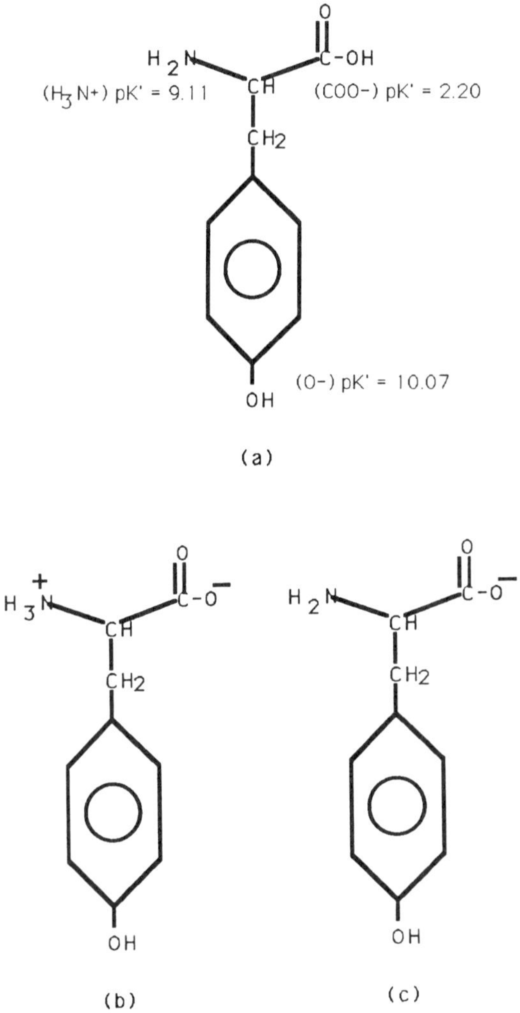

Figure 3. L-Tyrosine: (a) groups shown as un-ionized with pK' values; (b) zwitterionic form, typical at moderate pH levels; (c) single anion form, *e.g.,* at pH 10 (idealized case with no hydroxyl group ionization).

rates, (5) if the emulsion is to be broken for reuse, this should not be deleterious to the carrier, (6) the carrier should be as solute specific as possible, and (7) if possible, carrier size should be minimized to promote faster diffusivity of the ion-pairs. These criteria were established for unidirectional transport of the single solute, L-phenylalanine, with TOMAC as the carrier and chloride as the counterion. More complex LM enzyme systems require bidirectional transport of desired substrates and products with fairly rigid process parameters (e.g. pH and ion strength control). An essential criterion is emulsion stability, since any LM breakage would produce enzyme leakage. This laboratory is currently evaluating carriers and is using a rational approach to further develop anion and cation carriers suitable for LM enzyme systems.

ANION CARRIERS

Anion carriers, particularly quaternary ammonium salts and secondary or tertiary amines, have been the most widely studied carriers for LM biotechnology applications. These carriers have been adapted from solvent extraction reagents. Amberlite LA-2, a fatty secondary amine, was found by Mohan and Li (18) to be useful for transport of nitrate and nitrite ions through LMs. Scheper et al. (19) used Amberlite LA-2 to transport penicillin G because of this carrier's stability in the LM compared to e.g. quaternary ammonium salt carriers. Tertiary amines have been shown to be useful for the extraction of acids by LMs (20); these extraction systems are limited by the requirement of a low pH in order to have the carrier and/or the acid protonated, hence preventing their exploitation in LM enzyme systems using moderate pHs. This problem is precluded by incorporating a quaternary ammonium salt in the LM, particularly TOMAC, trioctylmethylammonium chloride--Aliquat 336 (Henkel Corp.) or Adogen 464 (Sherex Corp.), which is ionized at all pHs and can freely exchange with aqueous anions. TOMAC has been applied in LM systems for both extraction of amino acids and for LM enzyme systems. Extraction of amino acids can be accomplished, for example, by using excess counterion in the internal aqueous phase (9). Enzymatic reactions of amino acids or other anions are carried out by using LM encapsulated enzymes: the substrate is transported by TOMAC to the internal aqueous phase of the LM where subsequent enzymatic reaction occurs (19).

Transport of L-tyr through LMs containing TOMAC has been studied in this laboratory. The emulsions are made according to Figure 4(a) with either plain buffer solution or buffer with chloride counterion. Then the emulsion is dropped into an external aqueous solution containing dilute L-tyr, Figure 4(b). Table I shows the transport of L-tyr through LMs containing TOMAC for various external aqueous phase pHs (the internal aqueous phase was buffered at pH 5.4). At pH 8.6, where only 24% of the amino group is deprotonated, no transport of L-tyr was observed. This fully demonstrates that anion carriers cannot transport zwitterions. At pH

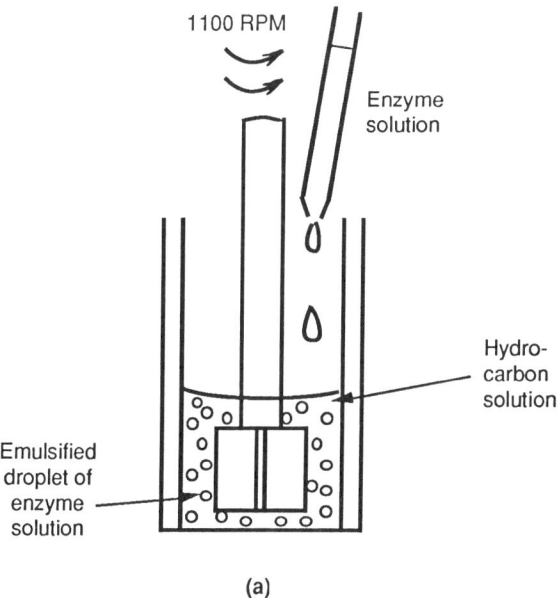

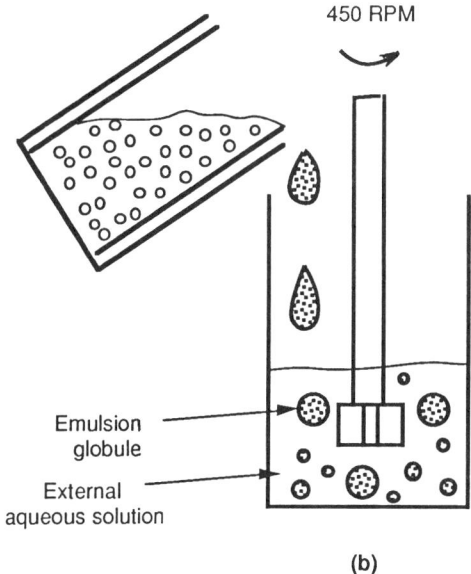

Figure 4. LM formulation: (a) Emulsification under high shear; (b) Dispersion of emulsion in external aqueous solution under low shear.

10.0, where 89% of the amino group is deprotonated and 46% of the hydroxyl group is deprotonated, L-tyr is readily transported through the LM. Including chloride counterion in the internal aqueous phase further improves transport (Table I).

Table I. Percent of L-tyrosine transported into the LM from the external aqueous solution (%T) vs. time for a batch reactor

external: Time, min.	pH 8.6 % T	pH 10.0 %T	pH 10.0 + NaCl %T
0	0	0	0
30	0	59	64
60	0	56	62
120	0	42	50
240	0	26	35

L-tyrosine can be transported into the LM via anion carriers; however, the transport is not generally reversible. For example, the conversion of L-tyr to L-DOPA by the LM encapsulated enzyme tyrosinase is favorable at neutral or lower pHs (as discussed later): any L-DOPA formed would therefore be trapped as a zwitterion in the internal aqueous phase. A non-zwitterionic form of L-tyr is thus needed to allow for bidirectional transport of the substrate and products, and this can be accomplished by blocking the amino or carboxyl group. The N-substituted N-formyl-L-tyrosine (Figure 5(a)) is anionic at moderate pHs and can therefore be transported readily without the need for basic conditions (Table II). There is some effect of pH on the anion transport of N-formyl-L-tyr, possibly due to ionization of the tyr derivative at pH 7.0. Comparing Table II to Table I shows that, without the trapping of the zwitterionic form in the internal aqueous phase, the concentration of N-formyl-L-tyr tends more toward the equilibrium value of 17% transported estimated for this particular system (Simmons, D. K. PhD Thesis, Georgia Institute of Technology, in preparation). Transport of N-formyl-L-tyr is reversible between the internal and external aqueous phases.

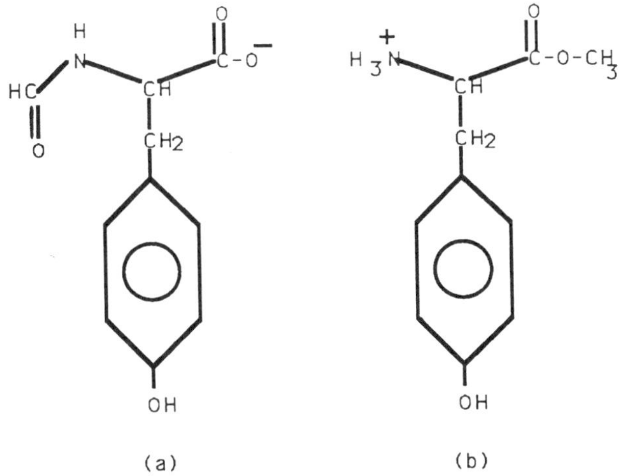

Figure 5. L-Tyrosine derivatives: (a) N-formyl-L-tyr, anionic form at moderate pH levels; (b) L-tyr methyl ester, cationic form at moderate pH levels.

Table II. Percent of N-formyl-L-tyrosine transported (for TOMAC carrier) or L-tyr methyl ester transported (for POPPA carrier) into the LM

CARRIER:	TOMAC			POPPA
external:	pH 5.4	pH 7.0	pH 7.0 + NaCl	pH 5.4
Time, min.	%T	%T	%T	%T
0	0	0	0	0
30	10	15	24	32
60	15	16	25	40
120	20	14	23	44
240	22	15	19	44

The use of TOMAC in LM enzyme systems is not without problems. This carrier has a relatively high HLB, e.g. about 12 (21), which is evidenced by the fact that it destabilizes the emulsion, generating turbidity in the external aqueous phase and leading to LM breakage. A cosolvent such as n-decyl alcohol will improve the stability of LMs containing this carrier (9). However, new problems arise in that even high quality decyl alcohol (Aldrich 99+ % Gold Label) contains phenols. These impurities leach out into the internal and external aqueous phases, not only causing contamination and interference with UV spectral measurements of the external aqueous phase but also producing unwanted side reactions with enzymes (e.g. tyrosinase) in the internal aqueous phase.

CATION CARRIERS

Carriers for cation transport in LMs have been demonstrated only very recently. Yagodin et al. (22) has proposed the use of di-2-ethylhexylphosphoric acid (D2EHPA) for the LM extraction of amino acids. Lorbach et al. (23) and Draxler et al. (24) have recommended bis-(2-ethylhexylthio)phosphoric acid for LM extraction of metals including zinc. These diesters of phosphoric acid contain a single ionizable proton which can exchange with other cations as shown in Figure 2(b).

Development of cation carriers is currently underway in this laboratory, fueled by the desire to produce the prodrug L-DOPA methyl ester from L-tyrosine methyl ester in an LM containing coencapsulated tyrosinase and ascorbate. The strongly acidic dinonylnapthalenesulfonic acid is unacceptable for use in LMs due to its requirement of aromatic cosolvents (which cause contamination and tend to separate from the aliphatic HC phase). Di-2-ethylhexylphosphoric acid (DEHPA) produces very unstable LMs, while the more hydrophobic di(p-octylphenyl-)phosphoric acid (POPPA) yields reasonably stable LM formulations. The polyamine surfactant (Paranox 100), which interacts with the acidic carrier, is replaced with Lubrizol #3702, originally suggested as an alternate membrane strengthening agent by Mohan and Li (18). Transport of L-tyrosine methyl ester (L-tyrOMe), Figure 5(b), by POPPA is considerably greater than that observed for anion transport of N-formyl-L-tyr, as shown in Table II. Comparison of Tables I and II shows that transport of L-tyrOMe by POPPA more closely approaches that for L-tyr (where the zwitterion is trapped inside the LM) than the transport observed for N-formyl-L-tyr (where transport is reversible). Since L-tyrOMe transport is reversible, a preliminary explanation is that POPPA remains bound with considerable amounts of L-tyrOMe in the HC phase. Attempts at using POPPA with a tyrosinase/ascorbate LM system have resulted in gross enzyme deactivation. The comparatively less acidic carboxylic acids are therefore being studied for application as cation carriers.

LM ENCAPSULATED ENZYMES

Enzyme Immobilization. The effective utilization of enzymes in industry requires that some form of immobilization technique be applied (25). Immobilization of enzymes provides numerous operational advantages -- ease of product separation; the possibility of continuously recycling the enzyme catalyst; flexibility in reactor design; facility for carrying out sequential reactions; and, sometimes, enhanced stability or selectivity of the enzymatic catalyst. Multienzyme systems, which demand mutual interaction of protein molecules during the catalytic cycle, render simple immobilization techniques such as polymer attachment unsuitable. Alternate immobilization techniques, for example, physical containment methods such as entrapment within crosslinked polymers (e.g. polyacrylamide) or within semipermeable membrane devices (e.g. hollow fiber or ultrafiltration cells) are therefore needed for such systems (26-28). However, these noncovalent attachment techniques can lead to leakage from the polymer network or through the membrane, and diffusional restrictions or non-specific adsorption complications can also occur. Encapsulation within "permanent" microcapsules whose membranes are formed by interfacial polymerization or coacervation of preformed polymers has been pioneered and subsequently developed by Chang (29). Microcapsules are suitable for multienzyme systems, allowing wide latitude in the relative concentrations and absolute amount of protein immobilized.

Encapsulation of enzymes in LMs offers further improvements for immobilization of complex enzyme systems, as the enzymes / cofactors, etc. are situated in aqueous droplets surrounded by a stable liquid hydrocarbon film (Figure 1). Instead of the physical pores present in microcapsules, the HC barrier, which has a diffusion thickness of about 0.1-1.0 µ, effectively blocks all molecules except those which are oil-soluble or transportable by the selected carriers. Encapsulation of enzymes in LMs is accomplished simply by emulsifying aqueous enzyme solutions. Hence, LMs offer many advantages over other systems used for separation and enzyme immobilization: they are inexpensive and easy to prepare; they promote rapid mass transport; they are selective for various chemical species; they can be disrupted (demulsified) for recovery of internal aqueous solutions; gradients of pH and concentration (even of small molecules) can be maintained across the HC barrier; multiple enzyme / cofactor systems can be coencapsulated; and enzymatic reaction and separation can be combined. Some of the potential disadvantages of LMs for enzyme encapsulation have been discussed earlier.

Previous studies. Only a few select groups have investigated LM enzyme encapsulation. May and Li (30, 31) developed the LM enzyme encapsulation concept and demonstrated its utility for a number of enzymes. Phenolase (tyrosinase) was used to show that enzyme activity is retained inside the LM with negligible enzyme leakage. To verify that enzyme denaturation was not occurring in the LM, encapsulated trypsin was recovered by demulsification: active site titration indicated only a small loss of enzyme activity (30). Analysis of kinetic data for the hydrolysis of urea by LM encapsulated urease indicated that diffusional restrictions are present, *i.e.* transport of the substrate (urea) to the enzyme-containing internal aqueous phase by the passive transport mechanism is slower than the diffusion of substrate to enzyme expected when both are present in a single aqueous solution. May and Landgraff (32) first demonstrated that LM encapsulation of a multienzyme-cofactor system is possible: the cofactor NAD^+ was regenerated inside the LM by a coupled YADH/NADH / diaphorase system.

Significant research on LM enzyme encapsulation systems has also been conducted at the University of Hannover, West Germany. Scheper et al. (19) proposed the use of LMs to resolve racemic D,L-phenylalanine methyl ester with encapsulated chymotrypsin. This enzyme cleaves the ester bond of the L-isomer only. The process employed Adogen 464 (TOMAC) as an anion carrier, but the pHs used were such that any L-phenylalanine formed would be zwitterionic: LM transport of zwitterions would be expected to be poor. Further work has included development of an LM enzyme reactor for detoxification of blood (33), reductive amination of α–ketoisocaproate by L-leucine dehydrogenase with a coencapsulated enzyme for regeneration of the cofactor NADH, and conversion of penicillin G to 6-aminopenicillinic acid (6-APA) in a continuous countercurrent reactor system (19). The last study demonstrated the important effect that the carrier manifests on LM stability and enrichment of the product (due to the fact that transport rate is a function of carrier concentration in the LM).

Tyrosinase/ascorbate system. This laboratory is currently investigating LM encapsulation of mushroom tyrosinase (Simmons, D. K. PhD Thesis, Georgia Institute of Technology, in preparation). Tyrosinase is a monooxygenase enzyme which is widely distributed in nature (34). The most readily available form of tyrosinase is extracted from the common mushroom *Agaricus bisporus* (35): this enzyme oxidizes phenol and various p-substituted phenols to o-catechol derivatives (o-diphenols), which are in turn oxidized to o-quinones (Figure 6). Since o-quinones are highly labile, especially in aqueous environments, they undergo a series of both enzymatic and non-enzymatic reactions to form colored polymers, melanins, which are the pigments observed in plants and animals. Oxidation by tyrosinase occurs rapidly under mild conditions in the presence of oxygen; the hydroxylation of phenols is regiospecific, as o-catechols are always formed. Thus, it would be facile to prepare L-DOPA (Figure 6, R = -CH_2-CH-(COOH)NH_2), a valuable drug used for treatment of Parkinson's disease (36), from L-tyrosine (a readily available amino acid) if not for the fact that the catechols are short-lived intermediates.

Various reducing agents can be incorporated into the reaction mixture to shuttle the o-quinone back to the o-catechol. Vitamin C (ascorbate) has been shown to be particularly effective in reducing o-quinones such as dopaquinone to L-DOPA such that substantial amounts of L-DOPA can accumulate (37). Ascorbate reduces the o-quinone back to the o-catechol and is itself oxidized to dehydroascorbate (Figure 6). A steady-state concentration of the o-catechol is maintained until the ascorbate is consumed (38), and then the reaction proceeds with o-quinone and subsequent melanin formation (39). Recently, it was demonstrated that tyrosinase isolated from frog epidermis could be immobilized on a solid support and used to produce L-DOPA in a variety of reactor configurations (40). Ascorbate was used as the reducing agent, and it was seen that enzyme inactivation was lessened as the ascorbate concentration was increased. One problem with using L-tyrosine as a substrate is its low solubility in aqueous solutions. To process a large amount of substrate in dilute aqueous solution would therefore require a large amount of ascorbate in order to maintain a high concentration in the substrate/product solution. This would not only be expensive but would also entail a separation process to free the L-DOPA from the ascorbate.

An LM enzyme system, which can combine reaction and separation as well as maintain a concentration gradient between internal and external aqueous phases, is thus being investigated for the L-tyrosine to L-DOPA conversion reaction. This combined enzyme reactor / separation system

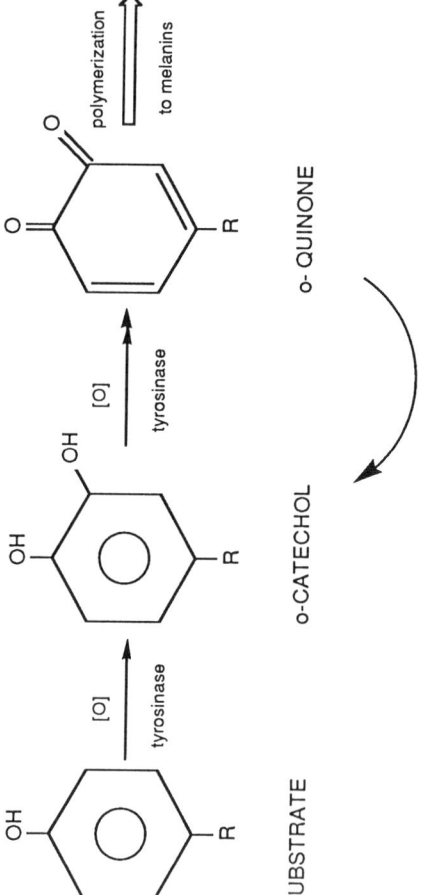

Figure 6. Oxidation of phenols catalyzed by tyrosinase showing the two enzymatic oxidations, the melanin pathway, and the ascorbate shuttle.

has the potential to separate L-DOPA from the enzyme as it is formed, preventing further enzymatic oxidation of the intermediate to useless products. High concentrations of ascorbate can be maintained in the internal aqueous phase to slow the overall oxidation of L-DOPA and thus improve the opportunity for L-DOPA to transport out and be recovered. If a suitable carrier is provided, then the L-tyrosine can diffuse into the internal aqueous phase and the intermediate L-DOPA can diffuse back out to the external aqueous phase. The overall system is shown in Figure 7.

Model substrates. To demonstrate the feasibility of the tyrosinase/ ascorbate LM system for isolation of reactive intermediate oxidation products such as L-DOPA, preliminary work has been carried out with model substrates. Phenol was used to demonstrate the enzyme activity (30), emulsion properties, and transport properties of this LM system. A tyrosinase-containing emulsion was made according to Figure 4, with the enzyme emulsion being dropped into an external aqueous solution containing dilute phenol. The rate of stirring was varied to observe the degree of dispersion of the LM globules in the external aqueous solution. Use of excessive shear caused disruption of the emulsion, as evidenced by gross enzyme leakage (rapid formation of melanins in the external aqueous phase). Normal reaction of phenol and the encapsulated enzyme was evidenced by a gradual browning of the emulsion during the time course of the experiment (typically four hours). The external aqueous solution was periodically sampled and monitored for phenol depletion and melanin formation by UV spectrophotometry. Lack of enzyme leakage was verified by methods similar to those developed by May and Li (31).

It is not feasible to isolate o-catechol from phenol even when ascorbate is incorporated in the LM, since the o-benzoquinone (Figure 6, R = -H) formed is highly labile and readily forms melanins. Decreasing the pH slows the formation of melanins as a result of lowered hydroxide ion concentration (41), but the enzyme's stability and activity begin to suffer (42): pH 5.4 was found to be approximately the lower limit. Since any p-substituted phenol yields an o-quinone having only one conjugate addition (Michael acceptor) site, it is expected that nucleophilic attack on such an o-quinone would be much slower than that occurring on o-benzoquinone itself. The model substrate p-hydroxyphenethanol (Figure 6, R = -CH_2-CH_2OH) has been found to freely diffuse into the LM and react with the enzyme/ascorbate system. Concentrations of coencapsulated tyrosinase and ascorbate were varied using this substrate until appropriate concentrations were found that yielded the catechol intermediate, 3,4-dihydroxyphenethanol, without accumulating any melanin products in the external aqueous phase over the course of the experiment (Figure 8). Without coencapsulated ascorbate, melanins are detected in the external aqueous solution (Figure 8(a)), while Figure 8(b) shows formation of product (the shoulder at 282 nm, verified by HPLC) and no melanin formation. This demonstrates the isolation of the reactive intermediate oxidation product from tyrosinase: only the substrate and the catechol

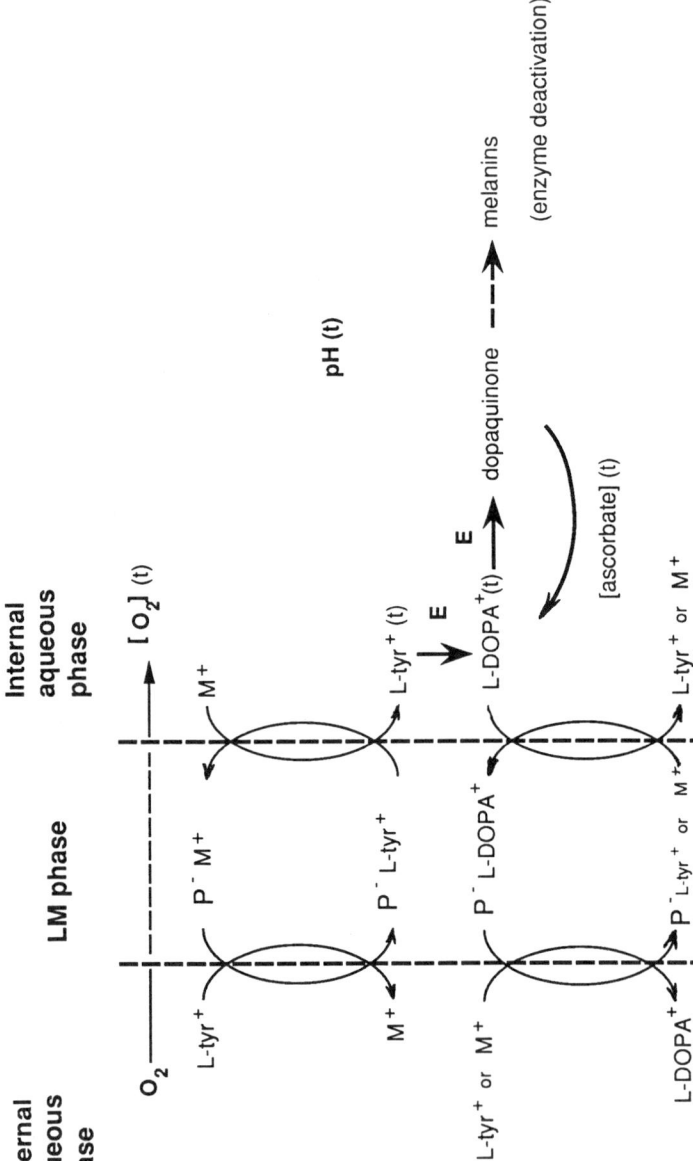

Figure 7. Schematic representation of the complete tyrosinase / ascorbate system in a Liquid Membrane showing diffusion of oxygen, carrier-facilitated transport of substrate and product through the LM, and the reactions occurring in the internal aqueous phase.

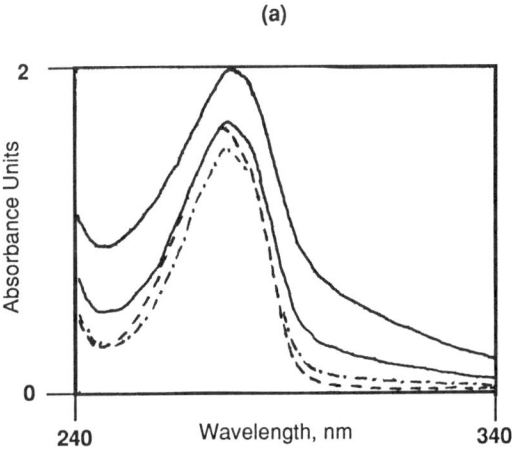

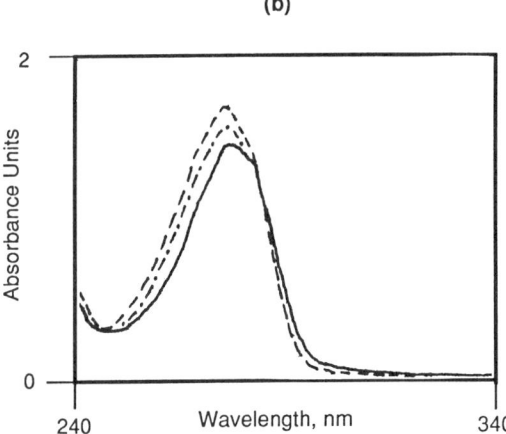

Figure 8. Isolation of intermediate 3,4-dihydroxyphenethanol from the enzymatic reaction of p-hydroxyphenethanol with LM encapsulated tyrosinase / ascorbate. Overlayed UV spectra: (a) no coencapsulated ascorbate: – – – 0 minutes-substrate peak; –·–·– 60 min.; —— 120 and 240 min., showing melanin buildup (b) coencapsulated ascorbate: – – – 0 min.; –·–·– 30 min.; —— 240 min.

product are present in the external aqueous solution (along with buffer ions and a very low concentration of ascorbate/EDTA). The enzymes and concentrated reducing agent are confined to the interior aqueous phase of the LM.

LM Reactor Configurations. The choice of reactor configuration to be employed in an LM system involves several considerations. The foremost among these is whether (1) the LM system is being used for separation applications alone where substances are accumulated in the internal aqueous phase or (2) the LM system is being used to conduct a reaction, for example, enzymatic conversion of a substrate in the internal aqueous phase to yield a high value product. The reactor operation is governed by whether this product remains in the internal aqueous phase or diffuses back to the external aqueous phase. A second consideration is the type of application. For example, in clinical or biomedical applications, a batch reactor may suffice; in industrial scale operations, a continuous flow reactor operation would be more appropriate. Yet another consideration involves the desirability of regenerating the enzyme and/or cofactor contained in the internal aqueous phase. In any reactor, the desirable characteristics also include the need to maintain emulsion stability as well as a large driving force for transport of all desired species across the liquid membrane.

Since the advent of LM systems, the major focus of their industrial application has been in separation processes. Only very recently have these systems been considered for simultaneous reaction and separation processes. Previous studies have involved potential applications which include (1) water treatment to remove organics, inorganics, metal traces, etc. (10); (2) uranium recovery from wet process phosphoric acid (43); (3) extraction of copper from acid leachates (44, 45); (4) detoxification of blood (33); (5) recovery of amino acids (46); (6) catalytic synthesis of acetaldehyde from CO and ethylene using the Wacker process (47); and (7) production and separation of L-leucine with continuous cofactor regeneration (19). The above-mentioned applications are not exhaustive but serve to illustrate the diversity of potential LM applications. Furthermore, the reactor configurations employed in separation processes can still provide guidance in developing reactor operations which involve both reactions as well as separation. Below is a brief summary of some of the reactor configurations used with LM systems.

The simplest mode of LM operation available is a batch reactor. The LM emulsions containing enzymes or cells are kept agitated in a vessel by using impellers. Periodic samples of external aqueous phase can be taken to monitor the reaction rate; it is much more difficult to monitor the conditions in the internal aqueous phase. Impeller speed becomes an important parameter in batch operation, as it determines the size of the emulsion globules and therefore influences the emulsion stability and the mass-transfer resistance (46).

The operation of a well-mixed reactor in a continuous flow mode (CSTR) provides some advantages over a batch operation in that the interpretation of reaction kinetics and mass-transfer resistance becomes less cumbersome. Two variations of CSTR have been employed. Scheper et al. (19) and Völkel et al. (33) have used a CSTR fitted with a membrane filter. This arrangement permits the continuous flow of external aqueous phase stream, while the emulsion phase remains in the reactor in a batch mode. Such a configuration is especially appropriate in those instances where there is little or no accumulation of products in the emulsion. The system used by Völkel et al. (33) for blood detoxification consisted of phenol in bovine blood or plasma as the exterior phase and a liver enzyme and cofactor in the interior phase. These investigators designed special experiments to compare a mass-transfer limited separation (where the interior reagent is base, thus having essentially instantaneous conversion of phenol to phenolate) to that of a kinetic limitation (plain water in the interior phase). It was observed that as the enzyme concentration in the interior phase was increased, the system moved from a kinetically-limited regime to a mass-transfer limited regime.

The second mode of CSTR operation is that used by Thien (17) and by Li and Shrier (10). Here, both the external phase and the LM emulsion are in a continuous flow mode. The reactor effluents are sent to gravity settlers where the exterior phase is separated from the emulsion phase. The emulsion phase is then demulsified to recover the product followed by remulsification and recycle back to the reactor. Hatton and Wardius (48) have developed the advancing front model for the analysis of such staged LM operations. Thien (17) employed this scheme to remove the amino acid L-phenylalanine from simulated fermentation broth (dilute aqueous solution).

CONCLUSION

The limited work with LM encapsulated enzyme systems shows that the method is indeed viable for enzyme immobilization. The ability to encapsulate multiple enzyme / cofactor systems (32) while maintaining a selective permeation barrier is a distinct advantage of LMs. Since enzymes in emulsions can associate with a hydrocarbon layer, there is some degree of hydrophobicity present in the enzymes' microenvironment. A hydrophobic environment is known to be beneficial for the stability of many potentially useful enzymes which deactivate in aqueous solution. Another advantage of LM encapsulation is that the enzyme can be protected from conditions and/or reagents present in the bulk external aqueous phase which might be deleterious to its activity.

Even with these potential advantages, LMs have been only slowly developed as useful industrial vehicles. While many theoretical modelling studies of LMs as separation tools have been performed in the past decade, it has been only recently that industrial processes utilizing LM separations have been realized (24, 49). Practical concerns associated with LMs such as membrane breakage and swelling, coupled with the novelty of the LM concept as an engineering process, have contributed to this problem. The ability to combine enzymatic reaction(s) and separation may lead to further development of LMs for biotechnology processes.

It has been shown that carrier-facilitated transport of amino acid ions plays a central role of development of LM enzyme systems. Better anion carriers and new cation carriers are needed to exploit enzymatic processes where amino acids (or derivatives) and their products must transport readily through the LM. Facile transport with little or no enzyme deactivation is required. Design of new carriers tailored specifically for LM processes will help pave the way for the industrial development of liquid membrane enzyme reactors.

ACKNOWLEDGMENTS

This work has been supported in part by a Georgia Tech Biomedical Research Support Group Grant and in part by a National Science Foundation Graduate Fellowship (DKS).

LITERATURE CITED

1. Li, N. N., U. S. Pat. 3,410,794 , Nov. 12, 1968.
2. Li, N. N., Cahn, R. P., and Shrier, A. L., U. S. Pat. 3,779,907, Dec. 18, 1973.
3. Davies, J. T., Proc. 2nd International Congress Surface Activity, London, 1957, 1, 426.
4. Becher, P., Emulsions: Theory and Practice, 2nd ed., Reinhold, NewYork, 1965.
5. Kriechbaumer, A., and Marr, R., ACS Symp. Ser. 272, 1985, pp. 381.
6. Sutheim, G. M., Introduction to Emulsions, Chemical Publishing Co., New York 1946.
7. Stroeve, P., and Varanasi, P. P., J. Coll. Int. Sci., 1984, 99, 360.
8. Tadros, T. F., and Vincent, B., Encyclopedia of Emulsion Technology, Becher, P., ed., Dekker, New York, 1983, Vol. 1.
9. Thien, M. P., Hatton, T. A., and Wang, D. I. C., In Separation, Recovery, and Purification in Biotechnology, Asenjo and Hong, eds., ACS Sym. Ser. 314, 1986, American Chemical Society, Washington.
10. Li, N. N., and Shrier, A. L., In Recent Developments in Separation Science, Li, N. N., ed., CRC Press, Cleveland, 1972, Vol. 1, 47.
11. Cahn, R. P., and Li, N. N., Separation Science, 1974, 9, 505.
12. King, C. J., In Handbook of Separation Process Technology, R. W. Rousseau, ed., 1987, Wiley, New York.
13. Noble, R. D., Way, J. D., and Bunge, A. L., Ion Exchange and Solvent Extraction, Marinsky, J. A., and Marcus, Y., eds., Dekker, New York, 1988, Vol. 10.
14. Behr, J. and Lehn, J., J. Am. Chem. Soc., 1973, 95, 6108.
15. Greenstein, J. P., and Winitz, M., Chemistry of the Amino Acids, Wiley, New York, 1961.
16. Lehninger, A. L., Principles of Biochemistry, Worth, New York , 1982.
17. Thien, M. P., PhD Dissertation, Massachusetts Institute of Technology, 1988.
18. Mohan, R. R. and Li, N. N., Biotech. Bioeng., 1975, 17, 1137.

19. Scheper, T., Likidis, Z., Makryaleas, K., Nowottny, Ch., and Schügerl, K., Enzyme Microb. Technol., 1987, 9, 625.
20. Cowan, R. M. and Ho, C. S., "Separating Lactic Acid from Fermentation Media with Liquid Surfactant Membranes," presented at the 194th National ACS Meeting, New Orleans, August 30-September 4, 1987.
21. McCutcheon, J. W., Detergents and Emulsifiers Annual, J. W. McCutcheon, Inc., Morristown, New Jersey, 1988.
22. Yagodin, G. A., Yurtov, E. V., Golubkov, A. S., ISEC '86 International Solvent Extraction Conference Preprints, Dechema, Munich, Germany, Sept. 11-16, 1986, Vol III.
23. Lorbach, D., Bart, H. J., Marr, R., ISEC '86 International Solvent Extraction Conference Preprints, Dechema, Munich, Germany, Sept. 11-16, 1986, Vol III.
24. Draxler, J. , Furst, W., and Marr, R.,J. Membrane Sci., 1988, 38, 281.
25. Wiseman, A., ed., Handbook of Enzyme Biotechnology, Wiley, New York, 1985.
26. Butterworth, T. A., Wang, D. I. C. and Sinskey, A. J., Biotech.& Bioeng., 1970, 12, 615.
27. Fink, D. J. and Rodwell, V. W., Biotech.& Bioeng., 1975, 17, 1029.
28. Mosbach, K. and Larsson, P., Biotech. & Bioeng., 1970, 12, 19.
29. Chang, T. M. S., Biomedical Applications of Immobilized Enzymes and Proteins, T. M. S. Chang, ed., Plenum, New York, 1977, Vol. 1, Chapter 7.
30. May, S. W., and Li, N. N., In Enzyme Engineering, 1974, 2, 77.
31. May, S. W., and Li, N. N., In Biomedical Applications of Immobilized Enzymes and Proteins, T. M. S. Chang, ed., Plenum, New York, 1977.
32. May, S. W. and Landgraff, L. M., Biochem. Biophys. Res. Comm., 1976, 68, 786.
33. Völkel , W., Bosse, J., Poppe, W., Halwachs, W., and Schügerl, K., Chem. Eng. Commun., 1984, 30, 55 .
34. Robb, D. A., In Copper Proteins and Copper Enzymes, Lontie, R., ed., CRC Press, Boca Raton, FL , 1984, Vol. II.
35. Lerch, K., In Metal Ions in Biological Systems, Copper Proteins, H. Sigel, ed., Dekker, New York, 1981, Vol. 13.
36. Calne, D. B., and Sandler, M., Nature, 1970, 226, 21.
37. Evans, W. C., and Raper, H. S., Biochem. J., 1937, 31, 2162.
38. Miller, W. H., and Dawson, C. R., J. Am. Chem. Soc., 1941, 63, 3375.
39. El-Bayoumi, M. A., and Frieden, E., J. Am. Chem. Soc., 1957, 79, 4854 .
40. Vilanova, E., Manjon, A., and Iborra, J. L., Biotech.& Bioeng., 1984, 26, 1306.
41. Dawson, C. R., and Tarpley, W. B., Ann. NY Acad. Sci., 1963, 100, 937.
42. Ingraham, L, L., In Pigment Cell Biology, Proceedings of the Fourth Conference on the Biology of Normal and Atypical Pigment Cell Growth, Gordon, ed., Academic Press, New York , 1959, p. 609.
43. Hayworth, H. C., Ho. W. S., Burns, W. A., and Li, N. N., Sep. Sci. Technol., 1983, 18, 493.
44. Cahn, R. P., and Li, N. N., U. S. Patent 4,086,163 (1978).
45. Igawa, M., Matsumura, K., Tanaka, M., and Yanabe, Nippon Kaysku Karshi, 625 ; Chem. Abstr. 1981 95, 49869r.
46. Thien, M. P., and Hatton, T. A., Separation Science, 1988, 23, 819.
47. Ollis, D. F., Thompson, J. B., and Wolynic, E. T., AIChE J., 1972, 18, 457.
48. Hatton, T. A., and Wardius, D. S., AIChE J., 1984, 30, 934 .
49. Cahn, R. P., and Li, N. N., "Commercial Applications of Emulsion Liquid Membranes," presented at The Third Chemical Congress of North America, Toronto, Ontario, Canada, June 5-10, 1988.

RECEIVED September 27, 1989

Chapter 7

Pilot-Scale Membrane Filtration Process for the Recovery of an Extracellular Bacterial Protease

John J. Sheehan[1], Bruce K. Hamilton[1], and Peter F. Levy[2]

[1]Washington Research Center, W. R. Grace & Company—Conn., 7379 Route 32, Columbia, MD 21044
[2]Amicon Division, W. R. Grace & Company—Conn., 17 Cherry Hill Drive, Danvers, MA 01923

> A two stage process consisting of crossflow microfiltration to remove bacterial cells at an initial dry weight concentration of 10-12 g/l (average flux of 25 liters/m^2-hr) followed by ultrafiltration for ten-fold concentration of an extracellular protease product at an initial broth concentration of 0.3-0.6 g/l (average flux 40-50 liters/m^2-hr) demonstrated consistently high recovery yield (>90%) of enzyme from 100 liter fermentation broths. High protease product yield (>90%) in the cell separation step, which involved transmission of the enzyme through the microfiltration membrane, was achieved only under conditions of low transmembrane pressure (<5 psi) and high crossflow recirculation rate (> 1 meter/sec). To maintain the required low transmembrane pressure with high recirculation rate, it was necessary to pressurize the permeate chambers of the hollow fiber cartridges used for the cell separation step. In addition, use of a pump on the permeate outlet to maintain a constant permeate flow rate during the run resulted in increased flux performance and stability, while keeping transmembrane pressure low. For the subsequent enzyme concentration step, a regenerated cellulose spiral ultrafilter achieved 100% recovery of the protease. Economic analysis of the cell separation step indicates that the membrane process is twice as cost effective as a centrifuge and equivalent to a precoat filter, on a basis of unit cost of enzyme product recovered.

A two stage pilot scale membrane process was developed to recover an extracellular protease from a bacterial fermentation. This process was first tested in the laboratory to establish an alternative to the use of a semi-continuous disk centrifuge which had been used in our pilot plant to remove cells from the fermentor broth. Centrifugation proved to be inefficient because of product yield losses, and because of the need to repeat centrifugation and filtration steps downstream to remove cells and other solids not removed in the primary separation step. The data presented in this paper are from 100 liter scale fermentation recovery runs in our pilot plant which were carried out to test the feasibility and economics of using membranes to remove the cells, as well as to produce enzyme for use in in-house research and development activities.

DEFINITIONS

Discussion of crossflow membrane filtration requires definition of a number of specific terms to describe operating conditions. **Crossflow** or **tangential** flow refers to the principal direction of process flow relative to the membrane surface (see Figure 1). When the fluid to be filtered flows tangential to the surface of the membrane, shear forces along the membrane mitigate fouling by sweeping retained species from the membrane surface. Tangential flow may be quantified either as **recirculation rate**, or as the **average linear velocity** through the retentate channel, or as **wall shear rate** at the membrane surface. The **retentate** side of the filter contains the fluid which does not pass across the membrane, while the **permeate** side contains the fluid which passes through the membrane's filtration barrier. For such systems, the pressure driving force across the filter is quantified as the **average transmembrane pressure** (TMP), defined (Figure 1) as:

$$\text{Avg. TMP} = [(P_i + P_o)/2] - P_{per} \quad (1)$$

Three basic quantities are defined to describe membrane performance. Flux is the permeate flow rate normalized to total membrane filter area. For protein recovery in the cell separation step, **instantaneous protein transmission** can be measured by determining enzyme concentration simultaneously on the retentate and permeate sides of the filter during cell concentration. Percent transmission is calculated as:

$$\text{Percent Instantaneous Transmission} = \frac{[\text{Enzyme}]_{permeate}}{[\text{Enzyme}]_{retentate}} \times 100 \quad (2)$$

Because conditions are not constant during the batch concentration of the cell broth, protein transmission usually changes during the run. Thus transmission is not a consistent parameter for comparison of different runs. For purposes of summarizing protein transmission during the cell concentration step, a protein passage efficiency is used, which is defined as:

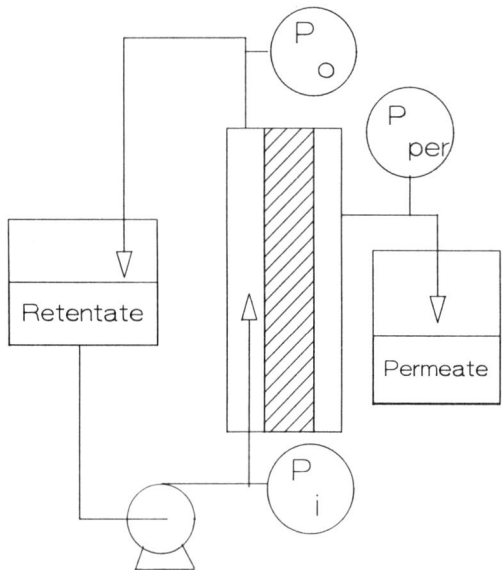

Figure 1. Generalized Schematic of Crossflow Membrane Filtration

$$\text{Passage Efficiency} = \frac{\frac{(\text{Grams of Enzyme})_{\text{final permeate}}}{(\text{Grams of Enzyme})_{\text{feed}}}}{\frac{(\text{Volume of Final Permeate})}{(\text{Volume of Feed})}} \times 100 \quad (3)$$

This is the weight fraction of enzyme in the feed actually recovered in the final permeate, normalized to the volume fraction of the feed collected as permeate. Unlike a simple yield calculation, this efficiency separates the effect of volume recovery from the effects of protein transmission. For example, for a five-fold volume reduction of the cell broth (before diafiltration), a protein passage efficiency of 100% corresponds to a yield of only 80%, because 20% of the protein originally fed is left behind in the retentate. For a constant passage efficiency, yield will go up or down with changes in the volume reduction of the feed. For computational purposes, equation (3) reduces to a simple ratio of enzyme concentrations in the feed and final permeates:

$$\text{Passage Efficiency} = \frac{[\text{Enzyme}]_{\text{final permeate}}}{[\text{Enzyme}]_{\text{feed}}} \times 100 \quad (4)$$

Physically, this simply means that, as passage efficiency drops, enzyme concentration will drop in the final permeate.

BACKGROUND

The use of ultrafiltration membranes for protein concentration and purification is well established, with the original work published almost 20 years ago (1, 6, 8, 15). Likewise, much has been published on the use of these types of membranes to concentrate bacterial suspensions (1, 2, 4-6, 10, 15, 17, 20, 22-25, 27, 28, 31-35), and, in some cases, to separate cells or cell debris from their products in solution (7, 9-13, 10, 17-19, 21, 23, 26, 30). The latter, however, is less well established for the recovery of protein products.

The ability to separate cells from a high molecular weight extracellular product is one of the touted advantages of using microporous membranes over ultrafiltration membranes. Microporous membranes compete with centrifuges and rotary vacuum filters for this type of recovery. High yield protein/cell separations with membranes, however, have not been well established in the literature. Rejection coefficients for extracellular proteins in bacterial cell broth have been reported to vary widely from 5 to 100% for microporous membranes of 0.1 to 0.6 micron nominal pore sizes (9, 10, 17-19), depending on the nature of the cell broth, size of the protein, and membrane characteristics (such as pore size). This separation is also problematic for protein/cell debris systems (7, 9-12, 19, 21, 26). A higher degree of success has been reported for the separation of proteins from whole cells for mammalian cell systems. Work in the area of plasmapheresis has, for example, demonstrated an ability to separate proteins from whole cells in blood with crossflow microfilters (30). High yield recovery of immunoglobulin proteins from mammalian cell suspensions has also been reported (23). Nonetheless, little data has been reported at the pilot scale (100 liters and above) which has demonstrated a consistent ability to recover an extracellular protein product from bacterial suspensions in high yield using crossflow microfiltration.

Finally, there is not a clear understanding of the relationship between membrane operating conditions and flux and protein passage. For example, different studies of the separation of cells or cell debris from a protein in solution have demonstrated conflicting results regarding the effect of transmembrane pressure on protein transmission through a microporous membrane. For cell lysate systems, Kroner et al. (19) observed an exponential decrease in protein transmission when TMP was increased from 0 to 8 psi. Quirk (26) reported a similar loss in enzyme yield at high (40 psi) TMP for a cell debris separation. Le et al. (21) and Gabler et al. (12), on the other hand, both report a strong increase in protein transmission with increasing TMP for cell lysate systems. The data reported in this paper support the former observation, showing a drop in protein transmission when TMP is increased to levels of only 10 psi.

In recent years, a new mode of operation has been suggested in which permeate flow is controlled at a constant rate, while transmembrane pressure is kept low. This approach was developed and tested for mammalian cells (23), where concern about cell damage due to compression on the membrane surface prompted an interest in controlling TMP and flux rate. The data reported in this paper demonstrate that the controlled permeate flow operating scheme also works well for bacterial systems, yielding more stable and improved flux as well as high protein passage.

THEORY

Theoretical analysis of crossflow filtration is often based on a steady state analysis of convective and diffusive transport of retained species between the membrane surface and the bulk fluid in the retentate. The build up of retained species at the membrane surface, which then acts as the major resistance to permeate flow, is known as concentration polarization. This analysis predicts that flux declines linearly with the log of the concentration of retained species (6):

$$J = K \ln(C_g/C_b) \tag{5}$$

where C_b is the bulk concentration of retained species, C_g is a "gel" concentration of retained species at the membrane surface, and K is a mass transfer coefficient describing backdiffusion of retained species into the bulk fluid. This model was developed to predict performance of ultrafiltration membranes used for protein concentration, and has been modified to predict membrane flux for cell concentration (6, 15, 25, 36, 37). Models based on polarization theory suggest that high recirculation rates (which affect the mass transfer coefficient, K) are needed to minimize fouling (1, 6, 15, 25, 36, 37). In addition, these models predict that, above a certain level, increased transmembrane pressure will not influence flux. Some researchers have suggested that, for cell systems, there is a strong dependence of flux on TMP, with no indication of a TMP-independent regime (6). Others researchers seem to have verified this idea experimentally (12, 35). Nevertheless, most researchers have observed a limiting TMP where flux appears independent of transmembrane pressure (1, 10, 21, 23, 28, 30, 32). More recent efforts to model the fluid mechanics of crossflow filtration have involved the use of standard cake filtration theory combined with a fluid mechanical analysis of suspensions in crossflow to predict flux performance (3, 14, 16, 28, 29, 36). Some of these models confirm the idea that, in order to reduce the fouling rate, it is important to keep transmembrane pressure as low as possible, while maintaining recirculation rate as high as possible (3).

DESCRIPTION OF PILOT SCALE ENZYME PRODUCTION AND RECOVERY PROCESS

One hundred liters of broth were produced in a 150 liter Chemap fermentor. In this fermentation, a marine bacterium (slightly smaller than E. coli) was grown to a dry cell density of 10-12 g/l in 12 to 14 hours. A complex nutrient medium (with about 0.5 g/l of insolubles) was used which contained a mixture of amino acids, proteins and trace minerals. SAG 4130 (Union Carbide), a silicone and polyglycol mixture, was used for foam control at dosages ranging from 100 to 500 ppm. Total protein at the end of the fermentation was roughly 6 g/l, of which 0.3 to 0.6 g/l was the protease product. The protease has a molecular weight of about 40,000 daltons. All other proteins are relatively small (less than 10,000 daltons).

A schematic and an equipment diagram of the membrane recovery system are shown in Figures 2a and 2b. The cell separation step consisted of two stages: a cell concentration step in which a five fold volume reduction of the cell broth was done, and a cell wash (or diafiltration) step in which four volumes of a salt solution were pumped into the cell concentrate at a rate equal to the permeate flow rate to wash out residual enzyme. These two steps were done using the same membrane system: two 23 square foot, 500,000 MWCO (1.0 mm ID) polysulfone hollow fiber units (AG Technology (Needham, MA), Catalog # UFP-500-E- 55). Though considered ultrafilters

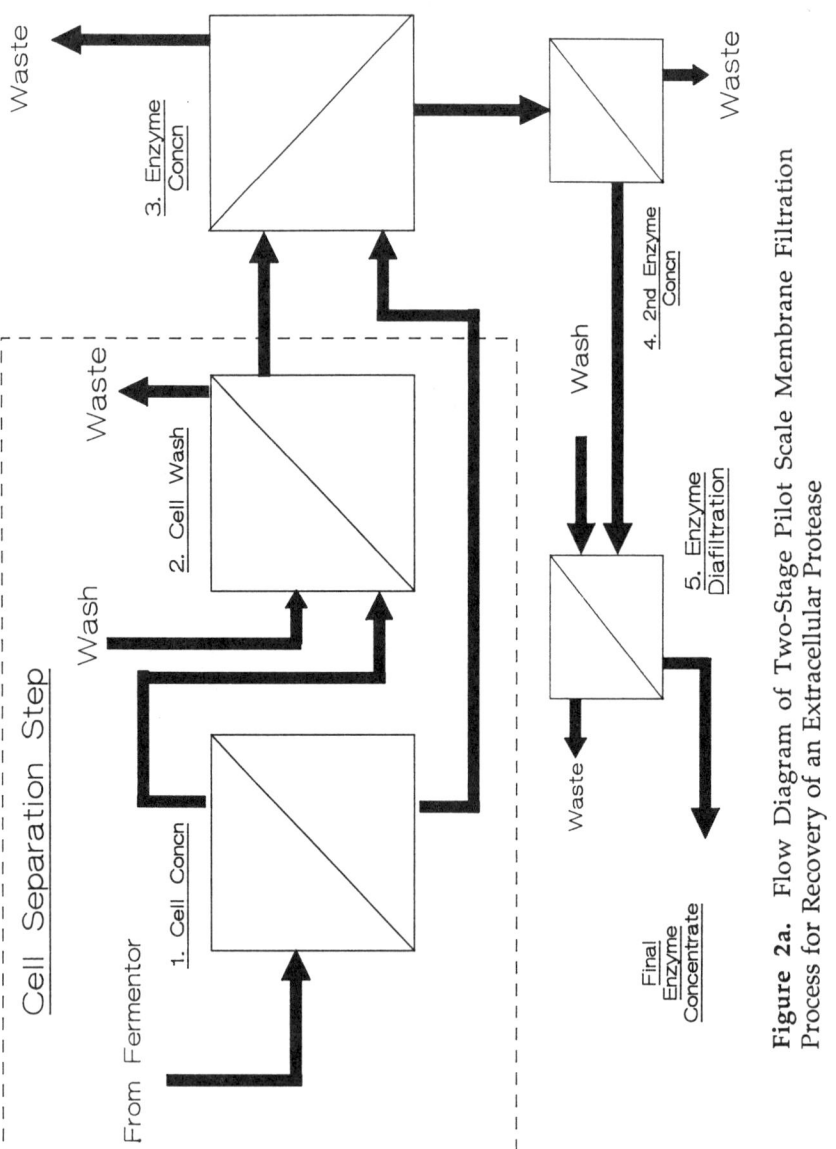

Figure 2a. Flow Diagram of Two-Stage Pilot Scale Membrane Filtration Process for Recovery of an Extracellular Protease

7. SHEEHAN ET AL. *Recovery of an Extracellular Bacterial Protease* 137

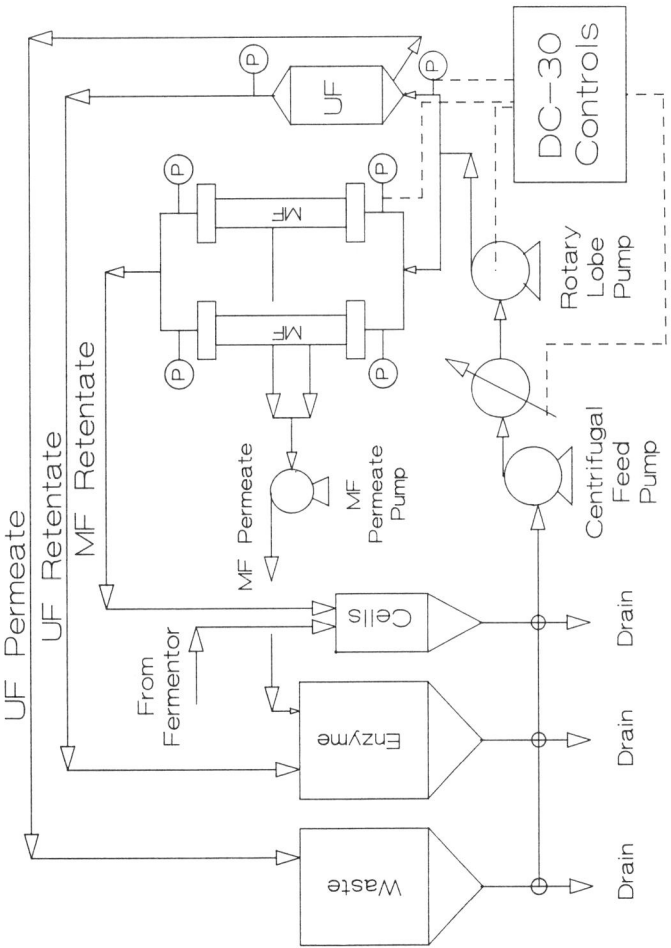

Figure 2b. Equipment Diagram for Two-Stage Pilot Scale Membrane Filtration Process for Recovery of an Extracellular Protease

by the manufacturer, these cartridges are microporous in nature, having nominal pore sizes of 0.02 microns (20).

The combined solutions containing the enzyme from the cell separation step (approximately 160 liters) were then concentrated using a 40 square foot, 10,000 MWCO regenerated cellulose S40Y10 Diaflo™ spiral membrane (Amicon Division of W. R. Grace & Co.- Conn., Danvers, MA). A roughly ten fold concentration of the enzyme was achieved in this step. Further concentration and diafiltration of the protein product solution were carried out at the lab scale.

An Amicon DC-30 pilot membrane system was used to supply recirculation pumps and controls (for pressure, flow, and temperature) for both the cell separation devices and the Amicon spiral ultrafilter. Temperature was controlled at 20°C using a heat exchanger on the pump inlet supplied with 4°C water on the shell side. A sanitary centrifugal pump (Tri-Clover, Kenosha, WI; model C114MF5GT-S) was used in a number of the tests as a feed pump to the DC-30 to prevent cavitation at the high recirculation rates of around 100 liters per minute.

As a substitute for the hollow fiber units, a Series "S" Ioplate™ plate-and-frame device (Amicon Division of W. R. Grace & Co. - Conn., Danvers, MA) was used in one series of tests to screen the effects of operating conditions on protein passage during cell concentration. This plate-and-frame device allowed sampling and measurement of individual permeate flows from each plate in series. Twenty one plates in series were used in these tests, each containing 0.7 ft² of Gelman 0.45 micron polysulfone membrane. The Ioplate™ plate-and-frame device was used to qualitatively evaluate the effect of transmembrane pressure on transmission of the protease, at constant recirculation rates. Again, the Amicon DC-30 system was used to supply pumping and controls.

RESULTS FOR CELL SEPARATION

Membrane Flux Performance for the Hollow Fiber Units.
Figures 3a and 3b show the results of a cell separation pilot scale run in which transmembrane pressure was maintained constant at 4.0 psi and linear velocity through the fibers was maintained at 1.0 m/sec (100 lpm). Transmembrane pressure was controlled with a backpressure valve on the combined permeates from the hollow fibers. Note that the flux vs. concentration curve for this run does not conform to theoretical predictions for concentration polarization (see Figure 3b). Flux on this semi-log plot does not decline linearly, but, rather, shows two phases of fouling. The initial phase (approximately 20 minutes in duration) is characterized by a rapid loss in flux, while the second stage (approximately 90 minutes in duration) is more gradual. The initially rapid decline caused an unacceptably low average flux of 10 l/m^2-hr.

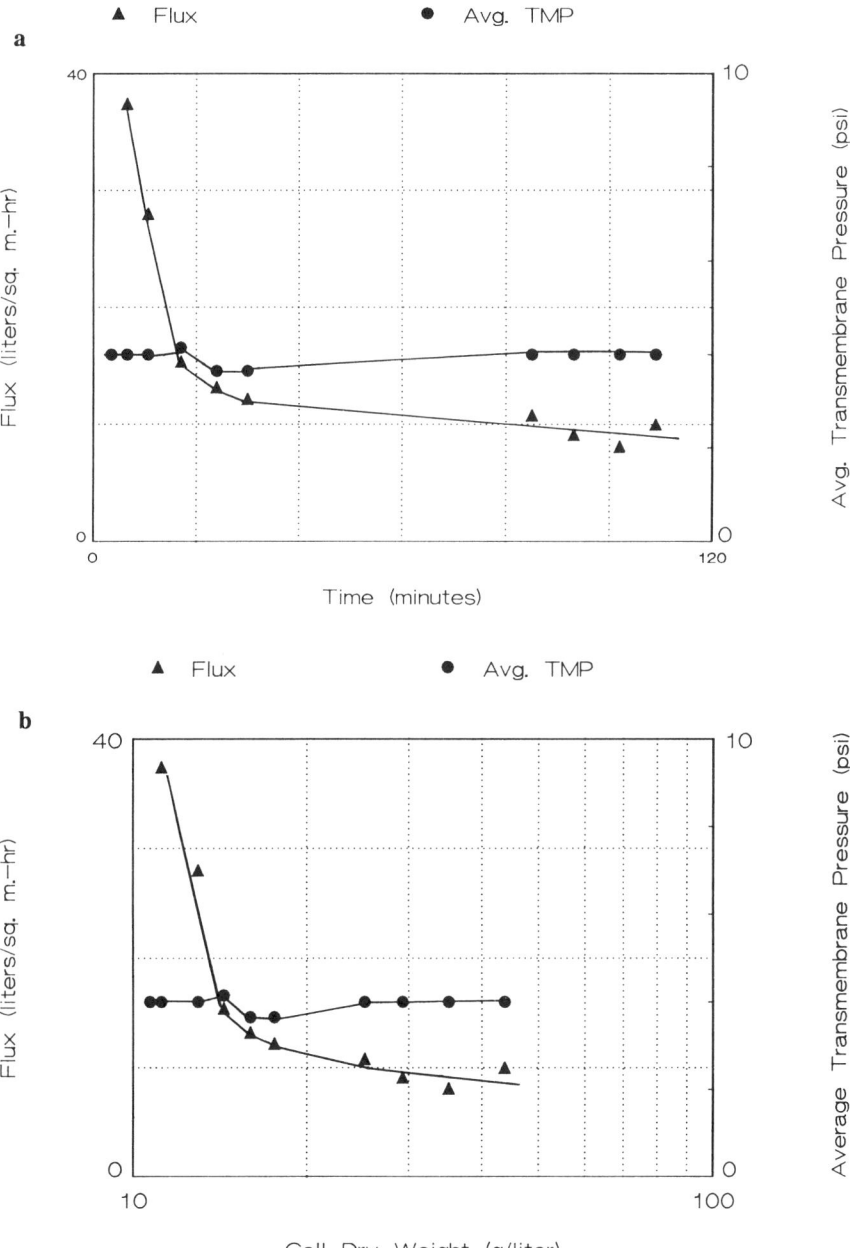

Figure 3. Concentration of Cells (AG Techn. 500K MWCO Hollow Fibers): **a,** Flux vs. Time Profile (Constant Transmembrane Pressure and Recirculation Rate); **b,** Flux vs. Cell Concentration Profile (Constant Transmembrane Pressure and Recirculation Rate)

In an attempt to alleviate the problems caused by the initial rapid fouling of the membrane during the first minutes of operation, a peristaltic pump was added to the combined permeate line of the two hollow fiber cartridges to control permeate flow. A constant permeate flow rate was then maintained throughout the run, while transmembrane pressure was allowed to increase. Figure 4 shows the effect (discussed below) of using permeate flow control, rather than permeate pressure control during the cell separation step. These results suggest that the controlled permeate flow approach stabilized membrane flux by avoiding the initially high fluxes which may cause the rapid fouling observed previously.

Transmembrane pressure in the permeate flow-controlled run (Figure 4) varied from 1.8 to 4.0 psi over the course of the run. Flux through the membrane was maintained at a constant 25 l/m^2-hr. This compares to an average flux of only 10 l/m^2-hr for the permeate pressure controlled run at 4.0 psi average transmembrane pressure (Figure 3b). Stabilizing permeate flow increased average flux performance by a factor of 2.5. Note that the observed increase in average transmembrane pressure during the early portion of this run (Figure 4) is the result of decreasing permeate

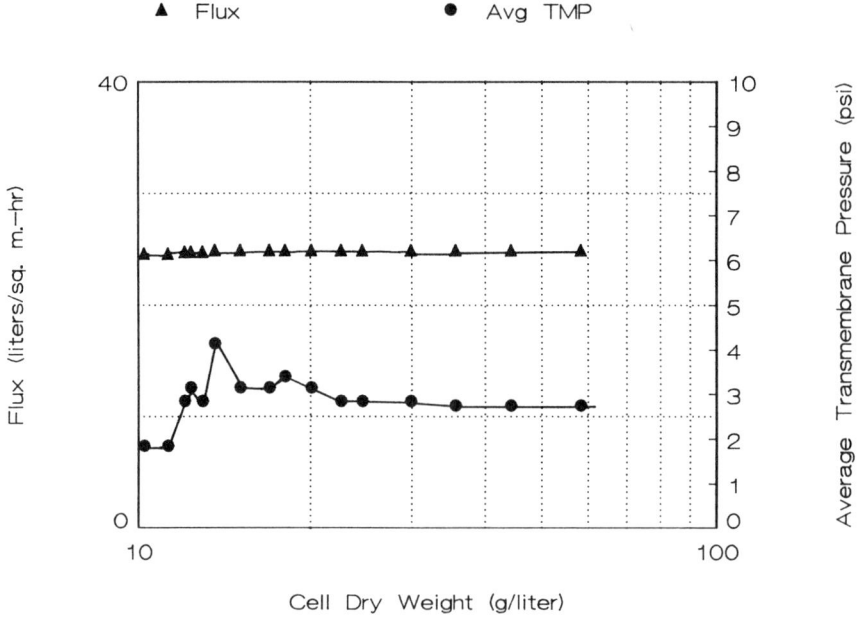

Figure 4. Concentration of Cells (AG Techn. 500K MWCO Hollow Fibers): Flux and Average Transmembrane Pressure Profiles for Constant Permeate Flow-Controlled Operation

backpressure, which occurs as the membrane fouls. The subsequent drop in average transmembrane pressure is the result of decreasing retentate pressure. Flow delivered by the centrifugal feed pump used in these tests typically dropped as fluid viscosity increased. The roughly 10% loss in flow is the cause of the drop in retentate pressure. In some runs, it was necessary to decrease permeate flow rate toward the end of the concentration step to avoid average transmembrane pressures in excess of 5.0 psi. Nevertheless, the 2 to 2.5-fold increase in flux is typical of most of the runs conducted with permeate flow control (see Table I).

Figure 5 shows the effect of recirculation rate on flux performance when permeate flow is controlled and transmembrane pressure is allowed to rise (up to a limit of 5.0 psi). Increasing recirculation rate from 60 to 100 liters/minute (corresponding to linear velocities of 0.6 to 1.0 m/sec) produced a roughly 50% increase in flux performance for the same transmembrane pressure profile. These results suggest that further improvements in flux performance might be obtained by increasing recirculation rates through the fibers. Pump limitations prevented testing higher recirculation rates.

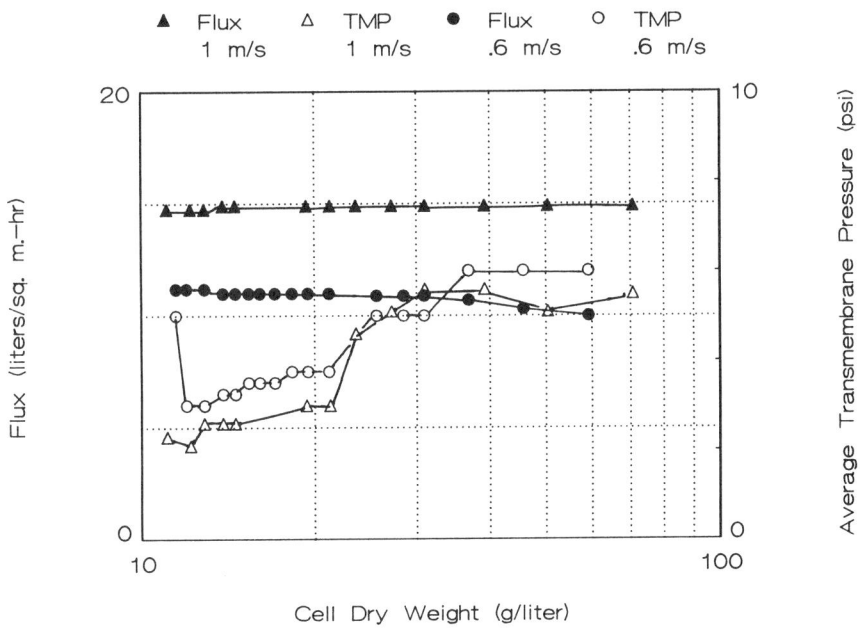

Figure 5. Concentration of Cells (AG Techn. 500K MWCO Hollow Fibers): Effect of Crossflow Velocity on Membrane Performance (Constant Permeate Flow-Controlled Operation)

Table I. Comparison of Permeate Flow Control vs. Transmembrane Pressure (TMP) Control

Run #	Control Scheme	Average Trans-Membrane Pressure (psi)	Recirc. Rate (lpm)	Time Averaged Flux (l/m^2-hr)
105	Const. TMP	3.95	100	10.0
108	Const. Permeate Flow	2.90	104	24.8
109	Const. Permeate Flow	3.38	97.3	24.6
110	Const. Permeate Flow	4.08	96.6	20.5
111	Const. Permeate Flow	4.86	99.9	19.7

Protein Product Passage and Yield Data. Protein product passage efficiency data for the hollow fiber units are summarized in Figure 6 for the cell concentration step. This efficiency, as discussed earlier, represents the ratio of the weight fraction of enzyme recovered in the permeate to the volume fraction of feed recovered in the permeate. The majority of the 100 liter scale runs achieved greater than 90% efficiency. Passage efficiencies greater than 90% were consistently achieved in all runs for which permeate flow control was used (runs 105 through 111).

The effect of transmembrane pressure (at a constant recirculation rate of 100 lpm) was demonstrated at the 100 l scale in a test carried out using an Amicon Series "S" Ioplate™ bench scale plate-and-frame crossflow filter device (shown schematically in Figure 7a). The advantage of using this device was that it allowed measurement of fluxes at each plate, which, because of the pressure drop across the series of plates, are all at different values of average transmembrane pressure. Because the permeate flow through the plates is much less than the recirculation rate through the stack, tangential flow is essentially constant for the entire stack. Thus, a single cell concentration test provides a controlled experiment on the effect of transmembrane pressure at fixed conditions of shear across the membranes, as well as an analysis of the effects of membrane fouling on protein transmission.

Figure 7b summarizes the results of this experiment. At the start-up (1x concentration) and at the two-fold concentration point in this 100 liter run, instantaneous transmission of the enzyme decreased dramatically as transmembrane pressure increased. The samples collected after two fold concentration showed the strongest trend, with transmission dropping

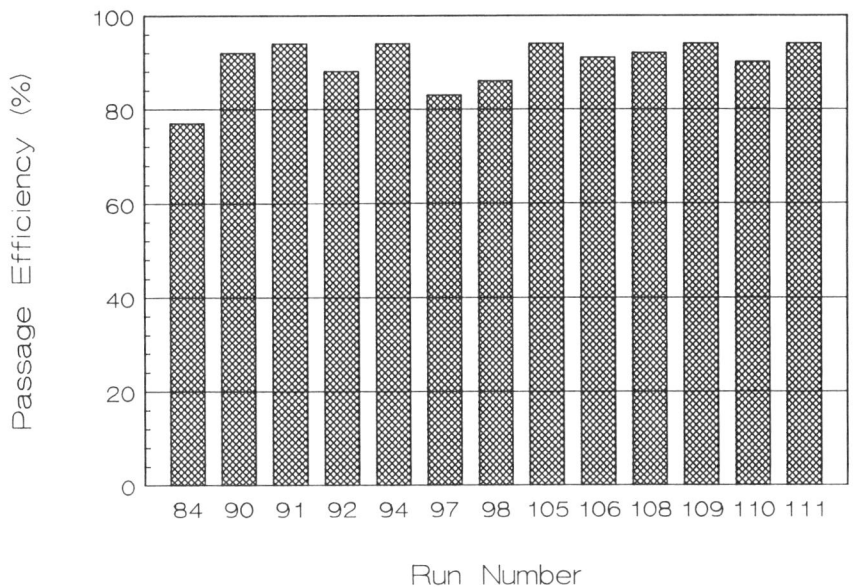

Figure 6. Concentration of Cells (AG Techn. 500K MWCO Hollow Fibers): Summary of Protein Passage Efficiencies for 100 liter Scale Runs

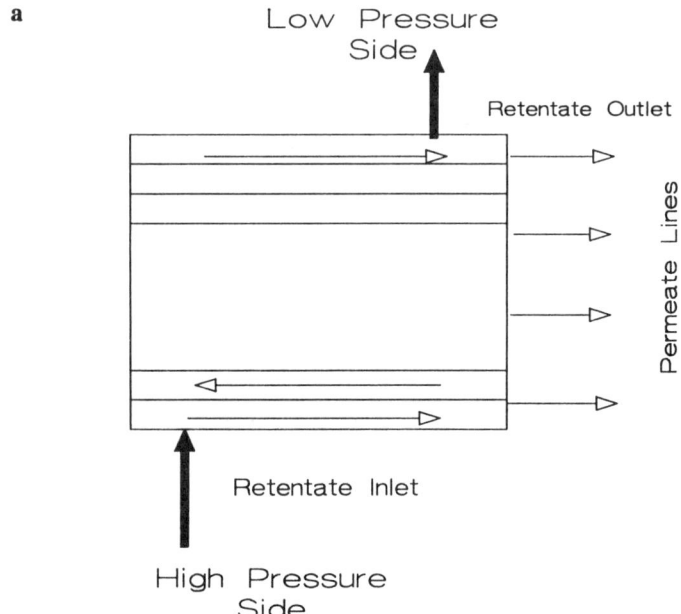

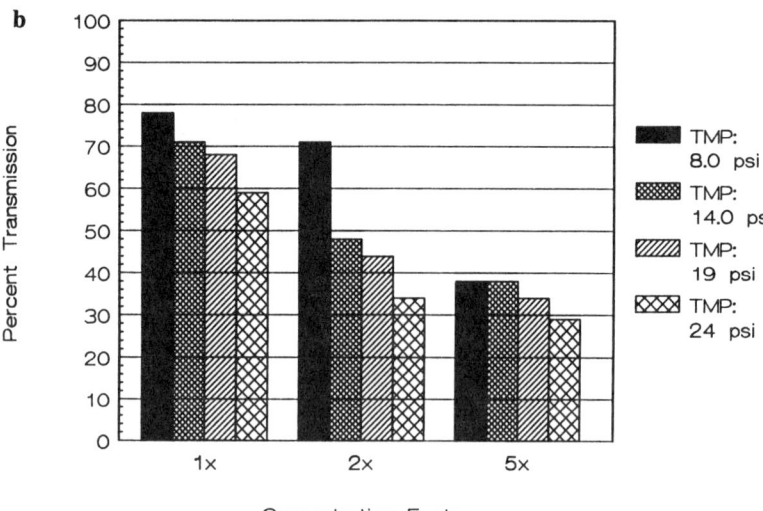

Figure 7. Effect of Transmembrane Pressure and Cell Concentration on Protein Transmission: **a,** Schematic of Amicon Ioplate Plate-and-Frame Device Used; **b**; Amicon Ioplate Plate-and-Frame Test Results

from 70 to 35% over the 8.0 to 24 psi range. At the end of the test, when the membranes became badly fouled (5x concentration), transmission was poor at all conditions.

Product Yield Data: Effect of Cell Wash (Diafiltration). Figure 8 summarizes yield data for the enzyme after the cell concentration and diafiltrations steps in the hollow fiber units. After concentration of the 100 liter broth to about 20 liters, approximately 80 liters of an isotonic salt solution (0.8% NaCl, 0.1% $CaCl_2$) were pumped through the cell concentrate. This wash step increased yield of the product from 60-80% to 90-100%, while diluting the final "cell-free" enzyme solution two-fold. The 100 liter pilot scale recovery runs demonstrated a consistent ability to achieve a high yield enzyme/cell separation using microporous crossflow membranes.

ENZYME CONCENTRATION: SPIRAL ULTRAFILTER PERFORMANCE

An Amicon Diaflo™ spiral ultrafilter (40 ft², 10,000 MWCO regenerated cellulose) was used to concentrate the approximately 160 liters of combined permeate from the two stage cell separation step. Typical flux profiles for various recirculation rates are shown in Figure 9. The average transmembrane pressure for all of these runs was approximately 30 psi. As expected, increasing recirculation rate through the spiral provided higher flux performance over the course of the enzyme concentration step. Fluxes of 42 l/m²-hr or greater were consistently achieved with this spiral device. In addition, the spiral membrane provided 100% rejection and yield of enzyme.

MEMBRANE CLEANING AND FLUX RECOVERY

Cleaning of both the polysulfone hollow fibers and the Amicon industrial spiral units was done using a combination of hypochlorite and detergent washes. The units were first flushed with DI water to remove solid debris and loosely attached foulants on the membranes. The water was pumped straight through the units, with no recycle. Separate washes consisting of 200 ppm sodium hypochlorite and 1% Terg-A-Zyme™, a protease-containing anionic detergent (Alconox, Inc, New York, NY), were then done by recirculating approximately 50 liters of the wash solution through the fibers with the permeate lines open (allowing some cleaning solution to pass through to the permeate side) for a minimum of one hour at room temperature. Cleaning solution in the permeate was recycled to the feed tank. These wash steps were used to try to remove more tightly bound or embedded foulants not removed by the water flush step.

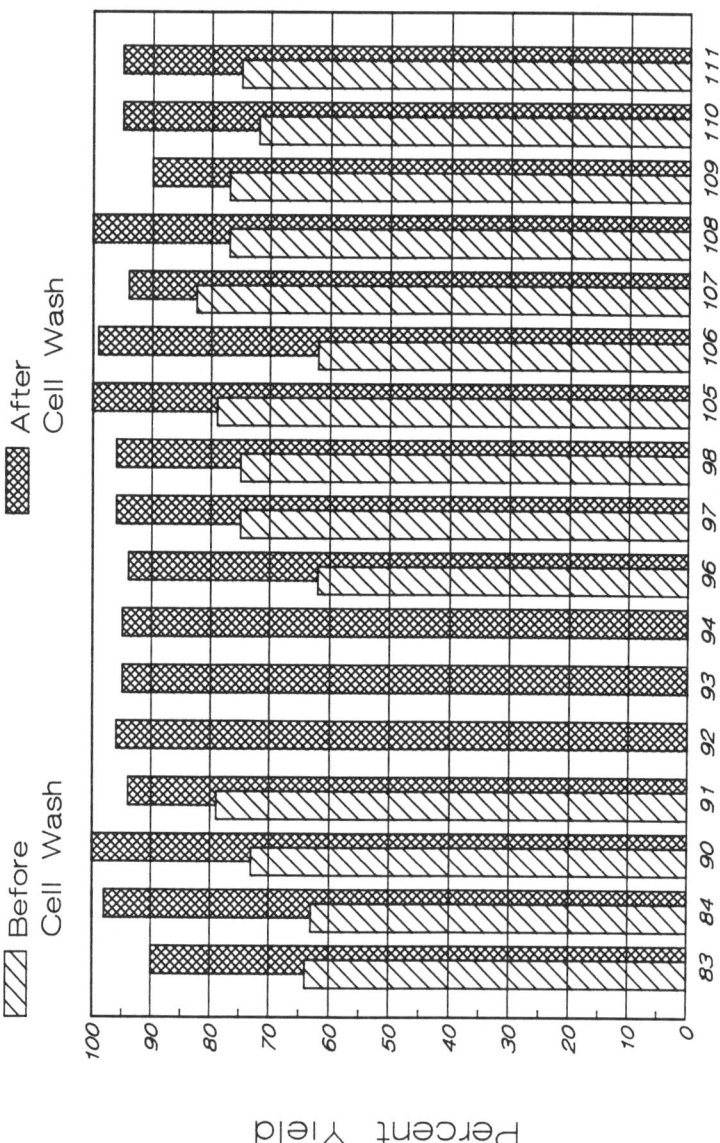

Figure 8. Cell Separation Steps (AG Techn. 500K MWCO Hollow Fibers): Summary of Product Yield Results Before and After Cell Wash Steps

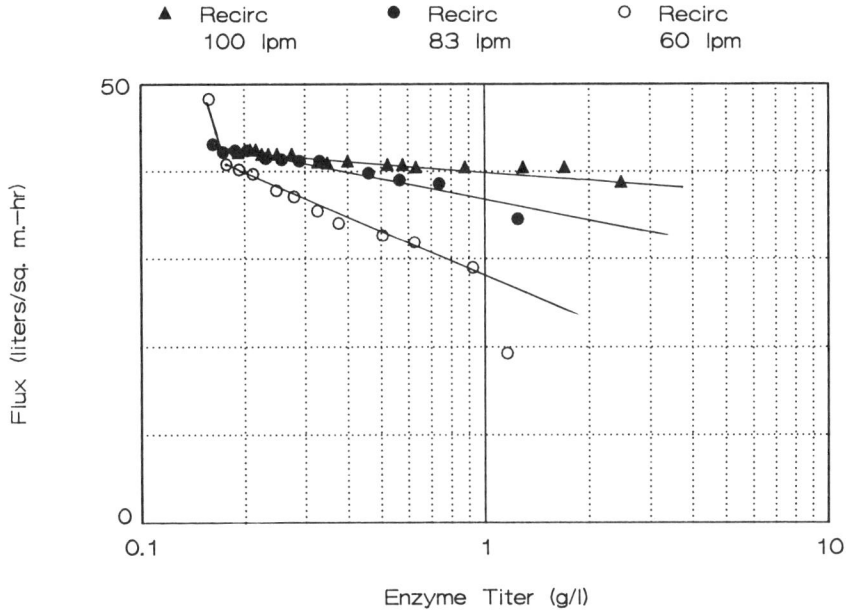

Figure 9. Enzyme Concentrations Step: Effect of Recirculation Rate on Flux Performance for the Amicon Diaflo S10Y40 Spiral Ultrafilter (10,000 MWCO, 40 ft^2) at Constant Average Transmembrane Pressure (30 psi)

Cell Separation Steps: 500,000 MWCO Polysulfone Hollow Fibers.
The hollow fibers used in the cell separation step were generally difficult to clean. This was due to: 1) the high solids process stream to which the membranes were exposed during the cell separation steps, and 2) the high degree of protein fouling commonly observed with polysulfone membranes. The condition of the membrane was measured by collecting data on water flux versus transmembrane pressure, and estimating the hydraulic permeability from the slope of this curve. These permeabilities are plotted in Figure 10a. This graph demonstrates the inconsistency of the effectiveness of cleaning. Periodically, apparent fouling became so severe that more rigorous cleaning had to be done (e.g., immediately before cycles 12 and 15). In these cases, repeated cleanings with the detergent and hypochlorite solutions were carried out. In addition, in these instances, pumping of the cleaning solution from the outside to the inside of the fibers was used. The latter step proved very effective.

Enzyme Concentration: Regenerated Cellulose Spiral Ultrafilter.
As indicated in Figure 10b, this membrane proved quite easy to clean. In contrast to the polysulfone membranes used in the cell separation step, the spiral ultrafilter showed 100% recovery of hydraulic permeability with a minimum effort to clean (i.e., no extra cleaning steps were ever used). Again, the ease of cleaning is related to both the nature of the process stream to which the unit was exposed, and the properties of the membrane itself. In the enzyme concentration step, solids load was very low (compared to the original cell broth). The membrane itself, on the other hand, exhibits much less tendency to adsorb protein than polysulfone because of the hydrophilic nature of regenerated cellulose.

ECONOMICS OF THE CELL SEPARATION STEP

A preliminary engineering economic analysis of the cell separation step was conducted to determine if the membrane-based cell separation process used in this pilot scale study is more cost effective than other alternative conventional technologies. Three different technologies were considered: precoat rotary vacuum filtration, continuous disk centrifugation, and crossflow membrane filtration. For the centrifuge and membrane case studies, two cases were considered: (1) a single stage ten fold concentration with no cell wash step (90% product yield), and (2) a two stage process consisting of a ten-fold concentration followed by a cell wash step in which four volumes of wash solution are used (99% yield). The wash step for the centrifuge requires rediluting the cell concentrate and reprocessing this solution through the centrifuge. There was insufficient information for the precoat filter to assess the cost of the 99% yield case accurately.

The same plant design basis was used for all three separation technologies. The units were designed to produce 17,000 kg of enzyme per

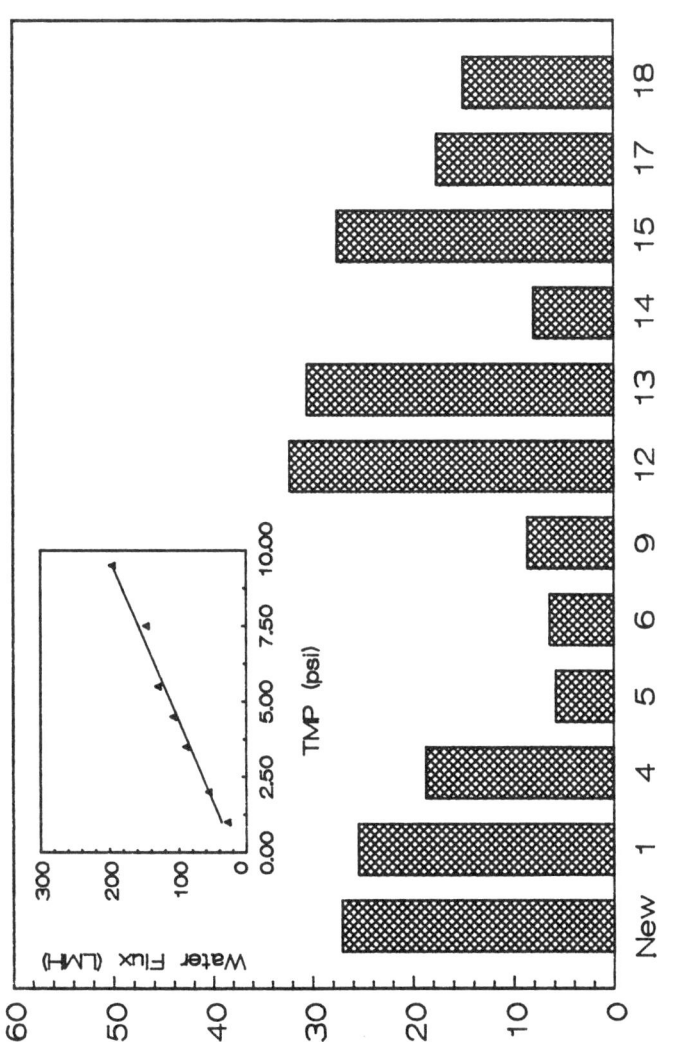

Figure 10a. Summary of Membrane Cleaning Results: Water Permeability Data for AG Technology 500,000 MWCO Polysulfone Hollow Fibers

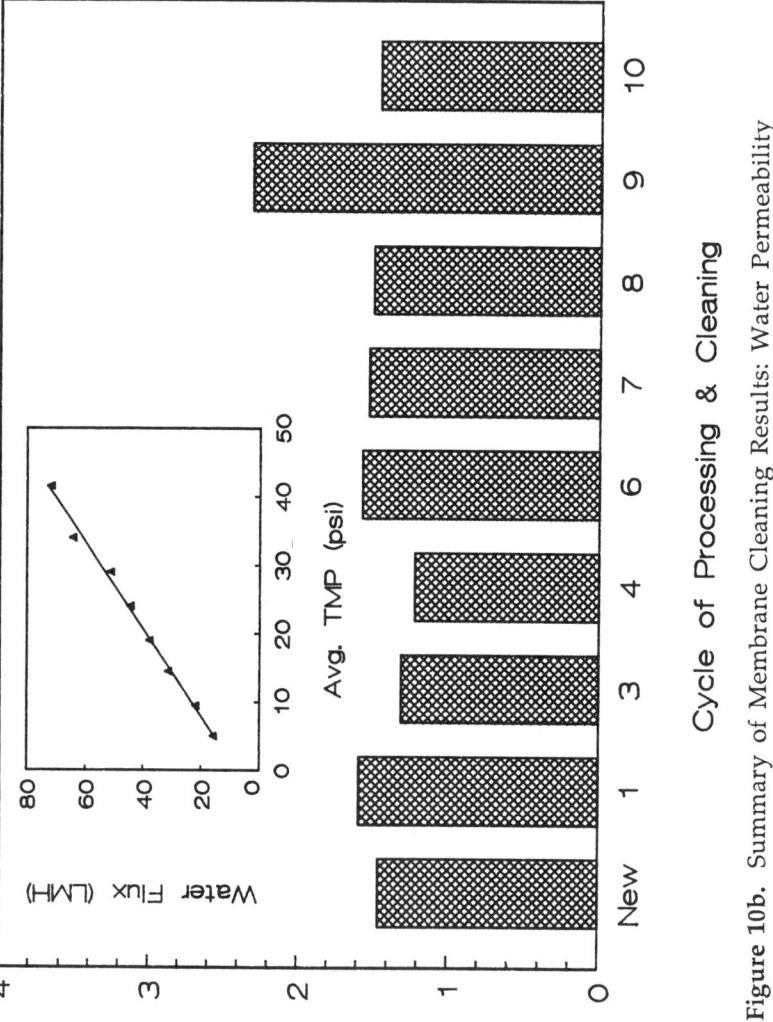

Figure 10b. Summary of Membrane Cleaning Results: Water Permeability Data for Amicon Diaflo S40Y10 Spiral Ultrafilter (10,000 MWCO, 40 ft^2)

year. The plant was assumed to operate 250 days per year, three shifts per day. To meet this annual output, two 17,000 liter fermentations are harvested per day, assuming that yield in the fermentation is increased to 2 grams of enzyme per liter. Finally, all of the systems were required to complete the cell separation step in four hours, to avoid cell lysis and loss of product. Design assumptions for each piece of equipment are shown in Table II.

The results of this analysis are summarized in Figures 11a and 11b. The capital cost analysis indicates that the membrane system is very attractive compared to either alternative technology. Capital cost for the continuous centrifuges is at least four-fold higher, while the precoat filter is two-fold more costly. The cost per kilogram of enzyme produced, which includes all operating costs and capital depreciation over ten years, indicates that the membrane system can achieve the same separation at half the cost for centrifugation.

The comparison of membranes versus precoat filters suggests that, on a cost per unit of enzyme basis, these two processes are roughly equivalent in cost. Final choice of a process would clearly boil down to a choice of the two filter technologies. Other issues, such as disposal of spent filter aid and consistent quality of the cell free product, then begin to play a significant role in the choice between these two systems (13).

CONCLUSIONS

Cell Separation.
1. This pilot scale study verified the ability to achieve consistently high yield and passage of an extracellular bacterial protein product using microporous membranes.

2. Fluxes of 25 l/m^2-hr were consistently achieved through permeate flow controlled operation of the hollow fibers. Further optimization of recirculation rates might lead to higher flux performance.

3. This study demonstrated the importance of maintaining low transmembrane pressure-to-recirculation rate ratios to achieve high flux and protein passage.

4. A preliminary economic analysis of the cell separation step indicates that membranes are two-fold more cost effective than centrifuges, and equivalent in cost to using a precoat vacuum filter for this application. Use of a precoat vacuum filter requires disposal of filter aid. This was not considered in the economic comparison, but could represent a significant incremental cost for precoat systems.

5. Based on measurements of water flux data, the polysulfone hollow fibers proved difficult to clean. Flux recovery from run to run was inconsistent, and the membrane occasionally required extra effort to clean.

Table II. Design Assumptions for Cell Separation

Scenario	Assumptions
Precoat Vacuum Filter (Petrides, D., MIT, personal comm., 1988)	• Flux: 200 l/m^2-hr • 0.1 kg diatomaceous earth per kg solid removed • Cake washwater sprayed on drum at a rate of 30% of feed flowrate gives 90% yield • Energy requirements: 0.12 Kw/m^2 • Capital cost: correlation[a] • Labor: 2 man-hrs/shift at \$20/hr • Maintenance: 5% of capital • Installation cost: 70% of Purchased Price • Indirect costs: 30% of Purchased Price
Continuous Disk Centrifuge (Petrides, D., MIT, personal comm., 1988)	• Throughput: 3,000 lph for a 233,300 m^2 area equivalent[b] disk centrifuge. • Energy: 40 kW per centrifuge • Capital cost: \$300,000 (Erickson, R., Alfa Laval, pers. comm., 1989) • Labor: 3 man-hrs/shift at \$20/hr • Maintenance: 5% of capital • Installation costs: 70% of Purchased Price • Indirect costs: 30% of Purchased Price
Membrane System:	• Hollow fiber flux: 25 l/m^2-hr • Recirculation rate: 1 m/sec • Capital cost: \$125/ft^2 (Pinto, S., Amicon Corp., personal comm., 1988 • Membrane replacement: \$35/ft^2 (Pinto, S., Amicon Corp., personal comm., 1988) • Membrane life: 1 year • Labor: 3 man-hrs/day (incl. cleaning) at \$20/hr • Maintenance: 5% of capital • Energy: recirc. pump • Installation cost: 10% of Purchased Price • Indirect costs: 30% of Purchased Price

[a] Hall, R. S.; "Current Costs of Process Equipment", Chemical Engineering, April 5, 1982.
[b] Area equivalent is the area required by a settling tank to produce the same separation efficiency as the centrifuge.

Enzyme Concentration.

1. An Amicon Diaflo™ spiral ultrafilter proved to be reliable in terms of flux and protein rejection. Fluxes of at least 42 l/m^2-hr were consistently maintained over repeated cycles of use.

2. Notable is the Diaflo™ ultrafilter's ease-of-cleaning. The spiral showed 100% flux recovery after cleaning in every cycle (up to 10 cycles) of use and cleaning.

ACKNOWLEDGMENTS

We thank William DeVane for technical assistance with microfiltration and ultrafiltration operations, and John Finch and the fermentation pilot plant staff for providing fermentation broths.

LITERATURE CITED

1. Amicon Division of W. R. Grace & Co.-Conn., "Laboratory Separation: Membrane Filtration/Chromatography"; Publication No. 716; Danvers, MA 1987.
2. Bailey, F.; Schulman, C.; Warf, R.; Maigetter, R. "Harvesting Microbial Cells Using Crossflow Filtration". Presented at ACS National Meeting September 25-30, 1988; Los Angeles, CA.
3. Belfort, G., et al. In Reverse Osmosis and Ultrafiltration; Sourirajan, S.; Matsuura, T., Eds.; ACS Symposium Series No. 281; American Chemical Society: Washington, D. C., 1985, 383-401.
4. Belfort, G.; Nagata, N. Biotechnology and Bioengineering, in press.
5. Bell, D.; Davies, R. Biotechnology and Bioengineering, 1987, 29, 1176-1178.
6. Blatt, W., et al. In Membrane Science and Technology: Industrial, Biological, and Waste Treatment Processes; Flinn, J., Ed.; Plenum Press: New York, 1972; 47-97.
7. Datar, R. Biotechnology Letters, 1985, 7, 471-476.
8. DeFilippi, R; Goldsmith, R. In Membrane Science and Technology: Industrial, Biological, and Waste Treatment Processes; Flinn, J., Ed.; Plenum Press: New York, 1972; 33-46.
9. Eriksson, A. Desalination, 1985, 53, 259-263.
10. Gabler, R. In Developments in Industrial Microbiology, 25; Society for Industrial Microbiology: Arlington, VA, 1984; 381-414.
11. Gabler, R., et al. "Separating Proteins from Cell Lysates with Cross Flow Filtration". Presented at Biotech '85 (Europe); Online Publications: Pinner, U.K., 1985.
12. Gabler, R.; Ryan, M. In Purification of Fermentation Products; LeRoith, D.; Shiloach, J.; Leahy, T., Eds.; ACS Symposium Series No. 271; American Chemical Society: Washington, D.C., 1985; 1-20.
13. Gravatt, D.; Molnar, T. In Membrane Separations in Biotechnology; McGregor, W. C.; Marcel Dekker, Inc.: New York, 1986; 89-97.
14. Green, G.; Belfort, G. Desalination, 1980, 35, 129-147.
15. Henry, J. D. In Recent Developments in Separation Science; CRC Press: Cleveland, OH, 1972; 202-225.
16. Ilias, S.; Govind, R. J. Membrane Science, 1988, 39, 125-141.
17. Kroner, K.; Nissinen, V.; Ziegler, H. Bio/Technology, 1987, 5, 921-926.

7. SHEEHAN ET AL. *Recovery of an Extracellular Bacterial Protease*

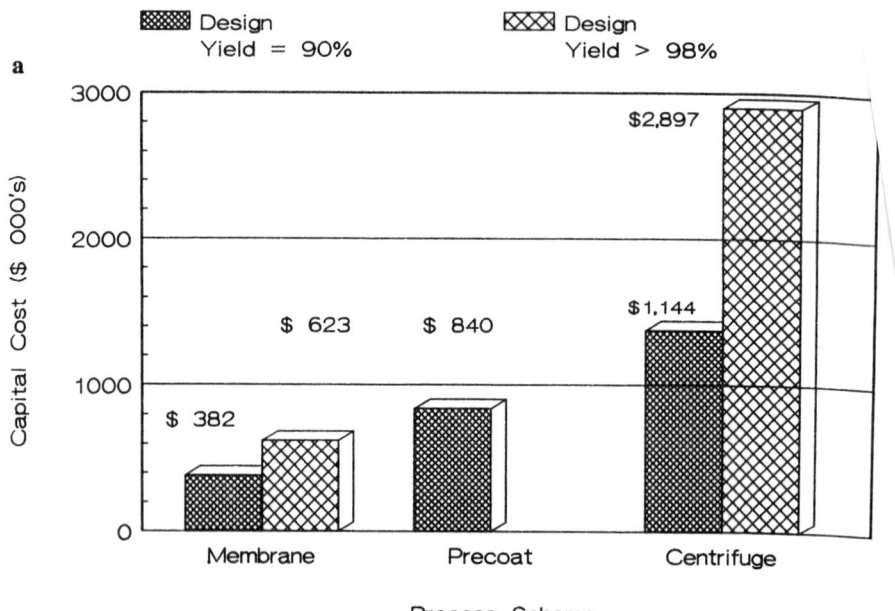

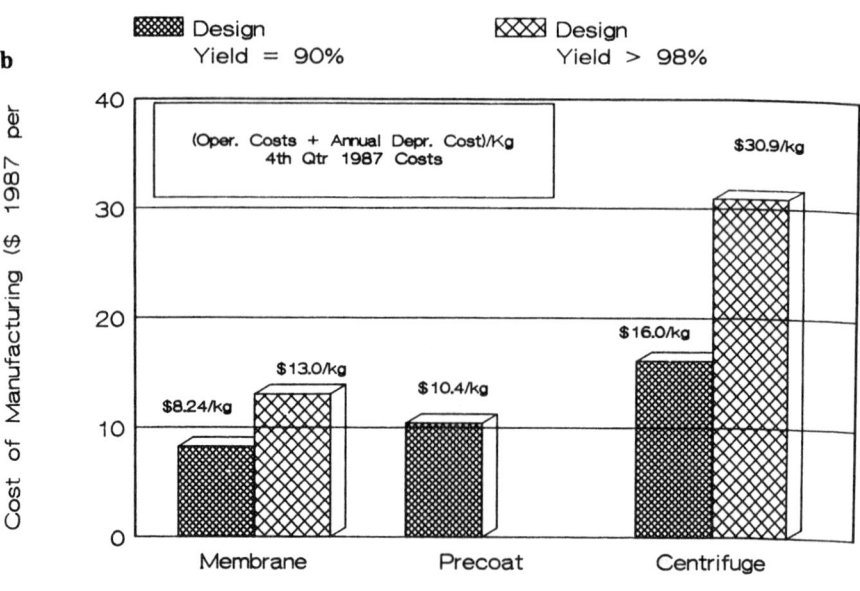

Figure 11. A Comparison of Crossflow Membrane Filtration, Precoat Rotary Vacuum Filtration, and Centrifugation. **a,** Capital Costs for Cell Separation Steps; **b,** Total Manufacturing Costs for Cell Separation Steps ($ per Kg of enzyme).

18. Kroner, K.; Schutte, H.; Hustedt, H.; Kula, M. Process Biochemistry, 1984, 19, 67-74.
19. Kroner, K.; Schutte, H.; Hustedt, H.; Kula, M. In Eur. Congress Biotechn., 3rd; Vol 3; Verlag Chemie: Weinheim, West Germany, 1984; 549-555.
20. Lasky, M; Grant, D. American Biology Laboratory, Nov/Dec 1985, 16-21.
21. Le, M.; Ward, P.; Atkinson, T. In Biotech '84 (USA); Online Publications: Pinner, U.K.; A97-A116.
22. Lindley, P.; Olson, W.; Faith, M. Polymer Science and Technology, 1982, 16, 173-190.
23. Millipore Systems Division. Downstream Process Update; Issue No. 2; Millipore Corporation: Bedford, MA, 1986.
24. O'Sullivan, T.; Epstein, A.; Korchin, S.; Beaton, N. Chem. Engr. Progress, April 1984, 68-75.
25. Porter, M. C. Ind. Eng. Chem. Res. and Dev., 1972, 234-248.
26. Quirk, A.; Woodrow, J. Biotechnology Letters, 1983, 5, 277-282.
27. Reid, D.; Adlam, C. J. Applied Bacteriology, 1974, 41, 321-324.
28. Riesmeier, B.; Kroner, K.; Kula, M. J. Membrane Science, 1987, 34, 245-266.
29. Romero, C.; Davis, R. J. Membrane Science, 1988, 39, 157-185.
30. Solomon, B.; et al. Trans. Am. Soc. Artificial Internal Organs, 1978, 18.
31. Tanny, G.; et al. Desalination, 1982, 41, 299-312.
32 Tutunjian, R. In Developments in Industrial Microbiology, 25; Society for Industrial Microbiology: Arlington, VA, 1984; 415-435.
33. Valeri, G.; Genna, G. Experientia, 1979, 35, 1535-1536.
34. Wolber, P; Dosmar, M; Banks, J. BioPharm, June 1988, 38-45.
35. Zahka, J.; Leahy, T. In Purification of Fermentation Products; LeRoith, D.; Shiloach, J.; Leahy, T., Eds.; ACS Symposium Series No. 271; American Chemical Society: Washington, D.C., 1985; 51-69.
36. Zydney, A. Ph.D. Thesis Digest, MIT, Cambridge, MA.
37. Zydney, A; Colton, C. Chem. Eng. Comm., 1986, 47, 1-21.

RECEIVED September 28, 1989

PROCESSES USING BIOSPECIFIC INTERACTION WITH PROTEINS

Chapter 8

Complexation Between Poly(dimethyldiallylammonium chloride) and Globular Proteins

Mark A. Strege, Paul L. Dubin, Jeffrey S. West, and C. Daniel Flinta

Department of Chemistry, Indiana University—Purdue University at Indianapolis, Indianapolis, IN 46223

> Studies of complexation between globular proteins and the synthetic strong polycation, poly(dimethyldiallylammonium chloride), were undertaken to gain insight into protein-polyelectrolyte complexation selectivity. Investigations of the protein-polyelectrolyte phase boundary, via turbidimetric titrations, suggest that phase separation may be a consequence of the saturation of protein binding sites on the polymer. Comparisons of the phase boundaries of various proteins reveal that the net protein surface charge density does not control phase separation, but rather suggests the importance of charge patches on the protein surface. Quasi-elastic light scattering measurements provide strong evidence for the existence of a stable soluble complex. Size exclusion chromatography, via the Hummel-Dreyer method, provides additional information on the binding equilibria for such soluble complexes.

Oppositely charged polyelectrolytes interact to form complexes. Depending primarily on the molecular weights and the linear charge densities of the polyelectrolytes involved, these complexes may be amorphous solids (1), liquid coacervates (2, 3), gels (4), fibers (4), or soluble aggregates (5-7). One particular case of inter-macroion complex formation involves synthetic polyelectrolytes and globular proteins. The formation of these complexes is generally evidenced by phase separation, where the denser, polymer-rich phase may be a liquid "complex coacervate" (8) or a solid precipitate. Examples of the former have been observed for gelatin and polyphosphate (9), and serum albumin and poly(dimethyldiallyl-ammonium chloride) (10). Systems that exhibit precipitation include hemoglobin and dextran sulfate (11), carboxyhemoglobin and potassium poly(vinyl alcohol sulfate) in the presence of poly(dimethyldiallylammonium chloride) (12), lysozyme and poly(acrylic acid) (13), and RNA polymerase and poly(ethyleneimine) (14).

0097-6156/90/0419-0158$06.00/0
© 1990 American Chemical Society

The ability of polyelectrolytes to remove oppositely charged proteins from solutions has been exploited through the incorporation of protein-polymer precipitation steps into a variety of protein purification procedures, wherein precipitated proteins are recovered from the insoluble complex aggregate via redissolution by pH or ionic strength adjustment (14-16). Furthermore, preferential complexation of polyelectrolytes with specific proteins has been substantiated (13). Although there are reports of the optimization of bulk complexation yield through the adjustment of solution parameters such as pH and ionic strength (17), very little has been accomplished in the optimization of the selectivity of complex formation.

The recovery of proteins through the formation of insoluble complexes with polyelectrolytes appears to be a very attractive non-denaturing separation process. After precipitating various enzymes with polyacrylic acids, Sternberg and Hershberger reported high recoveries of enzymatic activities, indicating that little or no denaturation had taken place during the separation (13). Other workers have reported the non-denaturing fractionation at slightly alkaline pH of intracellular proteins using synthetic polycations (14, 16, 17). Compared to other methods used for protein separation, such as gel filtration chromatography, the utilization of water-soluble, charged polymers as precipitating agents offers great economy with regard to materials and process, and, furthermore, is virtually unlimited in scale. Thus, an elucidation of the principles governing protein selectivity in polyelectrolyte separation would be of considerable applied significance.

In the present work, the mechanism of polyelectrolyte-protein interaction was studied using two approaches. To gain insight into the cooperativity of binding, and to determine whether complexes are intra- or interpolymer, complexes were characterized by size exclusion chromatography (SEC) and quasi-elastic light scattering (QELS). Size exclusion chromatography was also used to determine complex stoichiometry, i. e. the number of protein molecules bound per polyion. Such methods will also lead to the determination of the number of binding sites per polymer molecule, and the intrinsic association constant. Our second approach involved the analysis of plots of the ionic strength dependence of the critical pH (the pH at which an abrupt increase in turbidity is observed). The effects of protein:polymer stoichiometry and protein type on such phase diagrams were studied.

EXPERIMENTAL

Turbidimetric Titrations. Poly(dimethyldiallylammonium chloride) (PDMDAAC), a commercial sample "Merquat 100" from Calgon Corp. (Pittsburgh, PA) with nominal molecular weight 2×10^5 and reported polydispersity of $M_w/M_n \cong 10$, was dialyzed and freeze-dried before use. All proteins were obtained from Sigma Chemical Corp. Solutions were prepared as mixtures of PDMDAAC (0.05 - 1 g/l) and protein (0.25 - 25 g/l), corresponding to protein/polymer weight ratios (r) ranging from 0.25 to 200, at pH<4 in dilute (0.05 - 0.3 M) NaCl. The optical probe (2 cm path length) of a Brinkmann PC600 fiber optics probe colorimeter, and a pH electrode connected to an expanded scale pH meter (Orion 811 or Radiometer-Copenhagen 26), were both placed in the solution. Changes in turbidity were monitored as %T, relative to a blank (polymer-free) solution, as the pH was adjusted by the addition of dilute (0.01 - 0.10 M) NaOH.

Quasi-Elastic Light Scattering. Solutions for QELS were prepared by combining one volume of 3.0 g/l PDMDAAC with two volumes of 10.0 g/l bovine serum albumin (BSA) in 0.01 M NaOAc buffer. After pH adjustment with 0.10 M NaOH, the solutions were filtered (0.20 µm Millipore), and in some cases, centrifuged (2000 rpm, 15-20 min) in the 1 cm cylindrical sample cell, which was then placed in the toluene refractive index matching bath (25.0°C) of a Malvern RR102 spectrometer. The light source was a 20-mW He-Ne laser (Jodon); scattered light was collected at 90° through a photomultiplier frontal aperture of 0.5 mm. The photomultiplier output was analyzed with a Nicomp TC-200 computing autocorrelator.

Photon counts were acquired until computed distributions were stable, usually corresponding to a "fit error" of less than 5 and a "residual" (18) of less than 5, normally requiring the acquisition of 50,000 counts above the base line and a typical run time of 15 hours. The channel width was adjusted to encompass 2.0 decays. The autocorrelation function was analyzed using both the method of cumulants (19) and a technique similar to that of Provencher's (20) in that a nonlinear least squares, non-negatively constrained method is used to achieve the inverse Laplace transformation of the autocorrelation curve, but differing from CONTIN in the degree of smoothing and the degree of coupling of the individual exponentials (21). This technique produces the same bimodal size distributions as the non-negative constrained least-squares method of Morrison et al. (22) when applied to autocorrelation curves obtained for soluble complexes of PDMDAAC and mixed anionic/nonionic micelles (23).

Size Exclusion Chromatography. Exclusion chromatography was carried out on an apparatus comprised of a Minipump (Milton Roy), a model 7012 injector (Rheodyne) equipped with a 100-µl loop, an R401 differential refractometer (Waters), and a Model 153 UV detector, λ=254 nm (Altex). A Superose-6 column (30 cm x 1 cm OD)(Pharmacia) was eluted at 0.53 ml/min.. Column efficiency, determined with D_2O, was at least 12,000 plates/meter.

A mobile phase of 0.25 M NaOAc buffer has been found to sufficiently repress ionic adsorption effects, especially with regard to the polycation (35). The mobile phase also contained 0.10 mg BSA / ml.

Samples to be injected were prepared by dissolving 2.5 mg PDMDAAC per ml in the mobile phase, along with increasing amounts of BSA. All samples were filtered (0.20 μm Millipore) before injection. Injections were performed in mobile phases of various pH, all below $pH_{critical}$. To determine the stoichiometries of polymer-protein complexes, we used the Hummel-Dreyer method (24), as recently applied to dextran sulfate-hemoglobin complexes (25).

RESULTS

Turbidimetric Titrations. A typical set of turbidimetric titrations is shown in Figure 1, where solutions of 2.5 g BSA / l and 1.0 g PDMDAAC per liter have been titrated with NaOH at varying ionic strength (i.e. molarity of added NaCl), and turbidity is represented as 100-%T. Phase separation is evidenced by an abrupt rise in turbidity, and the pH at which phase separation occurs ($pH_{critical}$) is evaluated as shown. Plots of $pH_{critical}$ vs. ionic strength (phase boundaries) are shown in Figure 2 for BSA and PDMDAAC, over a range of r. In the case of the completely dissociated, univalent salt employed in these determinations, NaCl, ionic strength corresponds to salt molality.

Phase boundaries were also developed for β-lactoglobulin, chicken egg albumin, lysozyme, ribonuclease, and trypsin, all at r=100, a weight ratio at which polymer saturation appears to take place (see Discussion section). For each protein, $pH_{critical}$ was converted to net negative surface charge (Z_{pr}) per unit protein surface area (A), using potentiometric titration curves (26-31) and hydrodynamic radii (32) found in literature. Plots of surface charge density (Z_{pr}/A) vs. I are shown in Figure 3.

QELS. QELS results for mixtures of PDMDAAC and BSA at pH values near $pH_{critical}$ are presented in Figure 4. The multiexponential curve-fitting procedure yields a set of decay constants and corresponding coefficients, and, hence, an approximate distribution of apparent Stokes radii (33). This multiexponential algorithm broadens single-exponential peaks, so that QELS histograms in which only BSA contributes to the scattering look anomalously polydisperse (in solution, free polymer scatters too weakly to contribute to the autocorrelation function). However, in the interest of consistency, all comparisons were made by using a multiexponential fit, even when only one decay constant was observed.

SEC. The chromatograms obtained after injection of polymer into the mobile phase containing protein are displayed in Figure 5. The chromatograms reveal the presence of protein-polymer complex, followed by a negative or positive peak at the retention volume of the protein, which represents the amount of protein bound to the polymer.

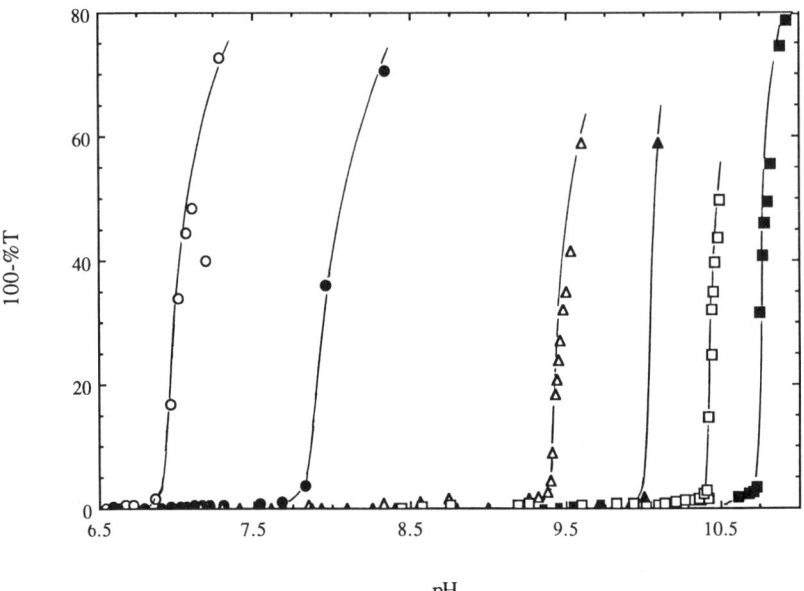

Figure 1. Turbidimetric titrations of solutions of BSA and PDMDAAC (r = 2.5) at various ionic strengths: (○) I = 0.05; (●) I = 0.10; (△) I = 0.15; (▲) I = 0.20; (■) I = 0.25; and (□) I = 0.30.

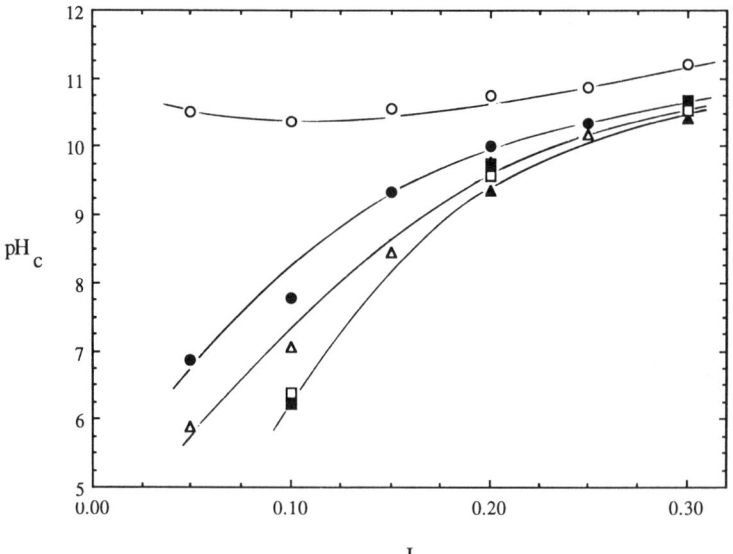

Figure 2. BSA–PDMDAAC phase boundaries over a range of r: (○) r = 0.25; (●) r = 2.5; (△) r = 25; (▲) r = 50; (□) r = 100; and (■) r = 200.

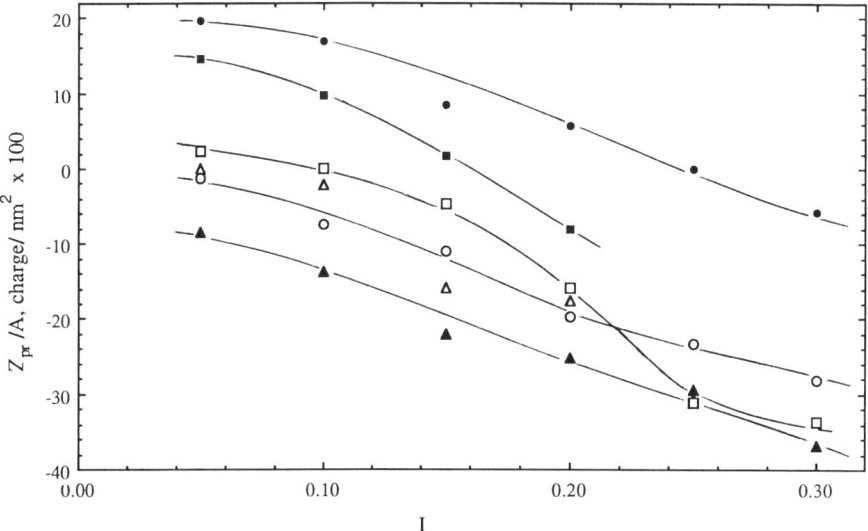

Figure 3. Phase boundaries (r = 100) of six proteins, plotted as net surface charge density (net charge/nm^2) vs. I (ionic strength): (○) bovine serum albumin; (●) lysozyme; (Δ) ribonuclease; (▲) chicken egg albumin; (□) β-lactoglobulin; and (■) trypsin.

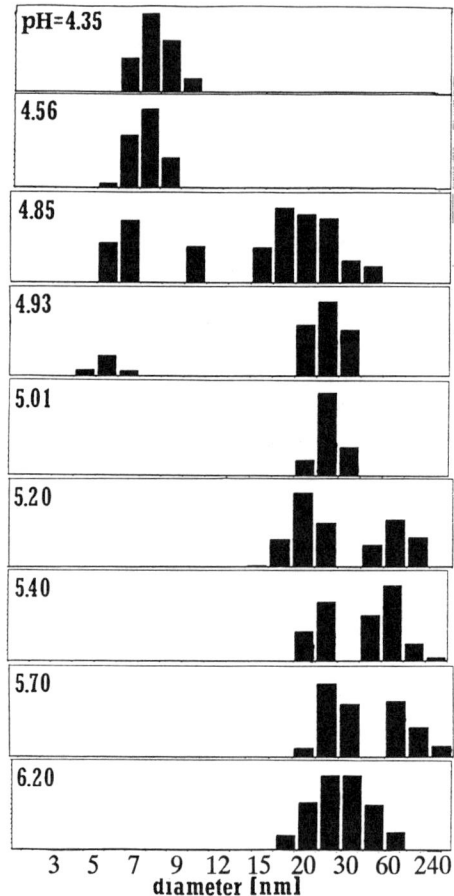

Figure 4. Apparent distributions of equivalent Stokes diameters, from quasi-elastic light scattering analysis of mixtures of 1.0 g/L PDMDAAC and 6.7 g/L BSA, in 0.01 M NaOAc buffer, at pH values shown.

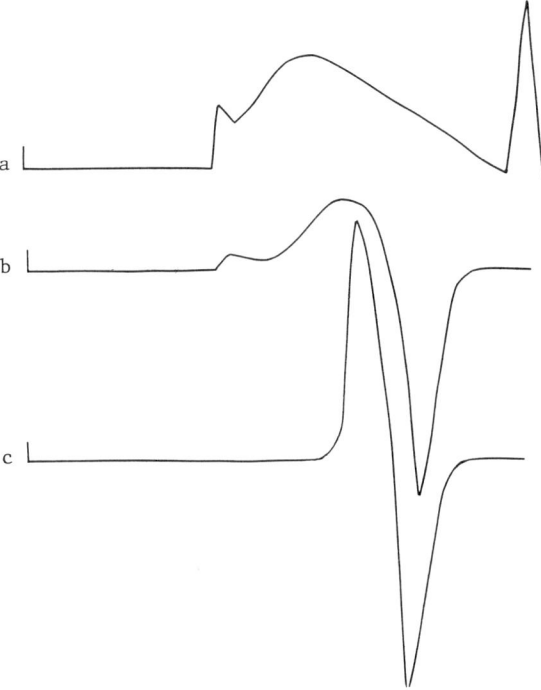

Figure 5. (a) Refractive index chromatogram of 2.5 mg PDMDAAC/mL, using a 0.4 M NaOAc buffer as the mobile phase. The peak at the low MW end is a salt imbalance peak. (b) UV chromatogram of 2.5 mg PDMDAAC/mL, using a 0.25 M NaOAc buffer, containing 0.1 mg BSA/mL, as the mobile phase, at pH 8.90, and (c) at pH 10.30.

By plotting the area of this peak versus the excess of protein, relative to that in the mobile phase, and by interpolating to zero, one may determine the excess of protein corresponding to the exact amount of protein bound to the polymer (24, 25).

DISCUSSION

Turbidimetric Titrations. The abrupt turbidity increases in Figure 1 correspond to phase separation of the polymer-protein complex. $pH_{critical}$ increases with I because the attractive Coulombic interactions between the protein and polyion are screened by added salt, so that a larger net negative protein charge is required for phase separation (10).

The phase diagrams of Figure 2 reveal a shift toward lower values of $pH_{critical}$, at constant I, as r is increased from 0.25 to 50. This result may be explained by assuming that hydrophobicity or dehydration of the complex, sufficient to permit phase separation, occurs when some critical fraction of polycation sites is electrically neutralized by binding the requisite number of protein-borne carboxylates. The conditions required for phase separation may be represented by:

$$Z_T = Z_P + nZ_{pr} = 0 \qquad (1)$$

where Z_T represents the net charge of the complex (which can be assumed to approach zero at phase separation), Z_P is the net positive charge of the polycation, Z_{pr} is the net negative charge of the protein, and n is the number of proteins bound per polycation. Since the charge of PDMDAAC results from the quaternary nitrogen, Z_P (and thus the conformation of the polymer) is independent of pH. At high values of r, i.e. "excess protein", the polycation sites may become saturated with proteins, so that n is large. Then Z_{pr} at critical conditions may be relatively small. At low values of r, the number of protein molecules bound per polycation may be decreased, and larger values of Z_{pr} - hence higher $pH_{critical}$ - may be required. If n increases with r, then $(Z_{pr})_{critical}$ decreases with r. Thus, $pH_{critical}$ decreases with increasing r. The shift of the phase boundaries, with increasing r, to lower values of $pH_{critical}$, appears to reach a limit at r=50. This occurrence may indicate saturation of the polymer molecule with a maximum number of protein molecules, i.e. one protein molecule per polycation binding site. Unfortunately, accurate calculations of complex stoichiometry at conditions of phase separation are hindered by the high degree of polycation polydispersity.

Using titration data and phase diagrams, one may observe whether a single charge parameter might control complex formation for all proteins. Phase boundaries of six different proteins, displayed in Figure 3 as plots of Z_{pr}/A vs. I, are observed to vary greatly. Especially striking is the fact that basic proteins, i.e. lysozyme and trypsin, form complexes with the polycation even while possessing a net positive charge, i.e. below their respective isoelectric points. It is believed that the existence of negative "charge patches" (36) on the protein surface, rich in acidic groups, are responsible for overcoming the net positive charge and initiating the electrostatic interaction between the basic proteins and the polycations.

QELS. Although there are often numerous hazards present in the literal interpretation of QELS measurements, especially when bimodal distributions are obtained, considerable evidence supports a literal interpretation of the data presented in Figure 3. The apparent Stokes radius from QELS of the polymer alone (ca. 35 nm) is in excellent agreement with the viscosity radius R_η of 33 nm (34), and the fastest diffusion mode in these distributions, with an apparent Stokes diameter of ca. 6 nm, compares well to results for BSA alone (or BSA in the presence of polymer below $pH_{critical}$). The observed distributions are stable with respect to run time, and are reproducible. Also, the coefficient of the slowest diffusion mode is diminished upon centrifugation, so this mode presumably corresponds to a large species.

The results of this investigation suggest the following; (1) there is no binding of proteins to polyion below pH ~ 4.6, (2) complexes (25 ± 5 nm) are in equilibrium with free protein in the range 4.6<pH<4.9; the sizes of these species are very nearly equal to that of free polymer, so they may well be highly solvated intrapolymer complexes, in which a number of protein molecules are bound to single polyion chains without collapse of the latter, (3) no free protein is observed above pH 5, so all protein is presumably bound, (4) at pH≥5.2, larger aggregates are formed.

The critical pH for bulk phase separation seems to be significantly larger than the pH at which intrapolymer complexes are first formed, and the aggregation of molecular complexes appears as an intermediate process. One may speculate that, near the isoelectric point of the protein (ca. 4.8 for BSA), each polyion chain binds some number of proteins which depends on polymer chain length and flexibility, and also upon protein dimensions. The net charge of this complex, however, remains positive until the pH substantially exceeds the isoelectric point. Under such conditions, intrapolymer complexes may begin to associate.

SEC. The UV chromatograms of Figure 5 reveal the presence of soluble BSA-PDMDAAC complexes, and the loss of protein from the mobile phase, as evidenced by the negative peak at the retention volume of BSA (ca. 16.5 ml). At pH = 8.90 and 9.60, the UV peak corresponding to the complex resembles, in shape and position, the refractive index trace of polymer alone (35), suggesting that the complexes existing at these conditions may be solvated, intrapolymer species, consisting of one or more proteins bound per polymer, in accord with the QELS results discussed above. As pH approaches $pH_{critical}$ (ca. 10.60), the size of the complex shifts toward that of a lower MW species, approaching the elution volume of the free protein. This collapse in size may occur as the proteins acquire increasing negative charge in response to the increasing pH, resulting in the binding of one protein to multiple polycation binding sites. A collapse of the complex was not evident in QELS measurements, where the BSA molecules possessed comparitively less negative charge at the lower pHs (<6.30) studied.

As stated earlier, we hope to determine complex stoichiometry through the use of the Hummel-Dreyer technique. A calculation of stoichiometry at the point of polymer saturation will provide the average number of binding sites per polymer molecule, from which an intrinsic dissociation constant may be estimated. Currently, calculations of complexation stoichiometries have been hampered by a chromatographic overlap of the protein peak with that of the complex, and also by the extremely high polydispersity of the PDMDAAC samples.

CONCLUSIONS

These results shed light upon the phase separation phenomenon. The information generated will be applied toward the optimization of complexation selectivity.

Analyses of phase boundaries reveal evidence for polymer saturation in the presence of excess protein. Phase boundaries also facilitate comparisions of the behavior of various proteins. The failure of net surface charge density as a universal parameter for protein-polyelectrolyte interaction is believed to be related to the existence of "charge patches" on the protein surface. The determination of a more realistic protein charge parameter possesses great importance, since the ionic interactions of proteins are exploited in a variety of applications, including protein purification via ion exchange liquid chromatography.

The results of QELS measurements provide strong evidence for the existence of a stable protein-polymer complex. Size exclusion chromatography via the Hummel-Dreyer method, provides one way to characterize such soluble complexes. Static light scattering will be employed to find complex stoichiometry, by measuring the average MW of the complex. A subsequent goal will be the measure of intrinsic binding constants.

ACKNOWLEDGMENT

Support from the Eli Lilly Corporate Research Center is gratefully acknowledged.

LITERATURE CITED

1. Michaels, A.S.; Miekka, R.G.; J. Phys. Chem. 1961, 65, 1765.
2. Veis, A.; Bodor, E.; Mussel, S.; Biopolymers 1967, 5, 37.

3. Polderman, A. Biopolymers 1975, 14, 2181.
4. Tsuchida, E.; Abe, K.; Honma, M.; Macromolecules 1976, 9, 112.
5. Kabanov, V.A.; Zezin, A.B. Macromol. Chem. Suppl. 1984, 6, 259.
6. Dauzenberg, H.; Linow, K.J.; Philipp, B. Acta Polymerica 1982, 33, 619.
7. Shinoda, K.; Sakai, K.; Hayashi, T.; Nakajima, A. Polymer J. (Japan) 1976, 8, 208.
8. Bungenburg de Jong, H.G. In "Colloid Science ; Kruyt, H.R., Ed.; Elsevier: Amsterdam, 1949, vol. II., Chapter X.
9. Lenk, T.; Theis, C. In Coulombic Interactions in Macromolecular Systems; Eisenberg, A., Bailey, F.E., Eds.; American Chemical Society: Washington, D.C., 1986, Chapter 20.
10. Dubin, P., Ross, T.D., Sharma, I., and Yegerlehner, B., in Ordered Media in Chemical Separations ; Hinze, W.L., Armstrong, D.W.,Eds.; American Chemical Society: Washington, D.C., 1987, Chapter 8.
11. Nguyen, T.Q. Makromol. Chem. 1986, 187, 2567.
12. Kokufuta, E.; Shimizu, H.; and Nakumura, I. Macromolecules 1981, 14, 1178.
13. Sternberg, M.; Hershberger, D. Biochim. Biophys. Acta 1974, 342, 195.
14. Jendrisak, J.J.; Burgess, R.R. Biochemistry 1975, 14, 4639.
15. Bell, D.J.; Hoare, M.; Dunnill, P. In Advances in Biochemical Engineering/Biotechnology ; Fiechter, A., Ed.; Springer-Verlag: New York, 1982, p.1.
16. Burgess, R.R.; Jendrisak, J.J. Biochemistry 1975, 14, 4634.
17. Jendrisak, J.J., In Protein Purification: Micro to Macro ; Burgess, R.R., Ed., Alan R. Liss, Inc., 1987, p. 75.
18. The "fit error" is defined in the NiComp coftware as the root-mean-square difference between each base line subtracted channel content and the value predicted from the fitted distribution; the "residual" is a measure of the amount by which the base line is raised to achieve the best fit and as such is a measure of large and erratic long-decay contributions to the autocorrelation function.
19. Koppel, D.E. J. Chem. Phys. 1972, 57, 4814.
20. Provencher, S.; Hendrix, J.; de Maeyer, L. J. Chem. Phys. 1978, 69, 4273.
21. Nicoli, D., private communication, December 1984.
22. Morrison, I.D.; Grabowski, E.F.; Herb, C.A. Langmuir 1985, 1, 496.
23. Dubin, P.L., Rigsbee, D.R., Gan, L.M., and Fallon, M.A. Macromolecules 1988, 21, 2555.
24. Hummel, J.P.; Dreyer, W.J. Biochim. Biophys. Acta 1962, 63, 530.
25. Barberousse, V.; Sacco, D.; Dellacherie, E. J. Chromatogr. 1986, 369, 244.
26. Tanford, C.; Swanson, S.A.; Shore, W.S. J. Am. Chem. Soc. 1955, 77, 6414.
27. Cannan, R.K.; Palmer, A.H.; Kilbrick, A.C. J. Biol. Chem. 1942, 142, 803.
28. Cannan, R.K.; Kilbrick, A.C.; Palmer, A.H. Ann. N.Y. Acad. Sci. 1941, 41, 243.
29. Tanford, C.; Wagner, M.L. J. Am. Chem. Soc. 1954, 76, 3331.
30. Tanford, C.; Hauenstein, J.D. J. Am. Chem. Soc. 1956, 78, 5287.
31. Duke, J.A.; Bier, M.; and Nord, F.F. Arch. Biochem. Biophys. 1952, 40, 424.
32. Cartha, G.; Bellow, T.; and Harker, D. Nature, 1967, 213, 826.
33. See, for example: Grabowski, E.F.; Morrison, I.D. In Measurement of Suspended Particles by Quasi-Elastic Light Scattering; B.E. Dahneke, Ed., Wiley, New York, 1983, Chapter 7.
34. Flory, P.J. Principles of Polymer Chemistry; Cornell University Press: Ithaca, NY, 1953, p.606.
35. Strege, M.A.; Dubin, P.L. J. Chromatogr. 1989, 463, 165.
36. Lesins, V.; Ruckenstein, E. Colloid Polymer Sci. 1988, 266, 1187.

RECEIVED September 27, 1989

Chapter 9

Protein Fractionation by Precipitation with Carboxymethyl Cellulose

Kathleen M. Clark and Charles E. Glatz

Department of Chemical Engineering, Iowa State University, Ames, IA 50011

The effects of pH, ionic strength, and polymer dosage on the precipitation of lysozyme and ovalbumin by Carboxymethyl cellulose (CMC) were investigated. Fractional precipitation was examined by performing the precipitations on binary mixtures of the proteins. Only proteins of opposite charge were precipitated by CMC, and fractionation was achieved through control of both the pH and dosage levels. Protein recovery was enhanced by low ionic strength or selection of pH at which the proteins were highly charged. While protein recovery initially increased with polymer dosage, beyond an optimum dosage level, protein recovery was reduced.

Interactions between protein and polyelectrolytes have been used to fractionate protein solutions (1-3), recover whey proteins (1, 3-6) and isolate serum glycoproteins (7) and recA protein (8). If the potential for protein recovery and purification from aqueous solution by precipitation with polyelectrolytes is to be fully exploited, several factors must be evaluated. The efficacy of the protein precipitation as well as the characteristics of the resulting precipitates depend on several variables.

Earlier work (9) has shown that the size of the precipitate, but not the protein recovery, depends on the method of addition of the polymer to the protein solution. Mixing conditions in the precipitation vessel also affect the precipitate size (10). The solubility of the protein-polyelectrolyte complex depends strongly on the solution conditions--pH, ionic strength, polymer dosage level, and the nature of the protein and polyelectrolyte. These factors are discussed below:

Protein. At the same solution conditions, different proteins display distinct solubilities with polyelectrolytes. The strength of the protein-polymer interaction depends on both the number and distribution of charged sites on the protein surface (11). At pH 3.4 and ionic strength of

0.01 M, carboxymethyl cellulose (CMC) precipitated nearly 90% of β-lactoglobulin but less than 30% of BSA (12).

This difference in solubility has been exploited to fractionate protein solutions. Polyacrylic acid was used to partially fractionate a mixture of four enzymes (2). Whey protein fractions have also been isolated by precipitation with CMC (3).

Polyelectrolyte. The nature of the polyelectrolyte plays a major role in the precipitation efficiency. Both the charge on the polymer as well as the charge density are important (5, 6, 12, 13). Steric factors, or the flexibility of the polyelectrolyte may also influence the effectiveness of the precipitation (6).

Solution pH. Several investigators (2, 12, 13) have indicated that the solution pH is an important determinant in the precipitation efficiency, and the optimum pH level will vary with both the protein (12) and the polyelectrolyte (5). However, the optimum pH for precipitation of protein by CMC did not change with the degree of substitution (DOS) of the CMC (12). This dependence is expected for the formation of an electrostatic complex. Changes in pH will affect the charge on the polyelectrolyte and the charge distribution on the protein.

Selective precipitation of whey protein has been accomplished by pH shifts of protein/polyelectrolyte mixtures. Beta-lactoglobulin is almost quantitatively precipitated by CMC at pH 4.0 but very little α-lactalbumin is precipitated at this pH. Following removal of β-lactoglobulin, the pH is reduced to 3.2 to allow recovery of α-lactalbumin (1).

Ionic Strength. The reduction in protein-polyelectrolyte interactions with increasing ionic strength has been noted by several investigators (1, 6, 11, 12), and gives further evidence that the complex results from electrostatic interactions. Hill and Zadow (5) noted that the decrease in precipitation varied with the polyelectrolyte used, and it has been reported that the effect of high ionic strength on precipitation of whey protein by CMC could be alleviated by using a highly substituted CMC (1, 6, 12).

Dosage. Protein Removal increases with polyelectrolyte dosage to an optimum, then decreases with further addition of polymer (6, 12, 13). The dosage requirement depends on the nature of the polyelectrolyte, the degree of ionization and the protein (1-3, 5, 12). Sternberg and Hershberger (2) have used the different affinities of proteins for polyacrylic acid to fractionate a mixture of four enzymes by successive addition of polymer.

The work presented here will show the effects of pH, ionic strength and polymer dosage on the precipitation and fractionation of proteins by

CMC. Two proteins were used: ovalbumin and lysozyme. Precipitation was performed on solutions of proteins, singly and in binary mixtures.

MATERIALS AND METHODS

Materials. The CMC was a commercial sample supplied by Hercules Inc. (Wilmington, DE) of average molecular weight 250,000 Da and a degree of substitution of 1.2. Lysozyme and ovalbumin were from Sigma Chemicals (St. Louis, MO).

Precipitation. The pH and ionic strength conditions for the precipitations are listed in Table 1. The pH levels were selected to allow investigation of the effect of protein charge on precipitation. The effect of pH on the charge of ovalbumin, lysozyme, and CMC is shown in Table 2. Solutions of single proteins and binary combinations were examined. For all runs, the total protein concentration was 1 mg/ml. Preliminary dosage screening experiments were performed to determine the effect of polymer dosage on protein removal. From this data an optimum polymer dosage level was determined. Precipitations were then repeated at the optimum dosage conditions. These optimum dosage levels are listed in Table 3. Identification of the optimum dosage levels is important since earlier work (9) had indicated that the largest particles are produced at optimum dosage levels.

The precipitations were performed in 32.4 ml (dosage screening runs) or 246 ml (optimum dosage runs) volumes. Carboxymethyl cellulose was added as a 0.1% solution to the protein solution. Mixing was provided by a magnetic stirrer. Each dosage screening sample was mixed for 3 minutes, then aged for 60 minutes in a shaker. The optimum dosage runs were mixed for 45 minutes. The slurries were then centrifuged at 20,000 g for 45 minutes. The supernatants were recovered.

Analysis. Protein concentrations of the supernatants from the binary precipitations were determined by ion-exchange-HPLC using an Aquapore CX-300 weak cation exchanger. The proteins were eluted by a linear ionic strength gradient from 0.05 to 0.65 M at pH 6.0, and detection was performed by 280 nm absorption.

Supernatants of precipitation slurries from single-component systems were analyzed by 280 nm absorbance.

Table 1. Summary of pH and ionic strength levels investigated in the precipitation of lysozyme and ovalbumin by CMC

Protein system	pH	Ionic Strength [M]
Ovalbumin	4.2	0.02
		0.04
		0.05
		0.06
		0.07
Lysozyme	4.2	0.02
		0.07
		0.10
		0.12
		0.15
	5.8	0.02
		0.07
	7.5	0.02
		0.07
Lysozyme-ovalbumin	4.2	0.02
		0.07
	5.8	0.02
		0.07
	7.5	0.02
		0.07

Table 2. The effect of pH on the net charge of ovalbumin, lysozyme and CMC

| pH | Net charge | | |
	Ovalbumin [14]	Lysozyme [15]	CMC [16]
4.2	+6	+11	-1047
5.8	-12	+7.5	-1489
7.5	-17	+5.8	-1550

Table 3. CMC dosage levels for the maximum total protein removals for the precipitation of ovalbumin and lysozyme as a function of pH and ionic strength

Protein	pH	Ionic strength [M]	Maximum total protein removal	CMC dosage g CMC / g protein	Extent of charge neutralization [%]
Ovalbumin					
	4.2	0.02	86.44	0.1222	22.52
		0.04	82.05	0.1421	18.38
		0.05	78.27	0.1448	17.21
		0.06	60.50	0.1447	13.31
		0.07	47.19	0.1366	11.00
Lysozyme					
	4.2	0.02	99.67	0.2241	81.69
		0.07	98.32	0.2267	79.66
		0.10	96.21	0.2370	74.56
		0.12	92.10	0.2287	73.97
		0.15	81.52	0.2238	66.90
	5.8	0.02	99.49	0.1566	55.94
		0.07	96.68	0.1847	46.09
	7.5	0.02	99.04	0.1389	46.65
		0.07	92.61	0.1707	35.49

RESULTS AND DISCUSSION

SINGLE COMPONENT SYSTEMS

Precipitations were performed on solutions of lysozyme and ovalbumin. For lysozyme, the effects of pH from 4.2 to 7.5 and ionic strength from 0.02-0.15 M were investigated. Ovalbumin precipitations were performed at pH 4.2, 0.02-0.07 M. For both systems, the polymer dosage levels were varied from 0.05 to 0.25 (g CMC/g protein). The results are discussed below.

Polymer Dosage. The effect of polymer dosage on the recovery of proteins by precipitation with CMC is evident in Figure 1. For both ovalbumin and lysozyme, protein removal increases with polymer level at low dosages to an optimum removal level. If the polymer dosage is increased beyond this optimal level for the ovalbumin precipitation, protein removal decreases. This is most likely the result of an electrostatically stabilized suspension. At high polymer levels excess polymer is incorporated into the protein-polyelectrolyte complex, resulting in a soluble complex with a significant net charge. Because of this charge, aggregation of the complexes into insoluble primary particles and flocs is hindered, and protein recovery is reduced.

This resolubilization is not observed for the precipitation of lysozyme at pH 4.2. However, at the higher pH level, where lysozyme has a lower net charge, the characteristic resolubilization is observed at high dosages.

pH. The results of lysozyme precipitation at pH 4.2, 5.8 and 7.5 are shown in Figure 2. As the pH is increased, decreasing the charge on the protein, the polymer dosage required for optimum recovery of protein is reduced. This is evidence that the precipitate formation is the result of electrostatic interactions.

The optimum dosage levels are also affected by the pH level. Although lysozyme recovery was nearly quantitative, the polymer requirements decreased with the net charge on the protein. Calculations, summarized in Table 3, indicate that for the precipitation of lysozyme at pH 4.2, a nearly electrostatically neutral complex is formed. Note that the extent of charge neutralization is reduced by increased ionic strength, and that this effect is most pronounced for proteins with low net charges.

The effect of protein charge, at constant pH, has been investigated in the recovery of recombinant β-galactosidase (17). At pH 5.7 only 30% of E. coli β-galactosidase was precipitated by polyethyleneimine. However, the addition of 6 and 11 aspartic acid residues to the protein increased the recovery by polyethyleneimine to 65 and 95%, respectively.

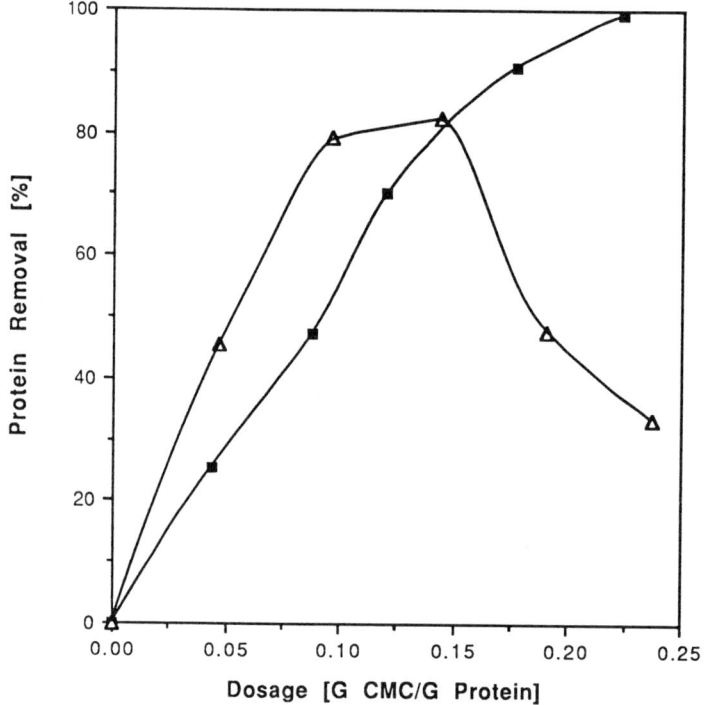

Figure 1. Effect of polymer dosage on the precipitation of ovalbumin and lysozyme by CMC at pH 4.2; 0.02 M.
■ Lysozyme; Δ Ovalbumin

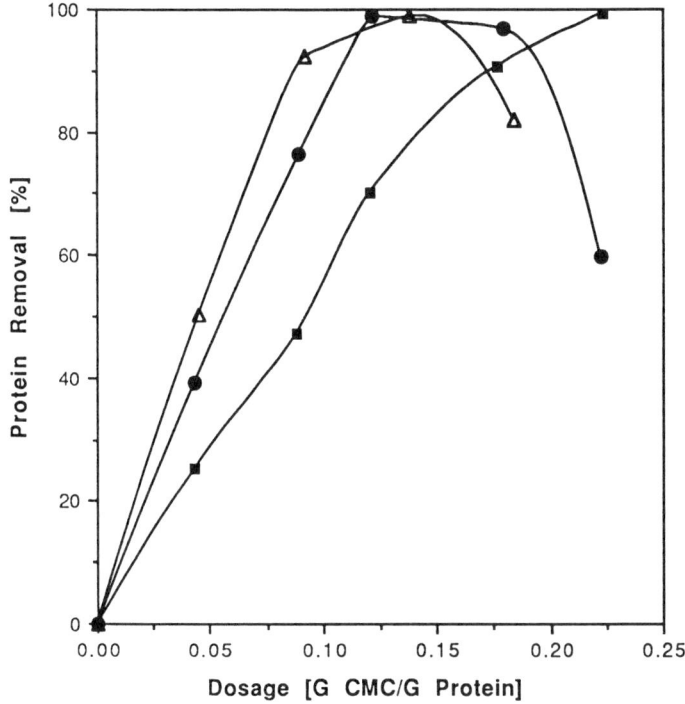

Figure 2. Effect of pH on the recovery of lysozyme by precipitation with CMC at 0.02 M.

■ pH 4.2;   ● pH 5.8;   △ pH 7.5

Ionic Strength. If the protein-polymer complex is formed as a result of electrostatic interactions, increased ionic strength should serve to reduce the attraction between the oppositely charged macromolecules, and decrease the precipitation efficiency. This is observed at pH 4.2 in Figures 3 and 4 for lysozyme and ovalbumin, respectively, and in Figure 5 for lysozyme at pH 5.8 and 7.5.

These data indicate an interaction effect between pH and ionic strength, as highly charged proteins are less affected by increases in ionic strength than proteins with a lower net charge.

As indicated in Table 3, increased ionic strength not only reduces the maximum precipitation possible, but also increases the polymer dosage requirement for optimum precipitation. This may be a result of the shielding effect attributed to high levels of counterions around charged macromolecules. A higher concentration of polymer-ionic groups is required to displace the counterions around the protein molecule.

BINARY SOLUTIONS OF PROTEINS

The effect of precipitation conditions on the selective precipitation of a target protein species (lysozyme) from a binary mixture of ovalbumin and lysozyme was investigated. The results are discussed below.

Polymer Dosage. Figure 6 shows the precipitation of protein from a binary solution of ovalbumin and lysozyme at pH 4.2 and ionic strength levels of 0.02 and 0.07 M. At this pH both proteins are positively charged; lysozyme has a net charge of 11, and ovalbumin, 6. Note that at very low polymer dosages, only lysozyme is precipitated. Only when virtually all the lysozyme is removed from solution does ovalbumin begin to precipitate. This indicates a selectivity of the polymer for the more highly charged protein species.

pH. The effect of pH on protein fractionation by polyelectrolyte precipitation has been noted by several authors (1, 2, 12), and is demonstrated in Figure 7 for precipitation at pH 5.8 and 7.5. At these pH levels, only lysozyme is positively charged. As expected, only lysozyme, is recovered in the precipitate, and no ovalbumin is removed.

One concern in using pH to control the fractionation efficiency of polymer precipitation, is that the recovery of the target proteins may be affected by the presence of the oppositely charged (non-target) protein species. For the conditions investigated in this work, no interference was noted, as shown in Figure 8. This may be a result of the low total protein concentrations investigated in this work, or simply the result of the lower charge density of the negatively charged proteins relative to the CMC.

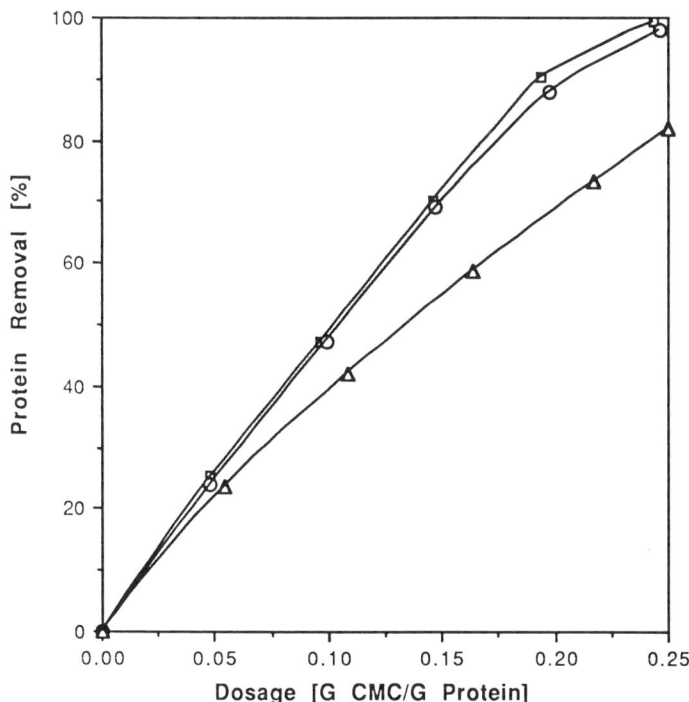

Figure 3. Effect of ionic strength on the precipitation by CMC at pH 4.2.

□ 0.02 M; o 0.07 M; Δ 0.15 M

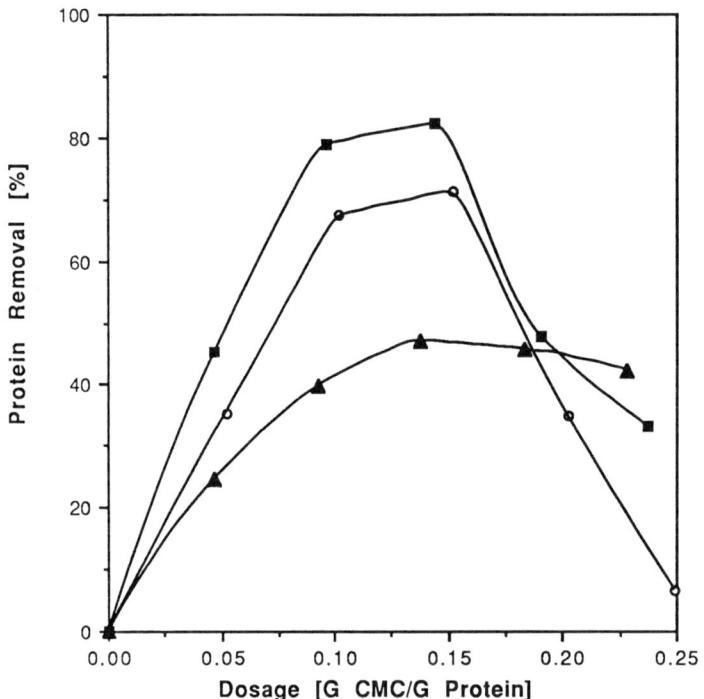

Figure 4. Effect of ionic strength on the precipitation by CMC at pH 4.2.

■ 0.02 M;   o 0.05 M;   ▲ 0.07 M

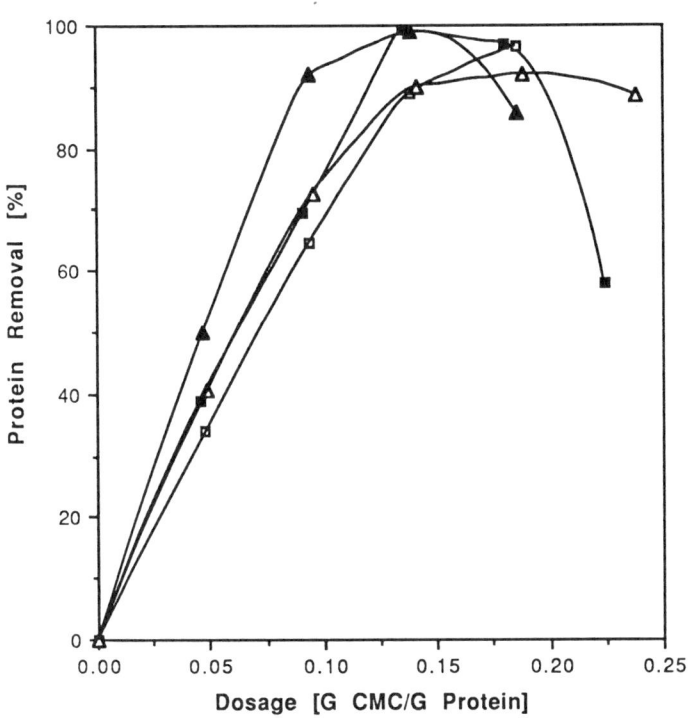

Figure 5. Effect of ionic strength on the precipitation of lysozyme by CMC at pH 5.8 and 7.5.

■ pH 5.8, 0.02 M; ▲ pH 7.5, 0.02 M
□ pH 5.8, 0.07 M; △ pH 7.5, 0.07 M

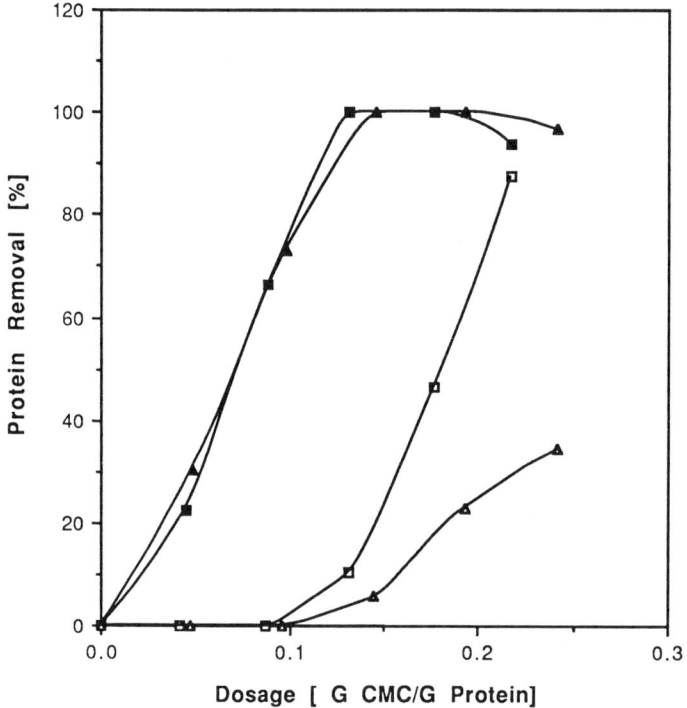

Figure 6. Precipitation of protein by CMC from a binary mixture of lysozyme and ovalbumin at pH 4.2.

■ Lysozyme 0.02 M;   ▲ Lysozyme, 0.07 M
❑ Ovalbumin 0.02 M;   △ Ovalbumin, 0.07 M

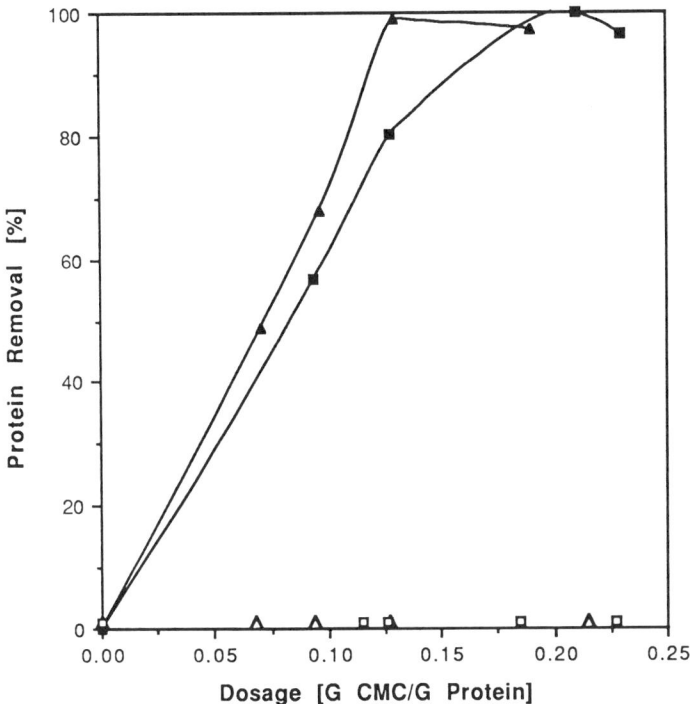

Figure 7. Precipitation of protein by CMC from a binary mixture of lysozyme and ovalbumin at 0.02 M and pH levels at which only lysozyme is positively charged.

■ Lysozyme pH 5.8;   ▲ Lysozyme, pH 7.5
□ Ovalbumin pH 5.8;   △ Ovalbumin, pH 7.5

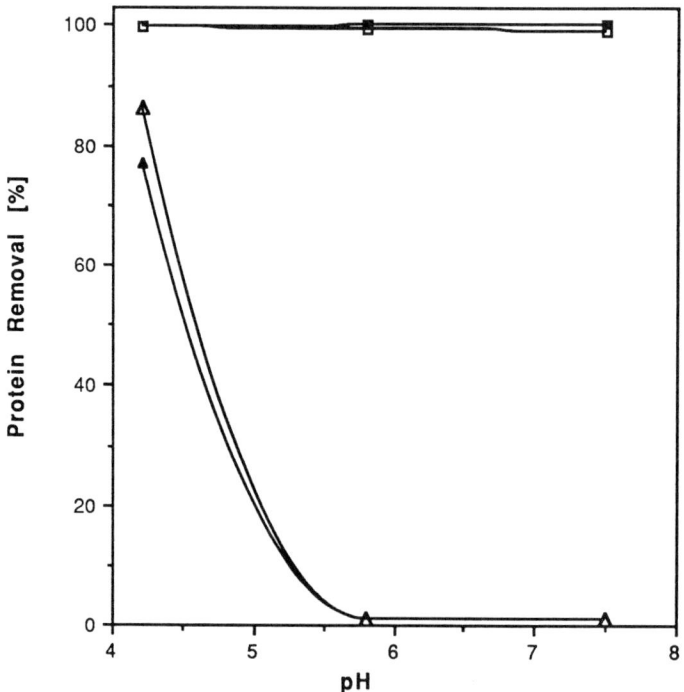

Figure 8. Effect of pH on the optimum precipitation of protein by CMC from single component and binary solutions at 0.02 M ionic strength. Filled symbols indicate the binary mixture of lysozyme and ovalbumin.

■ Lysozyme, Binary; ❏ Lysozyme
▲ Ovalbumin Binary; △ Ovalbumin

Ionic Strength. The effect of ionic strength on the fractionation efficiency for a binary mixture of proteins is the same as that for single component solutions of protein: increased ionic strength reduces the maximum precipitation possible. However, because the effect of ionic strength is less severe for highly charged proteins, this effect may be exploited to secure greater purification, as shown in Figure 6. Since the increase in ionic strength from 0.02 to 0.07 M served to reduce the ovalbumin recovery, but not that of the higher charged lysozyme, higher purification factors (Table 4) were achieved at the higher ionic strength level.

A model of protein-polyelectrolyte complexation as the result of electrostatic binding between charged protein and multiple binding sites of opposite charge on the polyelectrolyte has proven capable of describing most features reported here up to the point of optimum precipitation (18).

SUMMARY

From the empirical work presented above, several factors are apparent: The precipitation of proteins by oppositely charged polyelectrolytes is the result of electrostatic interactions, and is strongly dependent on the protein charge. Increased ionic strength reduces the efficacy of the binding, perhaps as a result of "competition" for macromolecular charges.

CONCLUSIONS

Several conclusions can be drawn about the effect of polymer dosage, pH and ionic strength on protein recovery and fractionation by precipitation with CMC:
1. Carboxymethyl cellulose precipitates only oppositely charged proteins, and preferentially precipitates proteins of higher charge.
2. Precipitation efficiency increases with protein charge.
3. At low polymer dosages, protein removal increases with dosage to an optimum. However, with additional polymer addition, protein recovery is reduced. This effect is alleviated if the protein is highly charged.
4. Increased ionic strength levels serve to increase the polymer dosage requirement, reduce the maximum precipitation possible, and reduce the effect of precipitation pH on protein recovery.
5. Fractional precipitation may be achieved through selection of precipitation pH or polymer dosage. Fractionation efficiency can be enhanced by high ionic strength levels, if the target protein is highly charged.

Table 4. Purification factors for the recovery of lysozyme from a 1:1 ovalbumin-lysozyme solution by precipitation with CMC at pH 4.2 and 0.02 and 0.07 M

Ionic Strength [M]	Dosage	Protein removals Lysozyme	Protein removals Ovalbumin	Purification factor	Purity of lysozyme in solid
0.02	0.0432	22.64	00.00	--	100.0
	0.0865	66.20	00.00	--	100.0
	0.1297	100.0	10.90	9.174	90.17
	0.1730	100.0	46.61	2.146	68.21
	0.2162	93.78	87.65	1.070	51.69
0.07	0.0478	30.52	00.00	--	100.0
	0.0956	72.95	00.00	--	100.0
	0.1433	100.0	8.96	11.16	91.78
	0.1911	100.0	23.07	4.335	81.25
	0.2389	96.82	34.15	2.835	73.93

ACKNOWLEDGMENTS

This work was supported by the Engineering Research Institute of Iowa State University through National Science Foundation Grant Nos. CPE-8120568 and ECE-8514865. The CMC was a gift from Hercules, Inc.

LITERATURE CITED

1. Hidalgo, J. and P. M. T. Hansen. 1969. Interactions between food stabilizers and β-lactoglobulin. J. Agr. Food Chem. 17:1089-1092.
2. Sternberg, M. and D. Hershberger. 1974. Separation of proteins with polyacrylic acids. Biochim. Biophys. Acta 342:195-206.
3. Hansen, P. M. T., J. Hidalgo, and I. A. Gould. 1971. Reclamation of whey protein with carboxymethyl cellulose. J. Dairy Sci. 54:830-834.
4. Sternberg, M., J. P. Chiang, and N. J. Eberts. 1976. Cheese whey proteins isolated with polyacrylic acid. J. Dairy Sci. 59:1042-1050.
5. Hill, R. D. and J. G. Zadow. 1978. The precipitation of whey protein with water soluble polymers. N. Z. J. Dairy Sci. Technol. 13:61-64.
6. Hill, R. D. and J. G. Zadow. 1974. The precipitation of whey proteins by carboxymethyl cellulose of differing degrees of substitution. J. Dairy Res. 41:373-380.
7. Anderson, A. J. 1967. The isolation of serum glycoproteins (seromucoid) by fractional precipitation with polyelectrolytes. Biochem. J. 104: 18-19.
8. Shibata, T., R. P. Cunningham, and C. M. Radding. 1984. Homologous pairing in genetic recombination: Purification and characterization of Escherichia coli recA protein. J. Biol. Chem. 256:7557-7564.
9. Clark, K. M. and C. E. Glatz. 1987. Polymer dosage considerations in polyelectrolyte precipitation of protein. Biotech. Prog. 4:241-247.
10. Fisher, R. R. 1987. Protein precipitation with acids and polyelectrolytes: The effects of reactor conditions and models of the particle size distribution. Ph.D. Thesis. Iowa State University, 165 pp.
11. Ledward, D. A. 1978. Protein-polysaccharide interactions. pages 205-217 in J. M. V. Blanshard and J. R. Mitchell. eds. Polysaccharides in food. Butterworths, London.
12. Zadow, J. G. and R. D. Hill. 1975. The precipitation of proteins by carboxymethyl cellulose. J. Dairy Res. 42:267-275.
13. Gault, N. F. S. and R. A. Lawrie. 1980. Efficiency of protein extraction and recovery from meat industry by-products. Meat Sci. 4:167-190.
14. Longsworth, L. G. 1941. The influence of pH on the mobility and diffusion of ovalbumin. Ann N. Y. Acad. Sci. 41:267-285.
15. Beychok, S. and R. C. Warner. 1959. Denaturation and electrophoretic behavior of lysozyme. J. Am. Chem. Soc. 81:1892-1897.
16. Chowdhury, F. H. and S. M. Neale. 1963. Acid behavior of carboxylic derivatives of cellulose: Part I. Carboxymethyl cellulose. J. Pol. Sci. Pt. A 1:2881-2891.
17. Zhao, J., C. Ford, S. Gendel, M. Rougvie, and C. Glatz. 1988. Genetic engineering of β-galactosidase with a poly-aspartate tail to enhance downstream product recovery by polyelectrolyte precipitation in K. Brew et al. eds. Advances in gene technology: Protein engineering and production. IRL Press, Washington, D. C.
18. Clark, K. M. and C. E. Glatz. In preparation. A multi-equilibrium binding model for the precipitation of proteins by carboxymethyl cellulose.

RECEIVED September 27, 1989

Chapter 10

Protein Separation via Affinity-Mediated Membrane Transport

Liese Dall-Bauman[1] and Cornelius F. Ivory[2]

[1]Crew and Thermal Systems Department, Lockheed Engineering and Sciences Company, C-70, 2400 NASA Road 1, Houston, TX 77058-3711
[2]Department of Chemical Engineering, Washington State University, Pullman, WA 99164

> Affinity-mediated transport is a form of facilitated transport in which a 'switch' monoclonal antibody is used as a highly selective protein carrier. The membrane's physicochemical environment is controlled so that the antibody exhibits a high binding affinity for its antigen at the upstream boundary and a significantly lower binding affinity at the downstream boundary. Hence, complexation is favored upstream and decomplexation is favored downstream.
>
> An affinity-mediated system in which a 'switch' monoclonal antibody is used to transport its antigen, human growth hormone, has been modeled. The affinity of the antibody for the hormone is dependent on local pH. In addition to the kinetic effect, macroscopic and microscopic electrochemical effects were considered. On the larger scale, modest induced and applied electric fields were found to exert considerable influence on fluxes of antibody, hormone, and complexes. The short-range effect of Donnan potential was found to enhance the flux of hormone into the membrane.

Carrier-mediated transport provides a means of increasing the flux of a selected permeant across a liquid membrane. As shown in Figure 1, the Fickian flux of permeant is augmented by a second mechanism in which the permeant is absorbed at the membrane's upstream boundary and reacts reversibly with a carrier molecule to form a complex. The complex then diffuses across the membrane to the downstream boundary, where decomplexation occurs, the permeant is desorbed, and the carrier is free to return to the upstream boundary so that the cycle can be repeated. The carrier and complex species are generally considered to be nonvolatile in the sense of Schultz et al. (1); that is, they are assumed to be confined to the membrane phase.

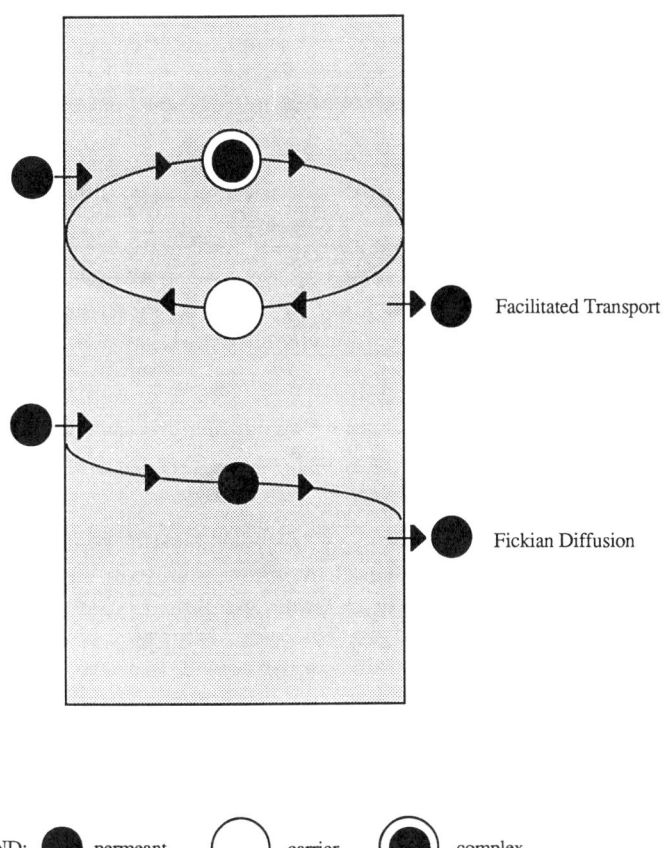

Figure 1. Schematic representation of facilitated transport and Fickian diffusion

There are certain benefits associated with controlling the direction of the complexation/decomplexation reaction. If complexation can be encouraged at the upstream boundary, more permeant can be introduced into and carried across the membrane. If the equilibrium is shifted away from decomplexation throughout the membrane, the permeant will be trapped in the membrane. More specifically, downstream decomplexation helps to ensure efficient use of carrier. Complexation and decomplexation can be favored in the appropriate regions if there is some difference between the physicochemical environments in the two regions.

For example, Jain and Schultz (2) found that hemoglobin/carbon monoxide complexation was favored in a dark environment, while decomplexation was favored in the presence of light. A greater flux of carbon monoxide was achieved through a partially illuminated membrane with darkness upstream and illumination downstream than through totally illuminated or non-illuminated membranes.

In this work, the consequences of controlling the chemical environment have been considered. Specifically, the effect of pH has been examined. A number of biochemical reactions are known to be pH-sensitive. For example, hemoglobin/oxygen binding tends to increase with increasing pH and most enzymes have a specific pH at which their substrate binding affinity is maximized. Here, a mathematical model describing the effect of a transmembrane pH gradient on transport of human growth hormone (hgh) is presented. 'Switch' monoclonal antibodies are considered as carriers.

'Switch' monoclonal antibodies were developed to improve the efficiency of immunoaffinity chromatography (3). The strength of an antigen-antibody bond is described in terms of the antibody's affinity for the antigen. In immunoaffinity chromatography, high-affinity antibodies are necessary for efficient adsorption of an antigen, particularly if the antigen is present in limited quantities in the feed mixture. Unfortunately, antigens are not easily desorbed from high-affinity antibodies and are sometimes denatured by the desorption process. 'Switch' antibodies can be transformed from high- to low-affinity states by relatively small changes in their environment (e.g. temperature, salt concentration, or pH). Hill and coworkers discussed two pH-sensitive antibodies, one of which had a high binding affinity for hgh at pH ~ 5 and a much lower affinity at pH ~ 4. The other underwent a similar change in the pH range 9.5-10.5. It was proposed that the pH change induced subtle changes in structure in the hormone or the antibody, similar to the way a change in pH influences the activity of an enzyme by affecting the enzyme's configuration (4).

Low selectivity is another problem that is encountered in immunoaffinity chromatography. Obviously, if an antibody shows little preference for one antigen over another, no separation will be achieved. By definition, a monoclonal antibody recognizes -- and binds to -- a single site and is therefore highly selective. As implied by their name, 'switch' monoclonal antibodies are selective and, under the right circumstances, are easily persuaded to release their antigens. For these reasons, they are promising candidates for use as carriers in facilitated transport of proteins.

SYSTEM DESCRIPTION

Virtually no information on the physical properties of 'switch' monoclonal antibodies is available and their reactions with antigens have not been described in detail. For the purpose of this work, the antibody was assumed to be a gamma globulin with two identical and independent binding sites for hgh. It was also assumed to have a binding site for hydrogen that could influence both hgh binding sites -- perhaps by slowing down the binding of the hormone and accelerating its release. This could occur if the bound hydrogen forced reorientation of the hgh binding sites with respect to each other or to the rest of the molecule. These assumptions imply the existence of several forms of complex. They, together with all other species present, are listed in Table 1.

Further assumptions were made concerning the physical parameters describing the antibody and the antibody-hormone complexes. The diffusion coefficients for all complexes were set equal to the value assigned to the antibody's diffusion coefficient. This is reasonable, since the molecular weight of the antibody is between 156,000 and 161,000 (5), while that of hgh is roughly 22,000 (6). The pH-dependence of the charge on the hormone was obtained by fitting the data read from a titration curve (6). The charge on the antibody was assumed to be a linear function of pH, with the exact formula varying from case to case. The charge on each complex species was calculated by summing the charges of the components forming the complex. The diffusion coefficients and charge profiles for all species are listed in Table 2.

The reactions occurring in the system are shown in Table 3. All of the rate and equilibrium constants are estimates. The rate constants tabulated by Steward (7) for a variety of antibody-antigen reactions range from 8.0×10^6 l mol^{-1} sec^{-1} to 6.2×10^8 l mol^{-1} sec^{-1} for the binding reaction and from 3.4×10^{-4} sec^{-1} to 6.0×10^3 sec^{-1} for the reverse reaction. These values were used as guidelines in a very limited parameter scan done at the beginning of this study. The rate constants appearing in the table were ultimately chosen for use throughout the remainder of the work. It must be stressed that these are not optimal values.

Examination of the rate constants shows that the rate constants for the binding of the first and second permeants have identical values, as do those for the release of the two permeants. This is true whether or not the carrier is protonated and is in accordance with the assumption of identical, independent sites. It can also be seen that the forward rate constants are lower for complexation of protonated species, while the reverse rate constants are lower for decomplexation of nonprotonated species. This is necessary, given the assumption that protonated forms are slower to bind and quicker to release hgh.

Table 1. Definition of Species Vector **s** for HGH - Antibody System

Index	Symbol	Description
1	S	human growth hormone (primary permeant)
2	M^+	supporting cation
3	X^-	supporting anion
4	H^+	hydrogen ion
5	A	antibody (carrier)
6	AS	single-hormone complex
7	AS_2	double-hormone complex
8	AH	protonated carrier
9	AHS	protonated single-hormone complex
10	AHS_2	protonated double-hormone complex

Table 2. Diffusion Coefficients and Charge Profiles for HGH - Antibody System

Species	D (10^{-5} cm^2/sec)	z
S	0.075	$f_0(pH)$
M^+	1.35	+1
X^-	1.35	-1
H^+	6.75	+1
A	0.040	$f_1(pH)$
AS	0.040	$f_0(pH) + f_1(pH)$
AS_2	0.040	$2f_0(pH) + f_1(pH)$
AH	0.040	$f_1(pH) + 1$
AHS	0.040	$f_0(pH) + f_1(pH) + 1$
AHS_2	0.040	$2f_0(pH) + f_1(pH) + 1$

Notes: Diffusion coefficient for antibody is from Sober (5).
Diffusion coefficient for hgh is an average of values given by Bewley and Li (6).
All coefficients are for diffusion of species through water at 25°C.
$f_0(pH)$ is obtained by fitting titration curve shown by Bewley and Li (6).
$f_1(pH)$ is a linear function, with formula varied for different cases.

The pH ranged from 5.0 at the upstream boundary to 4.2 at the downstream boundary. Hydrogen concentrations in the reservoirs were set to fix external pH at these values. The equilibrium constants for the protonation reactions were chosen so that protonated and nonprotonated carrier and complex could coexist in this pH range. The assignment of the same equilibrium constant to reactions 5, 6, and 7 implies a single hydrogen binding site which is independent of the binding of hgh.

The remaining species present, M+ and X-, are supporting electrolytes and are present as ions dissociated from the salt form of hgh, as products of the dissolved salt MX, or (in the case of X-) as counterions for H+. They are 'generic' ions and were given equal diffusion coefficients in the range of the coefficients for diffusion of ionic sodium, lithium, chlorine, and bromine in water (8). The permeating species are hgh (represented as S), H+, M+ and X-. The system is represented in Figure 2.

DONNAN EQUILIBRIUM

Donnan equilibrium is a well-understood phenomenon which is observed at any interface that prevents diffusion of at least one (but not all) charged species between two phases; a thin polymer membrane separating two liquid mixtures provides one example. Donnan behavior also occurs at other interfaces (9) and is especially important in ion exchange resins (10).

The requirement that the electrochemical potential of permeating species i be constant across the phase-separating interface and certain simplifying assumptions (11, 12, 13) can be used to obtain the following relationship:

$$\frac{C_i^{(1)}}{C_i^{(2)}} = \exp\left[\frac{z_i F}{RT}(\Phi^{(2)} - \Phi^{(1)})\right] \quad (1)$$

where $C_i^{(j)}$ is the concentration of species i in phase j, z_i is the electric charge, F is Faraday's constant, R is the gas constant, T is the absolute temperature, and $\Phi^{(j)}$ is the electric potential in phase j. This equation holds for each permeating species so that

$$\left[\frac{C_1^{(1)}}{C_1^{(2)}}\right]^{1/z_1} = \left[\frac{C_2^{(1)}}{C_2^{(2)}}\right]^{1/z_2} = \ldots = \left[\frac{C_P^{(1)}}{C_P^{(2)}}\right]^{1/z_P} = \exp\left[\frac{F(\Phi^{(2)} - \Phi^{(1)})}{RT}\right]$$
$$= \rho \quad (2)$$

Table 3. Reaction Network for HGH - Antibody System

RATE CONSTANTS

	Forward	Reverse
1. $A + S \leftrightarrow AS$	7.5×10^7 l mol^{-1} s^{-1}	7.5×10^{-1} s^{-1}
2. $AS + S \leftrightarrow AS_2$	7.5×10^7 l mol^{-1} s^{-1}	7.5×10^{-1} s^{-1}
3. $AH + S \leftrightarrow AHS$	7.5×10^6 l mol^{-1} s^{-1}	7.5×10^1 s^{-1}
4. $AHS + S \leftrightarrow AHS_2$	7.5×10^6 l mol^{-1} s^{-1}	7.5×10^1 s^{-1}

EQUILIBRIUM CONSTANTS

5. $A + H^+ \leftrightarrow AH$	3.2×10^4 l mol^{-1}
6. $AS + H^+ \leftrightarrow AHS$	3.2×10^4 l mol^{-1}
7. $AS_2 + H^+ \leftrightarrow AHS_2$	3.2×10^4 l mol^{-1}

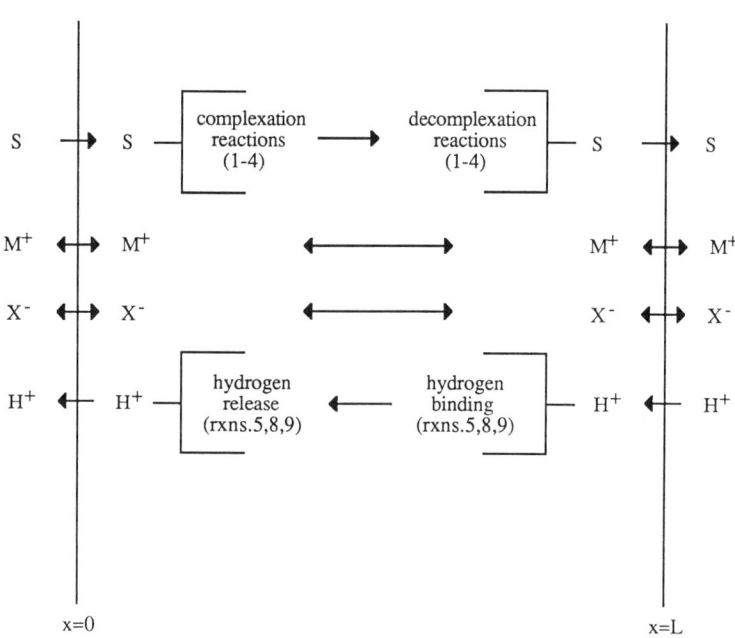

Figure 2. Schematic representation of hgh-antibody system

where P is the number of permeating species and ρ is the Donnan ratio (13). If concentrations are known in one phase then Equation (2) is, in effect, a system of P equations in (P+1) unknowns, with the Donnan ratio being the (P+1)st. The final equation describes electroneutrality in the other phase.

Ion exchange membranes and facilitated transport membranes are orders of magnitude thicker than the Debye length and can be considered phases in their own right. When such a membrane is placed between two solutions, there are two membrane-solution interfaces at which Donnan equilibrium is observed. In ion exchange membranes, the charged anion- or cation-exchange sites control the extent of Donnan inclusion and exclusion. The importance of the Donnan effect in ion exchange membranes is well-documented (14, 15, 16). On the other hand, when ion exchange membranes are used for carrier-mediated gas separations (17,18), there are no ionic permeating species present and thus Donnan equilibrium is not established.

The possible importance of Donnan equilibrium in facilitated transport membranes has not yet been explored. This is likely due to the applications considered to date. A number of gas-separating membranes (gas-liquid-gas systems) have used ionic carriers (19, 20, 21) and several examples of liquid-liquid-liquid systems in which membranes separate electrolytic solutions have been described (22, 23, 24). However, in G-L-G systems the reservoirs contain gases which are not electrically charged and the membrane solvents of choice in L-L-L systems have been organic (e.g. chloroform) and not hospitable to ionized species. The ionic species must be transported as ion pairs which have no net electrical charge. In either case, there exists no interface separating two charge-containing phases and the conditions required for Donnan equilibrium cannot be achieved.

If, however, a carrier-mediated transport membrane containing charged species -- in the form of either mobile ions or fixed sites -- were placed between two electrolytic mixtures, significant Donnan effects could be expected. For example, consider a membrane in which the carrier is a counterion to the permeant. The permeant would be expected to be preferentially included in the membrane phase. If significant inclusion were to occur, the use of simple first-kind boundary conditions would be inappropriate and could lead to underestimation of flux. On the other hand, if the permeant and carrier were coions, the permeant could be excluded and failure to account for exclusion could lead to overprediction of flux. Further complications would arise if the complex were charged or if other charged species were present, since the net charge density inside the membrane defines Donnan equilibrium conditions.

MODELING

The mathematical model describing the hgh-antibody system is based on the unidirectional continuity equation

$$\frac{dJ_i}{dx} = R_i \tag{3}$$

where J_i is the flux, x is the direction perpendicular to the membrane surface, and R_i is the net rate of production of i. The Nernst-Planck equation

$$J_i = -D_i \frac{dC_i}{dx} + D_i z_i C_i \frac{EF}{RT} \tag{4}$$

is used to define flux. The quantity D_i is the diffusion coefficient and E is the electric field strength. The Nernst-Planck equation together with the definition of current density

$$\frac{I}{F} = \sum_i z_i J_i \tag{5}$$

provides an expression for E:

$$E = \frac{RT}{F} \frac{\frac{I}{F} + \sum_{i=1}^{S} z_i D_i \frac{dC_i}{dx}}{\sum_{i=1}^{S} z_i^2 D_i C_i} \tag{6}$$

To avoid solving Equation (3) for each of the ten species present, the partial-equilibrium/ combined flux technique described by Gallagher et al. (25) is used to reduce the model to a set of seven differential and three algebraic equations. The algebraic equations describe the three (equilibrium) protonation reactions. The differential equations describe the conservation of the seven invariants, where an invariant is a linear combination of species that are independent of the equilibrium reactions (25). The invariants used here are (S), (M^+), (X^-), (H^++AH+AHS+AHS_2), (A+AH), (AS+AHS), and (AS_2+AHS_2). The first three are species which do not participate in equilibrium reactions. The fourth encompasses all protonated species, and each of the final three includes the protonated and nonprotonated forms of the carrier or of a complex.

The boundary conditions on the system are

$$x=0: \quad C_j^{(m)}(x=0) = C_j^{(r)}(x=0)\left[\frac{1}{\rho_0}\right]^{z_{j,0}} \quad j = 1,\ldots,4$$

$$\tilde{J}_i(x=0) = 0 \quad i = 5,6,7$$

(7a)

$$x=L: \quad C_j^{(m)}(x=L) = C_j^{(r)}(x=L)\left[\frac{1}{\rho_L}\right]^{z_{j,L}} \quad j = 1,\ldots,4$$

$$\tilde{J}_i(x=L) = 0 \quad i = 5,6,7$$

(7b)

where the index on the Donnan boundary conditions is the species index in s and the index on the zero-flux conditions is the index of the combined fluxes.

The number of zero-flux conditions to be replaced by integral constraints is

$$N_I = S - R - P = 10 - 5 - 4 = 1 \tag{8}$$

The number of reactions is 5 instead of 7 because the reactions are not all stoichiometrically independent; reaction 6 is a linear combination of reactions 1, 3, and 5 and reaction 7 is a linear combination of reactions 2, 4, and 6. The constraint chosen represents the conservation of carrier:

$$\frac{1}{L}\int_0^L (C_A + C_{AS} + C_{AS_2} + C_{AH} + C_{AHS} + C_{AHS_2})\, dx - C_{A,T} = 0 \tag{9}$$

where $C_{A,T}$ is the total amount of antibody present.

The model can be restated in dimensionless form through use of the dimensionless variables defined in Table 4. Here, $C_{ref} = C_{A,T}$ = total amount of carrier present, and $D_{ref} = D_S$.

The model was solved using orthogonal collocation on finite elements (OCFE). Orthogonal collocation on finite elements was developed by Carey and Finlayson (26) for solution of boundary layer problems. Carey and Finlayson used OCFE to solve the simultaneous heat and mass transfer equations describing a catalyst pellet and found the new method to be more efficient than finite difference techniques. They also showed that OCFE was applicable to boundary layer problems that could not be solved by global orthogonal collocation. Jain and Schultz (27)

Table 4. Dimensionless Variables and Dimensionless Groups

$$x^* = \frac{x}{L}$$

$$R_i^* = \sum_{j=1}^{R} \alpha_{ij} r_j^*$$

$$C_i^* = \frac{C_i}{C_{ref}}$$

$$r_j^* = k_{j,1}^* \prod_{i=1}^{s} (C_i^*)^{\alpha_{ij}^-} - k_{j,2}^* \prod_{i=1}^{s} (C_i^*)^{\alpha_{ij}^+}$$

$$D_i^* = \frac{D_i}{D_{ref}}$$

$$k_{j,m}^* = \frac{k_{j,m} L^2}{D_{ref}} C_{ref}^{(n-1)}$$

$$E^* = \frac{EFL}{RT}$$

$$K_j^* = K_j (C_{ref})^{-\sum_{i=1}^{s} \alpha_{ij}}$$

$$I^* = \frac{IL}{FD_{ref} C_{ref}}$$

Notation: $D_{ref} = D_{primary\ permeant}$
$C_{ref} = C_{A,T}$ = total amount of carrier present
$k_{j,m}$ = rate constant for j^{th} reversible reaction; m = 1 for forward reaction, m = 2 for reverse reaction
α_{ij}^-, α_{ij}^+ = stoichiometric coefficient for species i in reaction j; - signifies reactant, + signifies product
(n-1) = (number of molecules participating in reaction j,m) - 1

suggested that OCFE would be suitable for solving a wide range of membrane problems, including the frequently encountered boundary layer problem. They presented a development of the method as well as the results of solving models of three facilitated transport membranes using OCFE. Gallagher (28) has applied this method to his model of a CO_2/bicarbonate system with reaction boundary layers.

INDUCED FIELD EFFECTS

A FORTRAN program based on OCFE was used to solve the model for a number of zero-current cases. In each of these cases, the carrier loading was held at 10^{-5} M. Assuming a molecular weight of 150,000, this corresponds to 1.5 g/l of antibody, which is near the rule-of-thumb protein solubility limit of 1 g/l. This limit is only an approximation. Protein solubility is strongly dependent on local conditions such as temperature, pH, and ionic strength (29).

The function used to describe the pH-dependence of the antibody's charge was

$$z_A = -5 \text{ pH} + 20 \tag{10}$$

The fit of the titration curve data specified positive charges on hgh for the pH range of interest, with z_{hgh} varying from 2.5 to approximately 8.5. Upstream and downstream hgh concentrations were 5.0×10^{-8} M and 2.0×10^{-8} M respectively and the concentration of added salt (MX) ranged from 4.0×10^{-5} M to 3.0×10^{-4} M.

Facilitation factors are commonly used as measures of the effectiveness of carrier-mediated transport. Here, the facilitation factor f is defined as

$$f = \frac{J_{s,calc}}{D_s/L \, (C_{s,0} - C_{s,L})} \tag{11}$$

where the numerator represents the total calculated flux of hormone and the denominator is the Fickian flux that would be expected between the two reservoirs in the absence of reactions and electrochemical effects.

The facilitation factor predicted by the model is plotted against $\log_{10}$ (salt concentration) in Figure 3. The plot shows that the hormone flux decreases with increasing salt concentration. An explanation for this is offered by Figure 4, which shows the logarithm of the internal hgh concentration at $x = 0$ plotted against the logarithm of the salt concentration. The horizontal line corresponds to the hgh concentration in the upstream reservoir. The model predicts that the hgh concentration will be greater in the membrane phase than in the upstream reservoir; that

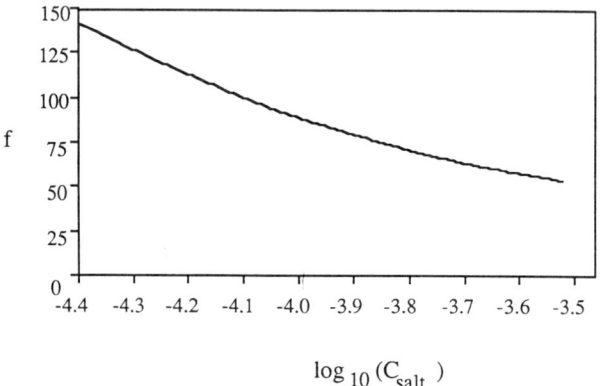

Figure 3. Facilitation factor as a function of salt concentration

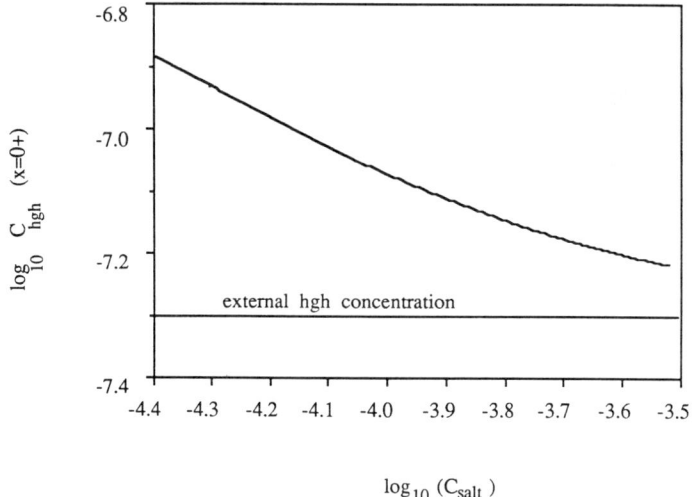

Figure 4. Hormone concentration at $x = 0+$ as a function of salt concentration

is, Donnan inclusion is predicted at the upstream boundary. Decreasing inclusion with increasing salt concentration is also predicted, in agreement with the literature pertaining to ion exchange membranes (10, 30). The internal and external hormone concentrations at several salt concentrations are shown in Table 5.

The fact that increased Donnan inclusion leads to increased flux is not surprising. If more hormone enters the membrane, more complexation can occur. In addition to the extra complexation, a higher hgh concentration near $x = 0$ implies a steeper concentration gradient and increased diffusion of free hgh. This is supported by Figure 5, in which the hgh concentration profiles corresponding to different salt concentrations are shown.

As expected, Figure 5 shows a much higher degree of upstream inclusion at low salt concentrations than at high salt concentrations. However, inclusion at the downstream boundary is slight and there is little difference between the inclusion at high and low salt concentrations. This is easily explained in light of the charge profiles of the carrier and permeant. The charges on all complex species are simply sums of charges of carrier, permeant, and (in some cases) hydrogen. Table 6 shows the upstream and downstream charges on all complex species for the case discussed here. It is seen that all species become more positively charged as they diffuse across the membrane. Some are negatively charged at $x = 0$ and positively charged at $x = L$. Even if the relative concentrations of the species are not known, it can be seen that there are a large number of negatively charged nonvolatile molecules at $x = 0$ and inclusion of the positively charged permeant is to be expected. On the other hand, nearly all species are highly positive at $x = L$ and a very low degree of inclusion -- or even exclusion -- of hgh is likely. When concentrations are taken into account, the reason for the difference in the degree of inclusion at the two boundaries is clear. To cite a specific example, when the salt concentration is 4.0×10^{-5} M, the sum of the charges due to nonvolatile species immediately inside the membrane is -5.057×10^{-5} M at the upstream boundary and -0.182×10^{-5} M at the downstream boundary; the net charge density due to nonvolatiles at $x = 0$ is nearly thirty times that at $x = L$.

The relative importance of the Donnan effect at high and low salt concentrations is further illustrated by the potential profiles shown in Figure 6. As can be predicted on the basis of Figure 5 and the definition of Donnan potential, low salt concentrations allow larger potential drops across the upstream interface. For all salt concentrations shown, the downstream Donnan potential is so small as to be negligible. Figure 6 also shows that the induced field, $E = -d\Phi/dx$, is positive for all salt concentrations. Thus, according to the Nernst-Planck equation, the presence of the field will assist the diffusion from 0 to L of such positively charged species as hgh. It will also accelerate the diffusion from L to 0 of the negatively charged carrier.

Table 5. Comparisons of Internal and External HGH Concentrations

Concentration of added salt	C_{hgh} (x=0-)	C_{hgh} (x=0+) predicted by model
4.0E-5 M	5.0E-8 M	13.08 E-8 M
5.0E-5 M	5.0E-8 M	11.65 E-8 M
1.0E-4 M	5.0E-8 M	8.501E-8 M
2.0E-4 M	5.0E-8 M	6.686E-8 M
3.0E-4 M	5.0E-8 M	6.081E-8 M

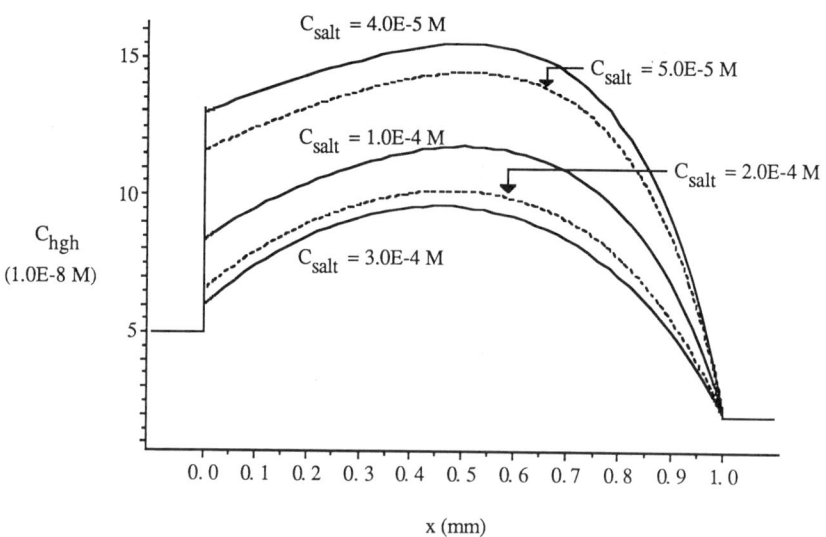

Figure 5. Concentration profile for hgh as a function of salt concentration, zero current case

Table 6. Upstream and Downstream Charges on HGH, Antibody, and Complexes

Species	z(x=0)	z(x=L)
S	2.5	8.5
A	-5	-1
AH	-4	0
AS	-2.5	7.5
AHS	-1.5	8.5
AS_2	0	16
AHS_2	1	17

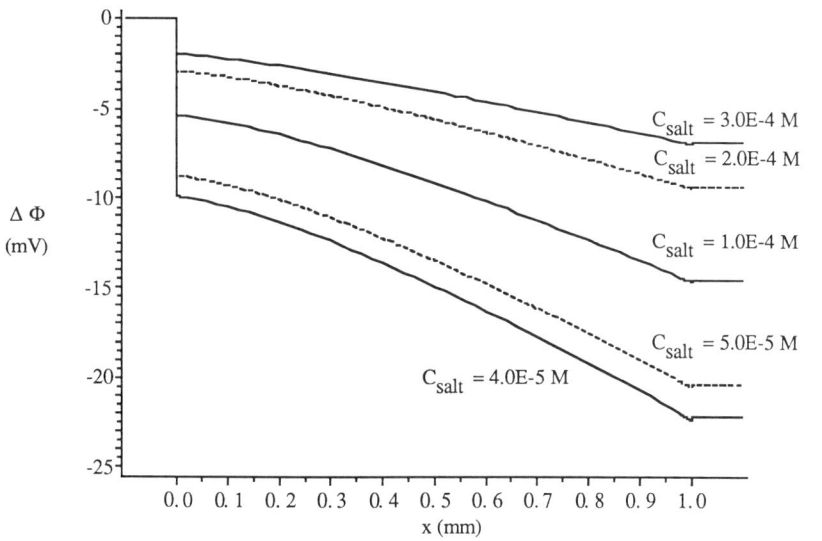

Figure 6. Potential profile as a function of salt concentration

EFFECT OF CURRENT ON PROTEIN FLUX

The model was again applied to some of the cases discussed in the previous section, this time with the dimensionless current density I^* varying from -750 to 500. As noted earlier, the present work is concerned with liquid-liquid-liquid systems, which do not require the electrodes to be in the immediate vicinity of the membrane. Thus, the boundary conditions need not be rewritten. If the current term is not set equal to zero, both applied and induced field effects are incorporated.

The total antibody concentration was again set at 10^{-5} M. The upstream hgh concentration was 5.0×10^{-8} M and the downstream concentration was 2.0×10^{-8} M. The pH was 5.0 upstream and 4.2 downstream, and the carrier charge again had the pH-dependence defined in Equation (10).

Facilitation factor is shown as a function of dimensionless current density and salt concentration in Figure 7. The model predicts increasing facilitation with increasing current density and indicates that a small degree of pumping (i.e. flux against the hgh concentration gradient) could occur at large negative current densities. High salt concentration has been shown to decrease permeant flux when an induced electric field is present. According to the data appearing in Figure 7, salt concentration has a similar effect when current is used.

While dimensionless variables are helpful for making comparisons and following trends, they are of little use in judging the feasibility of a process. Using current to increase flux is not likely to be popular if the necessary current is very high. In this study, the current density is given dimensions through multiplication by (F C_{ref} D_{ref}/L), which is equal to 7.236×10^{-9} amp/cm^2. After I^* has been transformed, it can be seen that the current density needed to achieve the fluxes shown in Figure 7 is on the order of microamps/cm^2. That is, a significant increase in flux can be achieved through use of a small current.

A current density measured in microamps/cm^2 would not significantly affect flux in most facilitated transport membranes. Gallagher and coworkers (31) modeled the effect of current on flux of CO_2 in a CO_2/bicarbonate system. The model predicted a near-linear increase in facilitation factor from roughly -200 to 320 as the dimensionless current density varied from 50 to -50. At first glance, these results appear to be comparable to -- if not better than -- those shown in Figure 7, where the facilitation factor in the hgh-antibody system was seen to increase from slightly less than zero to 400 as I^* was varied from -750 to 500. Comparing results based on dimensionless parameters can be misleading, however, if one does not verify that compatible scaling was used in defining the dimensionless variables. For example, the current scaling parameter used by Gallagher et al. was (F C_{ref} D_{ref}/L) = 1.052×10^{-2} amp/cm^2. This is six to seven orders of magnitude greater than the parameter used in this work.

The reference diffusivity used by Gallagher was $D_{HCO_3^-} = 1.09 \times 10^{-5}$ cm^2/sec and the reference concentration was 1.0 M (31, 28); in this work, $D_{ref} = 7.5 \times 10^{-7}$ cm^2/sec and $C_{ref} = 10^{-5}$ M. Rather than being on the order of microamps/cm^2, Gallagher's maximum current density was $\pm 5.26 \times 10^{-3}$ amp/cm^2.

There are two explanations for the response of the hgh-antibody system to extremely low current densities. First, it must be recalled that the hormone, antibody, and complex are multivalent. The extreme cases are the charge of -5 on the antibody near the upstream boundary and +17 on the protonated, two-hormone complex near the downstream boundary. In the CO_2/bicarbonate system, all species except CO_3^{2-} are monovalent. According to the Nernst-Planck equation, the influence exerted on a molecule by an electric field is directly proportional to the charge on that molecule. A very small applied field generated by a very small current can have a significant effect on the flux of a multi-charged species. A much larger current must be used to achieve similar results when the permeant, carrier, and/or complex are monovalent.

Another important difference between the hgh-antibody and CO_2/bicarbonate systems is the number of ions present. The most concentrated species in the former are the supporting electrolytes which have concentrations of 3×10^{-4} M or less. All other concentrations are at least one order of magnitude lower. In comparison, Gallagher specified an initial $NaHCO_3$ concentration of 1.0 M (28, 31). According to Newman (32), high electrolyte concentration increases conductivity and decreases resistance, and Ohm's law states that decreasing resistance leads to a decreasing field for a given current. If an identical current (e.g. 5 microamps/cm^2) were applied to Gallagher's CO_2/bicarbonate system and to the hgh-antibody system described here, the hgh-antibody system would be acted upon by a greater electric field and would respond accordingly.

The relative importance of the induced and applied electric fields can be judged by separating the field term into its induced and applied components. Values of these components at several different current densities are shown in Table 7. It is seen that at the highest negative current density, the magnitude of the applied field is roughly double that of the induced field. The same is true at the maximum positive current density. From the values given at $I^* = -350$ and $I^* = -300$, it can be concluded that the applied and induced fields will cancel each other at $I^* \sim -315$. When the field contributions cancel, the increase in flux is due solely to the chemical aspect of facilitated transport; there is no driving force other than the concentration gradients of the carrier, permeant, and complex. Examination of Figure 7 shows that at $I^* \sim -315$, the facilitation factor is approximately 50. In this case, simple facilitation is significant and is enhanced by the electric field.

Of course, the applied field affects the transport of all ionic species. Since the concentration gradient of hydrogen ions runs counter to that of hgh, an increase in the flux of positively charged hormone from $x = 0$ to

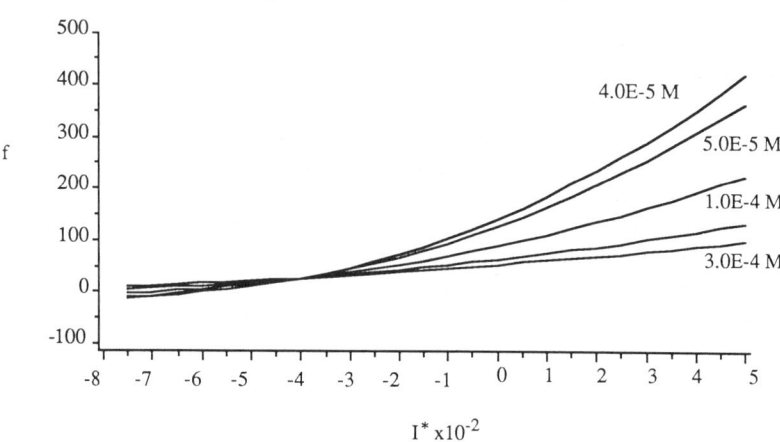

Figure 7. Facilitation factor as a function of current density and salt concentration

Table 7. Comparison of Applied and Induced Fields

I^*	I (amp/cm^2)	E_{total} (V/cm)	$E_{induced}$ (V/cm)	$E_{applied}$ (V/cm)
-750	-5.427E-6	-1.7704E-2	1.7150E-2	-3.4854E-2
-350	-2.533E-6	-0.1560E-2	1.4406E-2	-1.5966E-2
-300	-2.171E-6	0.0562E-2	1.4229E-2	-1.3666E-2
0	0	1.3542E-2	1.3542E-2	0
500	3.618E-6	3.5597E-2	1.3046E-2	2.2551E-2

$x = L$ implies a decrease in the flux of H^+ from $x = L$ to $x = 0$. This is illustrated by Figure 8, which shows the dimensionless flux of H^+ as a function of the current density and salt concentration. The plot indicates that hydrogen flux can be eliminated with sufficiently high current and that a large enough negative current will in fact accelerate the hydrogen flux from $x = L$ to $x = 0$ beyond the Fickian value. 'High' is a relative term; a current on the order of microamps/cm^2 is sufficient to eliminate the hydrogen flux.

Figure 8 also shows the effect of salt concentration on hydrogen flux. At a given current density, $J_{H^+}^*$ becomes increasingly negative with increasing salt concentration. Thus, a higher current must be used to achieve zero hydrogen flux in a high-salt system. The behavior of the hydrogen flux is not as easily explained as the relationship between the hormone flux and the salt concentrations. Recalling the definition of the Donnan ratio,

$$\left[\frac{C_i^{(m)}}{C_i^{(f)}} \right]^{1/z_i} = \rho \qquad i = 1,2,...,P \qquad (12)$$

and considering the fact that $z_{H^+} = 1$ while z_{hgh} is many times that, it can be seen that the concentration ratio for hgh will be much greater than the ratio for H^+. In other words, the multivalent hormone can be included to a much greater extent than the monovalent hydrogen ion and the reduction of the Donnan effect is not likely to alter hydrogen flux nearly as much as it does hgh flux. The increase in negative hydrogen flux is largely due to the previously noted fact that a high concentration of electrolytes reduces the electric field (32). As previously discussed, the positive field reduces flux of hydrogen from $x = L$ to $x = 0$ (i.e. makes the flux more positive); as the field is damped, a higher current must be used to achieve the same results.

CONCLUSIONS

The mathematical model described here has illustrated that electrochemical effects can significantly influence protein flux in an affinity-mediated transport system. The system considered consists of a supported liquid membrane containing a pH-sensitive monoclonal antibody as carrier and human growth hormone as permeant. On a microscopic scale, Donnan inclusion of the hormone can increase the flux of hormone into the membrane. This allows more complex to be formed and simultaneously generates a steep hormone concentration gradient which drives a greater flux of free hormone than would occur in the absence of inclusion.

Macroscopic electrochemical effects are also important. A positive induced electric field simultaneously enhances the flux of hormone from x = 0 to x = L and retards the diffusion of hydrogen ions in the opposite direction. A small (microamps/cm²), positive current provides an applied field which magnifies these effects. That is, hormone flux can be increased and hydrogen flux can be reduced to zero.

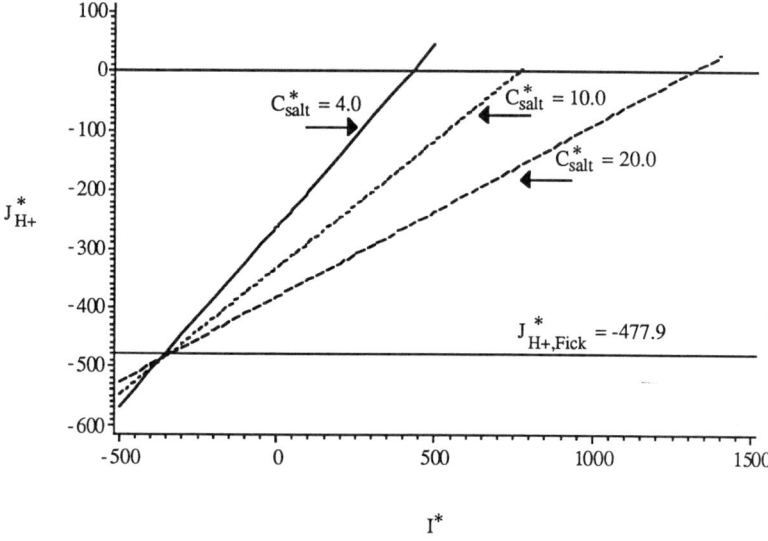

Figure 8. Hydrogen flux as a function of current density and salt concentration

SYMBOLS

C_i	concentration of ith species
D_i	diffusion coefficient of ith species
E	electric field strength
F	Faraday's constant
f	facilitation factor
I	current density
J_i	flux of ith species
K_j	equilibrium constant of jth equilibrium reaction
$k_{j,m}$	rate constant of jth reaction; m = 1 for forward reaction, m = 2 for reverse reaction
L	membrane thickness
N_I	number of integral constraints
P	number of permeating species
R_i	net rate of production of ith species
R	gas constant

R	number of reactions
r_j	rate of jth reaction
S	number of species in membrane
T	absolute temperature
x	distance coordinate
z_i	electric charge on i^{th} species
α_{ij}	stoichiometric coefficient for i^{th} species in j^{th} reaction
Φ	electric potential
ρ	Donnan ratio

LITERATURE CITED

1. Schultz, J. S.; Goddard, J. D.; Suchdeo, S. R.; "Facilitated Transport via Carrier-Mediated Diffusion in Membranes. Part I." *AIChE J.*, **1974**, *20(3)*, p. 417.

2. Jain, R.; Schultz, J. S.; "An Analysis of Carrier-Mediated Photodiffusion Membranes" *J. Membrane Sci.*, **1983**, *15*, p.63.

3. Hill, C. L.; Bartholomew, R. ; Beidler, D.; David, G. S.; "'Switch' Immunoaffinity Chromatography with Monoclonal Antibodies" *Biotechniques*, **1983**; 1(1), p.14 .

4. Bailey, J. E.; Ollis, D. F.; *Biochemical Engineering Fundamentals*, McGraw-Hill Book Company, New York, NY, 1977.

5. *Handbook of Biochemistry: Selected Data for Microbiology*; Sober, H. A. Ed.; Chemical Rubber Company, Cleveland, OH, 1970.

6. Bewley, T. A.; Li, C. H.; "The Chemistry of Human Pituitary Growth Hormone"; In *Advances in Enzymology and Related Areas of Molecular Biology*; Ed. Meister,A.; John Wiley and Sons: New York, NY, 1975, Vol. 42.

7. Steward, M. W.; "Affinity of the Antibody-Antigen Reaction and its Biological Significance" In *Structure and Function of Antibodies*, Glynn, L. E.; Steward, M. W., Eds.; John Wiley and Sons Ltd., Chichester, 1981.

8. Atkins, P. W.; *Physical Chemistry*, W. H. Freeman and Company: San Francisco, CA, 1978.

9. Lakshminarayanaiah, N.; *Transport Phenomena in Membranes*, Academic Press, New York, NY, 1969.

10. Helfferich, F.; *Ion Exchange*, McGraw-Hill Book Company, New York, NY, 1962.

11. Mackie, J. S.; Meares, P.; "The Sorption of Electrolytes by a Cation-Exchange Resin Membrane" *Proc. Roy. Soc. Lon. Ser. A*, **1955**, *232*, p. 485.

12. Boyd, G. E.; Bunzl, K.; "The Donnan Equilibrium in Cross-Linked Polystyrene Cation and Anion Exchangers" *J. Am. Chem. Soc.*, **1967**, *89*, p. 1776.

13. Lakshminarayanaiah, N.; *Equations of Membrane Biophysics*, Academic Press, New York, NY, 1984.

14. Blaedel, G. E.; Haupert,T. J.; "The Donnan Equilibrium through Ion Exchange Membranes: Analytical Applications" *Anal. Chem.*, **1966**, *38(10)*, p.1305.

15. Cox, J. A.; Gajek, R.; Litwinski, G. R.; Carnahan, J.; Trochimczuk, W.; "Optimization of Ion Exchange Membrane Structures for Donnan Dialysis" *Anal. Chem.*, **1982**, *54*, p. 1153.

16. Dasgupta, P. K.; Bligh, R. Q.; Lee, J. ; and D'Agostino, V.; "Ion Penetration through Tubular Ion Exchange Membranes" *Anal. Chem.*, **1985**, *57*, p. 253.

17. LeBlanc, Jr., O. H.; Ward, W. J.; Matson, S. L.; Kimura, S. G.; "Facilitated Transport in Ion Exchange Membranes" *J. Membrane Sci.*, **1980**, *6*, p. 339.

18. Way, J. D.; Noble, R. D.; Reed, D. L.; Ginley, G. M.; Jarr, L. A.; "Facilitated Transport of CO_2 in Ion Exchange Membranes" *AIChE J.*, **1987**, *33(3)*, p. 480.

19. Smith, D. R.; Lander, R. J.; Quinn, J. A.; "Carrier-Mediated Transport in Synthetic Membranes" In *Recent Developments in Separation Science, vol.III*, Li, N. N., Ed.; CRC Press Inc., West Palm Beach, CA, 1977.

20. Kimura, S. G.; Matson, S. L.; Ward III, W. J.; "Industrial Applications of Facilitated Transport," In *Recent Developments in Separation Science;* Li, N. N., Ed.; CRC Press Inc., West Palm Beach, CA, 1979, vol. V.

21. Athayde, A. L.; *"The Effects of Periodic Electric Fields on Carrier-Facilitated Membrane Transport"*; Ph.D. Thesis, University of Notre Dame, IN, 1985.

22. Schultz, J. S.; "Carrier-Mediated Transport in Liquid-Liquid Membrane Systems" In *Recent Developments in Separation Science, vol.III*, Li, N. N., Ed.; CRC Press Inc., West Palm Beach, CA, 1977.

23. Lamb, J. D.; Christensen, J. J.; Izatt, S. R.; Bedke, K.; Astin, M. S.; Izatt, R. M.; "Effects of Salt Concentration and Anion on the Rate of Carrier-Facilitated Transport of Metal Cations through Bulk Liquid Membranes Containing Crown Ethers" *J. Am. Chem. Soc.*, **1980**; *102 (10)*, p. 3399.

24. Lamb, J. D.; Izatt, R. M.; Garrick, D. G.; Bradshaw, J. S.; Christensen, J. J.; "The Influence of Macrocyclic Ligand Structure on Carrier-Facilitated Cation Transport Rates and Selectivities through Liquid Membranes" *J. Membrane Sci.*, **1981**, *9*, p.83.

25. Gallagher, P. M.; Athayde, A. L. ; Ivory, C. F.; "The Combined Flux Technique for Diffusion-Related Problems in Partial Equilibrium: Application to the Facilitated Transport of Carbon Dioxide in Aqueous Bicarbonate Solutions" *Chem. Eng. Sci.*, **1986b**, *41(3)*, p. 567.

26. Carey, G. F.; Finlayson, B. A.; "Orthogonal Collocation on Finite Elements," *Chem. Eng. Sci.*, **1975**, *30*, p. 587.

27. Jain, R.; Schultz, J. S.; "A Numerical Technique for Solving Carrier-Mediated Transport Problems" *J. Membrane Sci.*, **1982**, *11*, p.79.

28. Gallagher, P. M.; "*A Numerical Study of Steady State Electric Field Effects in Carrier-Mediated Transport*"; Ph.D. Thesis, University of Notre Dame, IN, 1986.

29. Lehninger, A. L.; *Biochemistry*, Worth Publishers, Inc., New York, NY, 1975.

30. Kesting, R. E.; *Synthetic Polymeric Membranes*, McGraw-Hill Book Company, New York, NY, 1971.

31. Gallagher, P. M.; Athayde, A. L. ; Ivory, C. F.; "Electrochemical Coupling in Carrier-Mediated Membrane Transport" *J. Membrane Sci.*, **1986**a, *29* p.49.

32. Newman, J. S.; *Electrochemical Systems*, Prentice-Hall Inc., Englewood Cliffs, NJ, 1973.

RECEIVED October 4, 1989

Chapter 11

Affinity Precipitation of Avidin by Using Ligand-Modified Surfactants

Roberto Z. Guzman[1], Peter K. Kilpatrick, and Ruben G. Carbonell

Department of Chemical Engineering, North Carolina State University, Raleigh, NC 27695−7905

The use of ligand-modified double-tailed phospholipid surfactants for selectively precipitating the tetrameric protein avidin from model and crude solutions is described. Dimyristoylphosphatidylethanolamine (DMPE) was derivatized by covalently attaching biotin, the specific ligand for the egg white protein avidin. The biotinylated surfactant (DMPE-B) was solubilized in aqueous buffer solution by the ethoxylated alcohol octaethyleneglycol mono-n-dodecylether (C12E8) at concentrations above the critical micelle concentration of the nonionic surfactant. The mixed surfactant solution of DMPE-biotin and C12E8 was then combined with protein solutions containing avidin which resulted in dilution of the nonionic surfactant below its critical micelle concentration. Upon binding of avidin to DMPE-B, the hydrocarbon tail groups of the phospholipid apparently aggregated with other DMPE-B molecules complexed to other avidin molecules. Because each avidin can bind four DMPE-B molecules, the result is a three-dimensional network of modified phospholipid-avidin complexes. This large aggregate grew until it precipitated from solution, as evidenced by gross turbidity which was monitored spectrophotometrically. The avidin-surfactant aggregates were then separated from solution by centrifugation and decantation of the supernatant. The avidin complex was resolubilized in a high concentration of (10^{-3} M) C12E8 in buffer solution, denatured by guanidinium chloride addition to debind the DMPE-biotin, and renatured after removal of the phospholipid by dilution and ultrafiltration. The technique was demonstrated with avidin solutions using lysozyme, bovine serum albumin, and myoglobin as model impurities. Greater than 85% recovery of avidin was achieved with no measureable coprecipitation of the model impurities. Avidin was also recovered from partially purified egg white solutions, which had been pretreated by ion exchange chromatography to remove hydrophobic and negatively charged protein impurities. Hence, greater than 90% of the avidin in the partially purified egg white fraction was recovered in pure active form, as evidenced by spectrophotometric assay and sodium dodecyl sulfate polyacrylamide gel electrophoresis (SDS-PAGE).

[1]Current address: Department of Chemical Engineering, University of Arizona, Tucson, AZ 85721

Precipitation is a commonly used purification step in protocols for isolating and recovering proteins from crude biological mixtures (1). The precipitation of proteins is commonly effected by adding a component which decreases the solubility of the desired biomacromolecule. The biomolecules in the resulting super-saturated protein solution rapidly aggregate to form seed particles for growth of larger precipitates. The differential solubility agents added to produce precipitation include, but are not limited to, electrolytes, organic solvents, and pH modifiers. These additives act to attenuate repulsive electrostatic interactions between the protein molecules. The resulting increased importance of hydrophobic attractions between the biomolecules leads to agglomeration. Pure protein precipitates are rarely obtained, partly because simple addition of differential solubility agents leads to regions of high local super-saturation. The resulting rapid aggregation of protein molecules tends to be non-specific. The result is simultaneous agglomeration and entrainment of different biomacromolecules and an impure precipitate.

Fisher et al. (2) attempted to control the local concentration of precipitating agent in protein solutions by using a dialytic membrane to control the rate of addition. The aim was to aggregate pure, dense crystals of the desired protein and ultimately produce a purer product. However, in a very complex mixture, containing several proteins with similar physical properties, it may be difficult to find a set of conditions for precipitation that will make the process very selective.

An alternative approach that can be used to impart greater selectivity to protein precipitation is to attach a ligand, which possesses specific affinity for the desired biomolecule, to a polymer which has functional groups which make it easy to precipitate. Schneider et al. (3) have successfully exploited this so-called affinity precipitation technique in the purification of the proteolytic enzyme trypsin from bovine pancreas. In their scheme, a competitive reversible inhibitor of trypsin, m-aminobenzamidine, was reacted with acryloyl chloride to form N-acryloyl-m-aminobenzamidine, one of the monomeric components of the polymer used in the precipitation. One of the other monomeric units was N-acryloyl-p-aminobenzoic acid, which has a pK in aqueous solution of about 5.5. These two monomers were polymerized with acrylamide to synthesize a substituted polyacrylamide. The m-aminobenzamidine group served as a specific ligand for trypsin while the benzoic acid substituent acted as a precipitation aid. At pH>6.0, the polymer was fully ionized and soluble while at pH<5.0 the polymer precipitated. After separating the precipitate from the supernatant, the precipitate was redissolved in water and the pH adjusted to a value of 2. This step desorbed the trypsin from the polymer, which then could be separated from the unbound trypsin by centrifugation. Yields obtained were in the range of 75% and the product consisted of 90% pure trypsin.

Schneider et al. (3) provided a concise list of the possible advantages of affinity precipitation in the large scale separation of biological molecules.

In addition to the fact that this process may be easily scaled-up, the high yields and specificity obtained in the preliminary experiments indicate that it would be a useful alternative to normal precipitation in the first stage of a purification process. Furthermore, since there are no particles carrying the ligand, diffusional limitations are negligible. Unfortunately, even though the polymer developed by Schneider et al. (3) was re-used several times, the synthesis is complex and therefore expensive. Furthermore, any technique that relies on pH changes for precipitation is also likely to induce the aggregation of other molecules in the suspension. This can complicate the precipitation of the polymer in other applications. Finally, the highly charged polymer may adsorb proteins of opposite charge non-specifically, and thus reduce the purity of the final product.

Larsson and Mosbach (4) and Flygare et al. (5) employed a different affinity precipitation technique which relies on the crosslinking of proteins that have multiple binding sites. In their scheme the bifunctional nucleotide N_2, N_2' - adipodihydrazido-bis-(N^6-carbonylmethyl-NAD), or Bis-NAD was synthesized. This molecule consists basically of a spacer arm separating two NAD moieties. Dehydrogenases are proteins that have a specific affinity for NAD and are also oligomeric, containing several NAD binding sites. In the presence of Bis-NAD, one protein molecule can bind to one of the NAD groups on the Bis-NAD, while a second protein can bind to the other group. If this proceeds, an aggregate of protein molecules can form, and if it is large enough, it will precipitate. This technique worked best with lactate dehydrogenase, a four-subunit protein. It was purified from ox heart extract using this method with about 90% recovery and a 50-fold purification. However, liver alcohol dehydrogenase, a two-subunit protein, did not precipitate because the aggregation process stopped when dimers were formed. Yeast alcohol dehydrogenase did not precipitate on its own until 0.2 M NaCl was used as a precipitating agent. Even though glutamate dehydrogenase did precipitate, it was difficult to redissolve the aggregate unless a high NADH concentration was used to remove the enzyme from the Bis-NAD complex.

Flygare et al. (5) mentioned that there were several additional difficulties to be surmounted in using this technique on a large scale. First, the kinetics of precipitation seem to be very slow, taking as long as 20 hours for the case of yeast alcohol dehydrogenase. Second, the recovery is very dependent on the ratio of ligand (Bis-NAD) to desired protein in solution. At ligand concentrations below the optimal ratio, the recovery drops because it is difficult to form large numbers of aggregates. At ligand concentrations above the optimal ratio the recovery also drops. This is likely because there is a much larger probability of completely binding each protein molecule with ligands before forming protein-protein complexes. Thus, finding the right aggregation conditions is important. Finally, in the design of ligands for the purification of other oligomeric proteins, the length of the spacer arm must be chosen carefully so as to prevent the two affinity moieties from binding to two sites on the same protein.

In this paper, we describe a new approach to affinity precipitation which offers several advantages over the existing methods. The method is based on the use of affinity surfactants, which are surfactants containing derivatizable polar head groups to which protein affinity ligands have been covalently attached (6, 7). The affinity ligand used in the affinity precipitation study reported here was biotin, which binds specifically to avidin with a very strong association constant of approximately 10^{15} M^{-1}. The affinity surfactant synthesized was dimyristoylphosphatidylethanolamine-biotin (DMPE-B), a phospholipid to which the ligand biotin has been attached. While DMPE-B is virtually insoluble in water, it can be solubilized in small aggregates with the water soluble surfactant C12E8 (octaethyleneglycol mono-n-dododecylether). The structural formulas of the two surfactants, DMPE and C12E8, are provided in Figure 1. Since the phospholipid is insoluble in water, its hydrocarbon tails are either incorporated in micelles formed with the non-ionic surfactant, or simply complexed to a small number of the solubilizing surfactants. Upon addition of this solubilized DMPE-B mixture to a solution containing avidin, precipitation occurs. No precipitation of any other protein has been observed. If the underivatized phospholipid is solubilized and added to the protein solutions, the avidin does not precipitate. As a result, the precipitation seems to be driven by the specific binding of avidin to the biotin in the phospholipid. The precipitation is probably caused by the fact that a single avidin molecule can bind four different phospholipids that either remain incorporated in their micelles or aggregates, or that get dislodged from the aggregates, and precipitate due to hydrophobic interactions between their hydrocarbon tails. In either case, the precipitation is specific enough that it has been used successfully to separate avidin from both model mixtures as well as from partially purified egg white solutions. A schematic illustration of the presumed network-like aggregates of avidin and phospholipid which form upon precipitation is provided in Figure 2.

The use of these affinity surfactants solubilized in solution offer several advantages over other affinity precipitation methods. The synthesis of the ligand-modified phospholipids is much simpler and cheaper than the synthesis of either a polymer as described by Schneider et al. (3) or the bifunctional ligand of Flygare et al. (5). The non-ionic surfactant used to solubilize the phospholipid does not carry any charges that can lead to non-specific interactions with undesired proteins. The degree of phospholipid aggregation, and to some extent the kinetics and yields of precipitation, may be controllable by varying the sizes of the hydrocarbon tail and the polar head group in the solubilizing surfactant. Only one affinity-modified phospholipid can bind to one of the binding sites on the desired protein and this therefore eliminates one of the possible difficulties with the bi-functional ligands. Finally, the aggregates may be micellar in size, or at least the size of several aggregated non-ionic surfactant molecules. This greatly reduces the likelihood of forming dimers which would stop the

$$CH_3-(CH_2)_{11}-(OCH_2CH_2)_8-OH$$

Octaethylene glycol mono-n-dodecyl ether, or $C_{12}E_8$

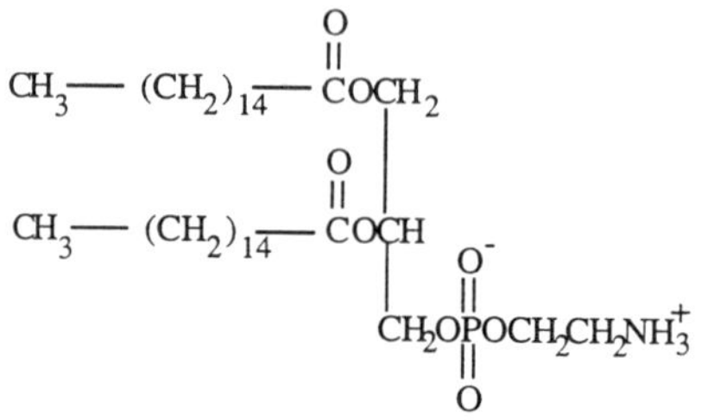

$$CH_3-(CH_2)_{14}-\overset{O}{\overset{\|}{C}}OCH_2$$
$$CH_3-(CH_2)_{14}-\overset{O}{\overset{\|}{C}}OCH$$
$$CH_2O\overset{O^-}{\overset{\|}{P}}OCH_2CH_2NH_3^+$$
$$\underset{O}{\|}$$

Dimyristoylphosphatidyl ethanolamine, or DMPE

Figure 1. Structural formulas of the ethoxylated alcohol and phospholipid surfactants used in the affinity precipitation of avidin. Octaethyleneglycol mono-n-dodecylether (C12E8) was used as a solubilizing surfactant and dimyristoylphosphatidylethanolamine (DMPE) was used as the insoluble surfactant to which biotin was covalently attached. (The structural formula of the derivatized phospholipid and a discussion of the reaction and purification scheme have been previously described by Powers et al. (7)).

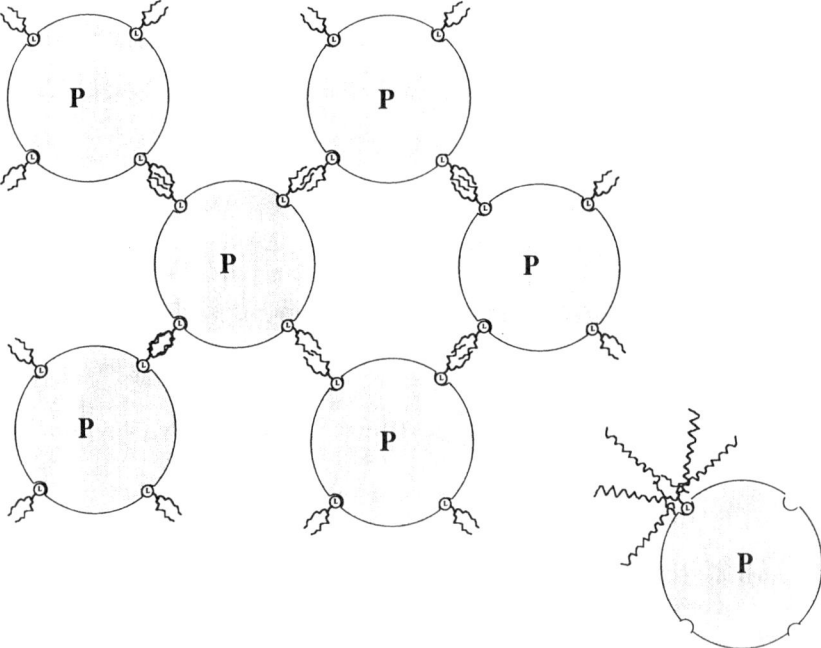

Figure 2. Schematic illustration of three-dimensional network formed by binding of affinity surfactant to multi-binding site biomolecule and subsequent hydrophobic aggregation of surfactant tail groups.

precipitation, as was observed by Flygare et al. (5) with the bifunctional ligands.

In what follows, we describe the synthesis and purification of the affinity surfactant, the study of its binding in aqueous solution to avidin, and the selective precipitation of avidin from model protein mixtures containing myoglobin, lysozyme, and bovine serum albumin. We also describe the selective purification of avidin from a crude natural source of avidin, hen egg whites. In this experiment, hydrophobic and aggregating impurities in egg whites were first removed by an ion exchange pretreatment of egg whites on carboxymethylcellulose. This partially purified egg white fraction, containing about 8% avidin on a total protein basis, was then purified to greater than 90% avidin by affinity precipitation with DMPE-B solubilized in small aggregates of C12E8. The characterization of this purified product by high performance liquid chromatography, SDS polyacrylamide gel electrophoresis, and biotin titration is described.

EXPERIMENTAL SECTION

MATERIALS

Dimyristoylphosphatidylethanolamine (DMPE), biotinyl-N-hydroxysuccinimide ester (BNHS), triethylamine, myoglobin, bovine serum albumin (BSA), lysozyme, guanidinium chloride, dimethylamino-cinnamaldehyde, acrylamide, ammonium persulfate, sodium dodecyl sulfate (SDS), and molybdenum blue reagent were all obtained from Sigma Chemicals and used without further purification as received. The compounds N, N, N', N'-tetramethylethylenediamine (TEMED), N, N'-methylene-bis-acrylamide, and Coomassie Brilliant Blue R-250 were from BioRad Laboratories. Avidin was obtained from Vector Laboratories with a quoted activity of 14 units/mg. All solvents were from Fisher Scientific. Water was passed through a Barnstead Nanopure system.

METHODS

Synthesis of Biotinylated Phospholipid. Biotin was covalently attached to dimyristoylphosphatidylethanolamine (DMPE) according to the procedure of Bayer et al. (8, 9). Our adaptation of their procedure has been described previously (7).

Determination of Avidin Activity. The ultraviolet spectrum of avidin in aqueous buffer solution is observed to shift to the red (hyperchroic effect) when biotin binds to one of the four avidin active sites (10). The difference between the normal spectrum of avidin in buffer and that of avidin bound to biotin is a characteristic of binding in which the maximum difference in the spectrum occurs at 233 nm (Figure 3). The difference

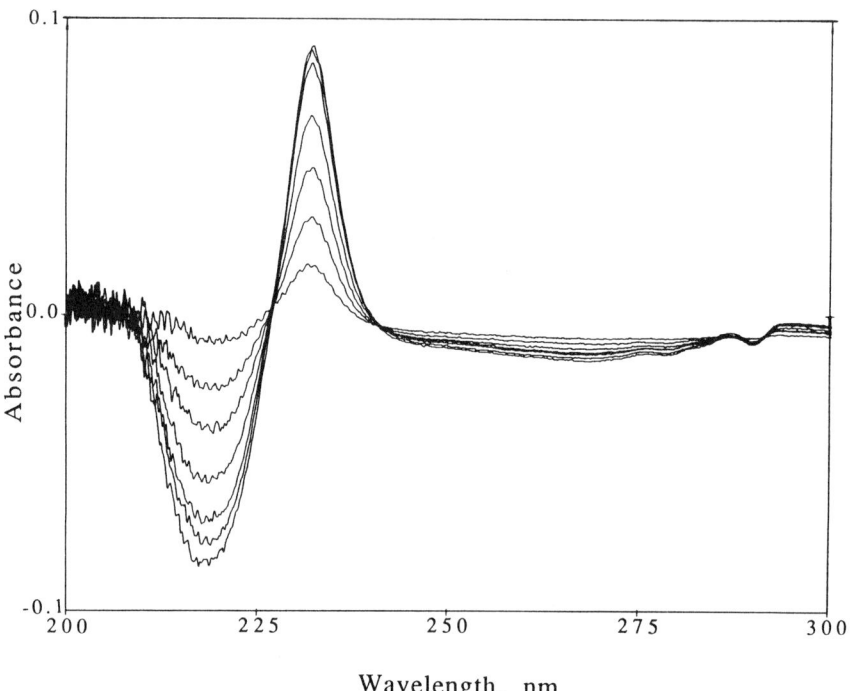

Figure 3. Difference in ultraviolet spectra of a standard avidin solution (10^{-6} M in 0.2 M ammonium carbonate buffer at pH 8.9) and a solution of avidin in which varying amounts of the specific ligand biotin have bound to avidin. The characteristic shift of the avidin spectrum to higher wavelengths upon binding to biotin is called the hyperchroic effect (Green (10)) and the maximum shift occurs at 233 nm as illustrated in this difference spectrum.

spectra in Figure 3 were obtained by placing standard avidin in buffer solutions in both the sample and reference cells of the spectrophotometer and zeroing at 233 nm. Aliquots of biotin in buffer solution were then added to the sample cuvette only and the change in the absorbance spectrum recorded upon addition of each aliquot. A plot of absorbance of an avidin solution at 233 nm as a function of amount of biotin added yields a titration plateau at the stoichiometric endpoint (four biotins per avidin molecule for completely active avidin). Absorbance measurements were made on a Shimadzu model UV-265 double-beam recording spectrophotometer equipped with a TCC-260 thermoelectric cell for temperature control. All measurements were made at 25°C, unless stated otherwise. Three ml of a standard avidin solution (typically 0.068 mg/ml, or 10^{-6} M) in 0.2 M ammonium carbonate buffer at pH 8.9 (standard buffer) were placed in the sample cuvette and this was zeroed against 3 ml of standard buffer solution at 233 nm. Aliquots of concentrated biotin in standard buffer solution were then added to both cuvettes and the increase in absorbance was determined for each aliquot until further addition of biotin resulted in no change in absorbance. The resulting titration plateau (Figure 4) was used to calculate the number of biotin molecules binding to each avidin molecule. The titration endpoint ratio of 3.89:1 is obtained, indicating our standard avidin is about 97% active.

Dissolution of Biotinylated Phospholipid in Nonionic Surfactant Solution. Solutions of the nonionic surfactant octaethyleneglycol mono-n-dodecylether (C12E8) at a concentration of 10^{-3} M in standard buffer were used to solubilize DMPE-B to obtain solutions of 5×10^{-5} M and 1.25×10^{-4} M of the modified phospholipid. The phospholipid was suspended in 100 ml of the C12E8 solutions and the mixture was vigorously stirred with heating to a uniform temperature of 60-70°C for about 45 minutes whereupon all of the DMPE-B was solubilized. Solutions of underivatized phospholipid in C12E8 were prepared in a similar fashion.

Binding Studies of DMPE-B to Avidin. In order to determine the stoichiometry of binding the modified phospholipid to avidin, standard avidin solutions (10^{-6} M in standard buffer) were titrated with DMPE-B solutions solubilized in C12E8. Just as with the avidin-biotin hyperchroic effect measurements, the change in the absorbance spectrum of a standard avidin solution was determined after addition of several aliquots of DMPE-B-C12E8 solution. The reference and sample cells in the spectrophotometer were both initially filled with an avidin solution (10^{-6} M) in standard buffer and unmodified phospholipid solubilized in C12E8 (3.6×10^{-6} M DMPE and 7.2×10^{-5} M C12E8) and zeroed at 233 nm. After zeroing, the solution in the sample cell was replaced by a freshly prepared avidin solution (10^{-6} M) containing modified phospholipid and C12E8 (3.6×10^{-6} M DMPE-B and 7.2×10^{-5} M C12E8) and the change in absorbance was then monitored in time. It was observed that the absorbance change at 233 nm far exceeded that due

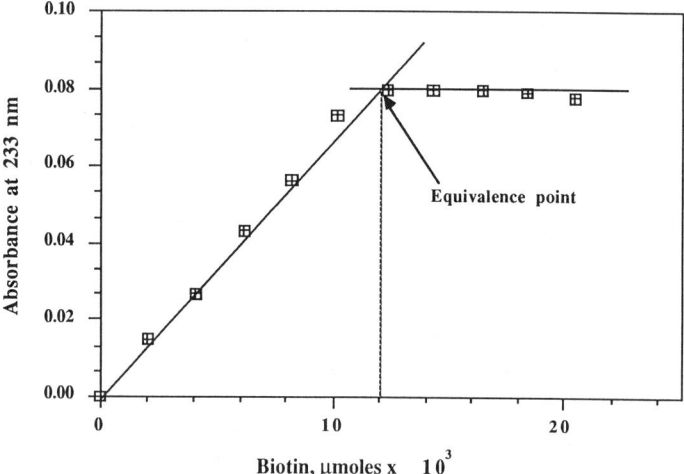

Figure 4. Absorbance difference at 233 nm between standard 10^{-6} M avidin solution in buffer (0.2 M ammonium carbonate at pH 8.9) and a 10^{-6} M avidin solution in which the protein is bound to varying amounts of biotin. The titration endpoint corresponds to a molar ratio of biotin to avidin of 3.89:1, in good agreement with the theoretical value for completely active avidin of 4:1.

solely to the hyperchroic effect and was in fact a combination of the hyperchroic shift and the turbidity caused by the formation of aggregates in solution. The major contribution to the absorbance was due to the turbidity. Thus, titrations of avidin with DMPE-B-C12E8 solutions gave turbidimetric rather than hyperchroic effect endpoints. A typical turbidimetric titration of avidin with a DMPE-B-C12E8 solution is shown in Figure 5.

Precipitation of Avidin from Model Protein Solutions. In order to test the efficacy of precipitating avidin from solution as a means of separating this protein from other protein impurities, a series of experiments were performed with model protein mixtures. Solutions of avidin (10^{-6} M) were prepared in standard buffer at pH 8.9, which is the pH of strongest avidin-biotin binding (10). Protein solutions were prepared consisting of avidin and myoglobin, avidin and BSA, and avidin, BSA, and lysozyme, in which all protein concentrations were 10^{-6} M in standard buffer. To these standard protein solutions, a sufficient aliquot of concentrated DMPE-B-C12E8 solution was added to yield a final concentration of 4 x 10^{-6} M DMPE-B and 7.5 x 10^{-5} M C12E8, i.e. a 4:1 stoichiometric ratio of DMPE-B to avidin. This was found to be the optimal ratio for precipitation of avidin, as will be discussed below. After aggregation of the avidin-DMPE-B complexes, which typically required from 1-2 hours, these standard solutions were centrifuged for 20 minutes in an International Centrifuge Centra 4 at 5000 rpm. The supernatants were analyzed by ultraviolet absorbance at 280 nm to determine residual protein concentration in solution. The solid pellet was typically redissolved in 10^{-3} M C12E8 solution and analyzed by UV absorbance. The avidin was dissociated from the DMPE-B by addition of 6 M guanidinium chloride at pH 1.5. The resulting avidin subunits were then reconstituted by a 15-fold dilution with standard ammonium carbonate buffer followed by ultrafiltration through a YM-5 membrane (5000 MWCO, Amicon). The avidin activity was then measured by the hyperchroic titration described above.

Purification of Avidin from Egg Whites. In order to test the applicability of the described precipitation procedure for purifying avidin from a naturally occurring crude biological mixture, a preliminary ion exchange purification of hen egg whites was performed. This was necessary as the crude egg white solution contained hydrophobic proteins which slowly precipitate from solution, even when the cloudy precipitated egg white solution was thoroughly clarified by filtration. Initially, 40 grams of carboxymethylcellulose (CMC) from Pharmacia (CM-Sephadex) was hydrated with deionized water at 80°C using a ratio of 10 grams of CMC per liter of water. The hydrated gel was washed several times by decantation to remove fines and then packed into a 100 cm x 5 cm glass column. The column was washed according to Melamed and Green (11) with a half liter of 0.5 NaCl, 0.5 N NaOH aqueous solution, followed by distilled water (4 liters), 1 mM EDTA solution (2 liters), and finally again with distilled water

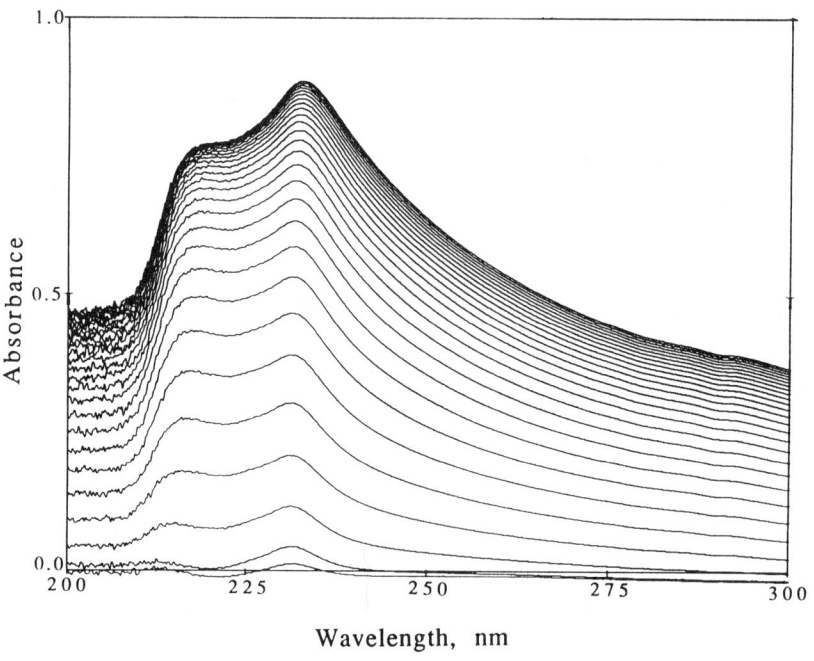

Figure 5. Difference in ultraviolet spectra of a standard avidin solution and a solution of avidin in which an aliquot of a DMPE-B-C12E8 solution (10^{-3} M C12E8, 1.25×10^{-4} M DMPE-B) has been added. Initially, a difference spectrum similar to that observed solely for the hyperchroic shift (Figure 3) is obtained. As time proceeds, the absorbance far exceeds that attributable to the hyperchroic shift, and the spectrum is typical of absorbance due to Tyndall scattering in which the magnitude of the absorbance decreases with the fourth power of the wavelength.

(4 liters). After washing, the gel was removed from the column and dried on a fritted glass funnel with gentle suction. The washing with water helps to remove small amounts of soluble CMC, which can complex avidin (10). The EDTA wash helps to remove metal ions which can denature avidin (12).

Egg whites from 15 eggs (ca. 0.5 liter) were gently homogenized with an equal volume of cold deionized water. A small amount of precipitate was removed by centrifugation at 4000 rpm for 30 minutes. The pretreated dried CMC (10 grams) was stirred into the homogenized egg white supernatant with a resulting suspension pH of 7.1. At this pH, only proteins in the egg whites with an isoelectric point (pI) exceeding 7.0 bind to the CMC. After the CMC had settled, the supernatant was siphoned off and a second fraction of the dried pretreated gel (10 grams) was added to the solution. Again the gel settled and the supernatant was removed, whereupon a third fraction of dried gel (10 grams) was added to the solution. The three gel fractions were combined and washed with five liters of 50 mM ammonium acetate buffer at pH 7.0. The gel was transferred to a fritted glass funnel and washed again with one liter of 50 mM ammonium acetate. These washings were intended to remove negatively charged proteins, proteins which tend to aggregate, and residual fines which could plug the column during elution of the proteins. The gel was washed a final time with 2 liters of 50 mM ammonium acetate, but now at pH 8.9, still below the pI of the desired avidin. The gel was then suspended in 3 liters of 50 mM ammonium acetate buffer at pH 9.0. The suspension was degassed and poured into the glass column, at the bottom of which the last fraction of the dry pretreated CMC was packed under hydrostatic pressure. The packed column was then washed with 1.5 liters of 50 mM ammonium acetate at pH 9.0 at a flowrate of 5.0 ml/min. The column was next washed with 0.8 liters of 0.3 wt% ammonium carbonate, pH 9.0, and the eluent collected until the absorbance at 280 nm was 0.04. The adsorbed proteins were then eluted with a gradient in ammonium carbonate concentration, pH 9.0, from 0.4% to 1.2 wt% in increments of 0.1%. With each discrete change in salt concentration, 500 ml of eluting buffer were applied, except for the case of 0.4% ammonium carbonate in which one liter of buffer was used. A uniform flowrate of 5.0 ml/min was maintained; at the end of the elution, the gel had compacted 25 cm. The elution profile in absorbance at 280 nm versus volume of effluent is shown in Figure 6. Assuming a void fraction of 0.4, the total elution time through the column was estimated to be 2.5 hours corresponding to about 750 ml of eluting buffer. Avidin, along with other globulin proteins, eluted primarily in the second peak (fraction B) in Figure 6, corresponding to an eluting buffer concentration from 0.5 to 0.8% ammonium carbonate. The eluting solutions were collected in 500 ml fractions and were concentrated by a factor of ten by ultrafiltration using a YM10 flat membrane (10,000 MWCO, Amicon).

The avidin content of the fractions eluted from the ion exchange column were determined by reverse phase HPLC analysis using an Alltech octyl-bonded silica column, a Perkin Elmer Series 410 quaternary solvent

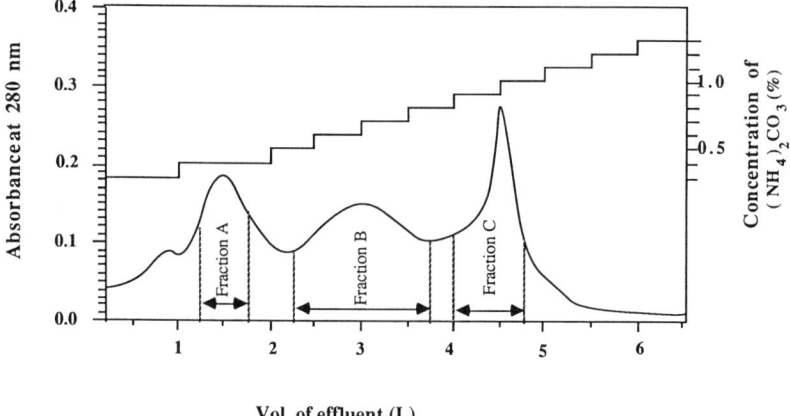

Figure 6. Optical density profile (280 nm) as a function of volume of eluting buffer during the ion exchange elution of solubilized egg whites from carboxymethylcellulose (details described in text). Also shown is the concentration of ammonium carbonate (in wt%) in the eluting buffer added at the top of the column in sequential 500 ml aliquots. Avidin eluted primarily in the second optical density peak (fraction B) corresponding to eluting buffer concentration of 0.5 to 0.8 wt% ammonium carbonate.

delivery system with environmental chamber, a Perkin Elmer model UV-95 ultraviolet-visible detector, and a Rheodyne injection valve. A standard gradient of 10 minutes for protein elution was applied from 90% water, 10% acetonitrile, 0.1% trifluoroacetic acid (TFA) to 100% acetonitrile, 0.1% TFA. Protein content was quantified by monitoring absorbance at 280 nm. After determination of the avidin content of a solution fraction, which for the fractions most concentrated in avidin was about 5-10% of the total protein, an aliquot of DMPE-B-C12E8 solution (typically 10^{-3} M C12E8 and 1.25×10^{-4} DMPE-B) was added to precipitate avidin such that the molar ratio of DMPE-B to avidin was slightly greater than the stoichiometric 4:1 ratio. Avidin was then unbound from the DMPE-B and reactivated in the manner described above.

The final purity of the reconstituted avidin was assessed by its activity according to a hyperchroic titration (as described above), its ultraviolet spectrum relative to pure avidin standards, and an SDS-PAGE analysis (13). Sodium dodecyl sulfate (SDS) polyacrylamide gel electrophoresis (PAGE) was carried out in gels of 15% acrylamide with 2.7% crosslinking in 8 x 10 cm slabs using a Bio-Rad Mini-Protean Gel II electrophoresis system. Electrophoresis was carried out at 150 V for 1.5 hours without cooling. Protein bands were visualized by staining for two hours with Coomassie Brilliant Blue following electrophoresis. The dye not bound to protein was removed from the gel using a destaining solution consisting of glacial acetic acid: methanol: water (1:1:8 volume ratio). After 12-18 hours and several changes of destaining agent, the gel background became colorless and protein bands remained colored blue. The resolution was increased with the use of a stacking gel (13).

RESULTS AND DISCUSSION

Titration of Avidin with Solubilized DMPE-B. In an effort to understand the mechanism and kinetics of binding affinity-modified phospholipid to avidin, hyperchroic and turbidimetric titrations of avidin solutions were performed by either adding small sequential aliquots of DMPE-B-C12E8 solution to avidin or by adding large single aliquots to the protein solution. In one experiment, 3.0 ml of an 8.82×10^{-7} M avidin solution in standard buffer (0.2 M ammonium carbonate at pH 8.9) was placed in the sample cell of the spectrophotometer and 3.0 ml of standard buffer in the reference cell, with the absorbance reading at 233 nm zeroed. Small aliquots (40 µl) of concentrated DMPE-B-C12E8 solution (5.0×10^{-5} M and 10^{-3} M, respectively) were then added to both reference and sample cell and the absorbance change monitored until a steady-state value was obtained. The results of this experiment are shown in Figure 7. Note that the maximum value of the absorbance reached, 0.682 at ca. 240 µl of solution added, far exceeds the absorbance value corresponding to the hyperchroic shift (shown in Figures 3 and 4) in a simple avidin-biotin titration, ca. 0.08-0.10 for this concentration of avidin. This maximum absorbance

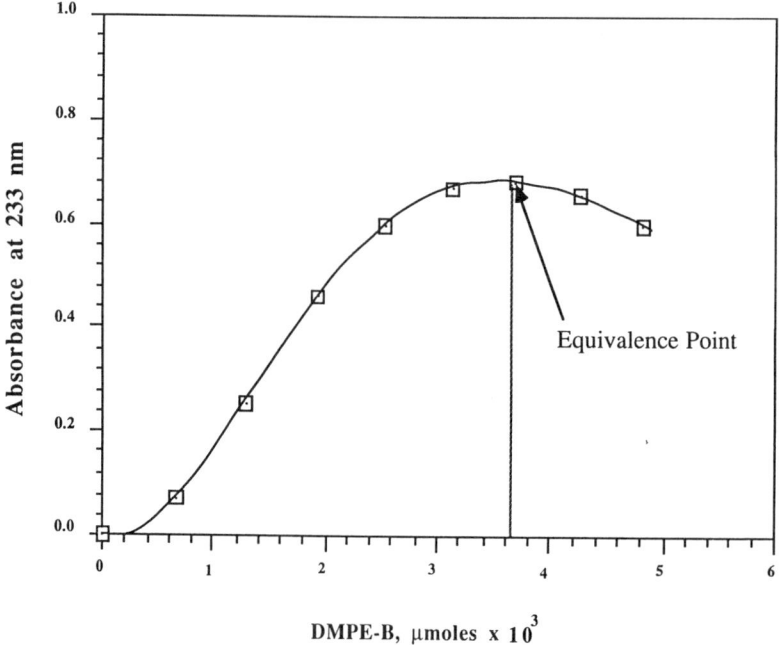

Figure 7. Absorbance difference at 233 nm between a standard avidin solution and varying amounts of a DMPE-B-C12E8 solution (10^{-3} M C12E8, 1.25×10^{-4} M DMPE-B). The turbidimetric endpoint (plateau value of absorbance) corresponds to a molar ratio of DMPE-B to avidin of about 4.5:1 which slightly exceeds the stoichiometric ratio corresponding to complete binding of biotin to avidin.

corresponds to a molar ratio of DMPE-B to avidin of 4.5 to 1, which slightly exceeds the stoichimetric ratio of 4:1 for complete binding of DMPE-B to avidin.

For these values of absorbance, the avidin solutions are visibly turbid indicating the presence of large super-molecular aggregates. Our interpretation of this observation is that upon binding of DMPE-B molecules to two different avidin molecules, hydrophobic interactions between the exposed surfactant tail groups lead to oligomer formation. Because avidin possesses four binding sites per molecule, each protein can interact simultaneously with tail groups on four other avidin molecules. The result is the formation of a three-dimensional network of avidin-DMPE-B complex which ultimately phase separates as a precipitate and can be sedimented or centrifuged. A schematic of the presumed aggregate types is shown in Figure 2.

In the titration experiment just described, the concentration of nonionic surfactant C12E8 at the absorbance maximum is about 8.0×10^{-5} M, just slightly below its critical micelle concentration of about 10^{-4} M (14). As additional aliquots of DMPE-B-C12E8 solution are added to the avidin solution, the absorbance values begin to decrease, suggesting sedimentation of the avidin-DMPE-B complexes. This is confirmed by the low value of the absorbance (ca. 0.08) observed for the solution when it was allowed to stand overnight with no agitation. The kinetics of aggregate formation were assessed by adding one large aliquot of concentrated DMPE-B-C12E8 solution to a standard avidin solution (8.82×10^{-7} M in buffer) such that the ratio of DMPE-B to avidin in the solution was 3.8:1. A plot of change in absorbance at 233 nm of the avidin-DMPE-B solution versus time is shown in Figure 8. Within a few seconds after addition, the change in absorbance of the avidin solution had reached a value of about 0.055, close to that of the hyperchroic titration endpoint. The absorbance remained at this value for more than 10 minutes whereupon apparent aggregation began to occur and the absorbance began to steadily increase over a 20-30 minute period. After several hours, the absorbance attained a steady asymptotic value of 0.666. This final absorbance is in good agreement with the value obtained by sequential addition of DMPE-B aliquots to this avidin solution.

Precipitation and Recovery of Avidin from Model Protein Solutions.
The exploitation of the observed large aggregate formation in avidin-DMPE-B mixtures for recovering avidin from protein solutions was first tested in model solutions of pure proteins. A standard avidin solution was mixed with concentrated DMPE-B-C12E8 solution so that the final concentration was 10^{-6} M avidin, 4×10^{-6} M DMPE-B, and 7.5×10^{-5} M C12E8. After two hours, the absorbance of the solution had reached a steady value of about 0.68 and the mixture was centrifuged at 5000 rpm for 20 minutes. The supernatant was analyzed by ultraviolet absorbance and found to contain 27% of the original avidin. The precipitated solid was resolubilized in 10^{-3} M C12E8 solution and found to contain ca. 70% of the original avidin. A more efficient separation of solids and supernatant could have

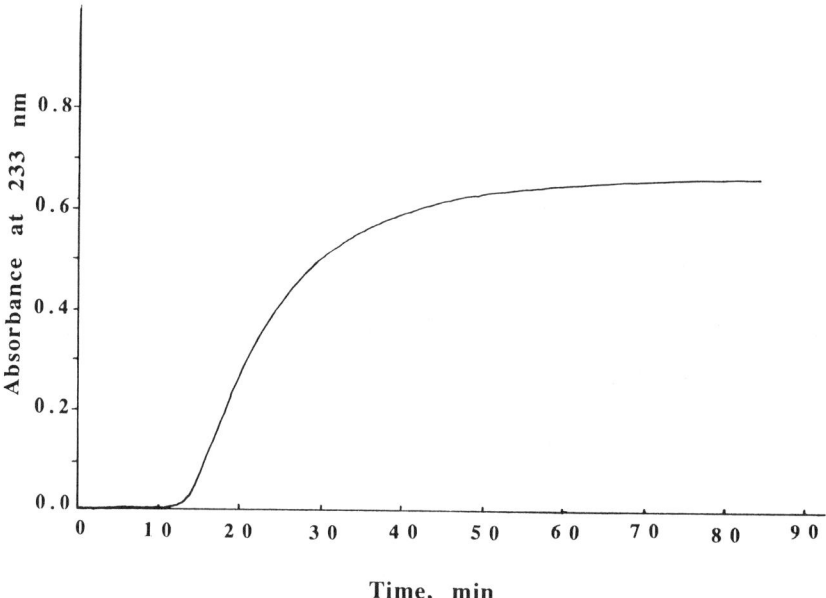

Figure 8. Absorbance difference at 233 nm as a function of time between a standard avidin solution (10^{-6} M in 0.2 M ammonium carbonate buffer at pH 8.9) and a solution of avidin of the same concentration with an added aliquot of a DMPE-B-C12E8 solution (10^{-3} M C12E8, 1.25×10^{-4} M DMPE-B) such that the final molar ratio of DMPE-B to avidin is 3.8:1.

perhaps resulted in a higher yield of avidin in the precipitate. Alternatively, a higher than stoichiometric (4:1) quantity of DMPE-B to avidin may be necessary to drive all of the avidin to aggregate and precipitate.

A protein mixture containing 10^{-6} M avidin and 10^{-6} M myoglobin in standard buffer was then prepared for affinity precipitation to test the specificity of the method. Myoglobin was selected as a test impurity because it possesses an absorbance maximum at 410 nm which does not overlap the UV-visible spectrum of avidin; thus the concentrations of the two proteins could be analyzed by simple absorbance. The UV-visible spectra of 10^{-6} M solutions of avidin and myoglobin in standard buffer are shown in Figures 9a and 9b, respectively. Upon addition of concentrated surfactant solution, so that the final concentrations were 4×10^{-6} M DMPE-B and 7.5×10^{-5} M C12E8, the onset of turbidity was observed. After four hours, the sample was centrifuged at 5000 rpm for 25 minutes and the solid separated from the supernatant. Analysis of the supernatant showed that at least 93% of the myoglobin remains in solution (Figure 9c) while the resolubilized solid in C12E8 solution showed no absorbance at 410 nm (Figure 9d), indicating that myoglobin was apparently not coprecipitated with avidin to any measureable extent. From the absorbance at 280 nm of the redissolved pellet, it was estimated that 68% of the avidin was recovered by precipitation. A second aliquot of DMPE-B was added to the supernatant (to give a DMPE-B concentration of 10^{-6} M) and the solution once again became turbid with time. After four hours, the solution was centrifuged, the solid separated from supernatant and resolubilized in C12E8 solution. Analysis of the resolubilized solid indicated that an additional 18% of the avidin was precipitated with no concomitant precipitation of myoglobin. Thus the two precipitation steps combined yielded 86% of the original avidin with no apparent co-precipitation of myoglobin.

Similar experiments were performed on solutions of avidin and BSA (both 10^{-6} M) and avidin, BSA, and lysozyme (all 10^{-6} M). In the avidin/BSA system, a total of 77% of the original avidin was recovered in resolubilized precipitate, while in the avidin/BSA/lysozyme system, 87% of the original avidin was recovered in resolubilized precipitate. As a control, solutions of myoglobin in buffer (10^{-6} M), BSA in buffer (10^{-6} M), and BSA plus lysozyme in buffer (both 10^{-6} M) were mixed with concentrated surfactant solution to obtain a mixture concentration of 4×10^{-6} M DMPE-B and 7.5×10^{-5} M C12E8. In each case, no turbidity developed in the solution and no precipitate was obtained upon prolonged centrifugation. Thus, it appears that precipitation is completely restricted to avidin, the protein to which the affinity phospholipid binds specifically.

Affinity Precipitation of Avidin from Partially Purified Egg Whites. Following the ion exchange fractionation of egg white solution, described in the experimental section above, fractions of eluted protein solution (obtained from the elution described in Figure 6) were mixed with

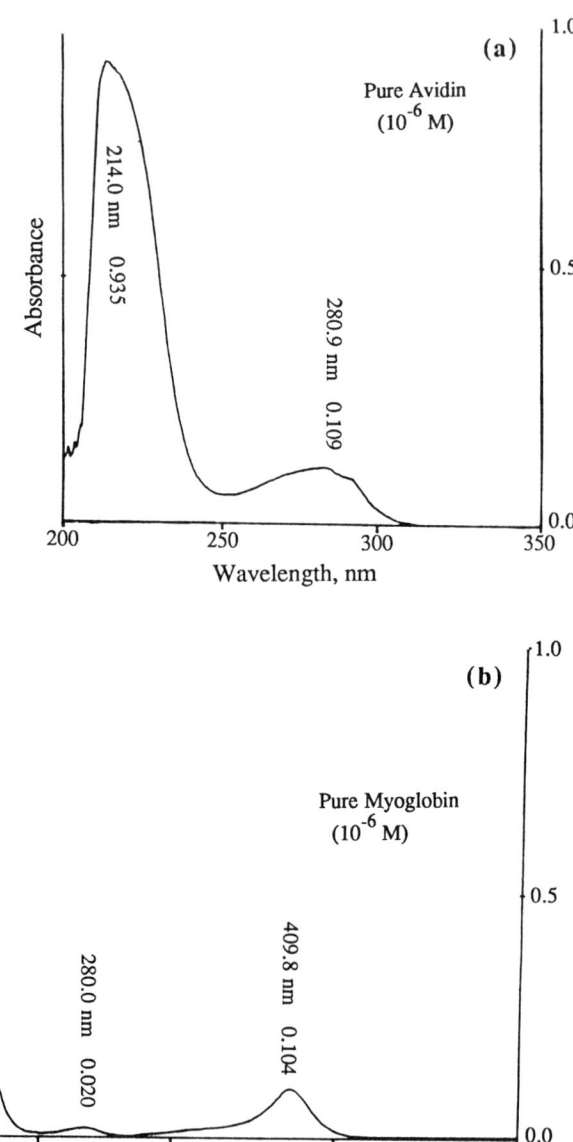

Figure 9. (a) Ultraviolet spectrum (zeroed at 350 nm) of a standard avidin solution (10^{-6} M in 0.2 M ammonium carbonate buffer at pH 8.9) relative to the buffer solution. (b) Ultraviolet–visible spectrum (zeroed at 550 nm) of a standard myoglobin solution (10^{-6} M in 0.2 M ammonium carbonate buffer at pH 8.9) relative to the buffer solution. *Continued on next page.*

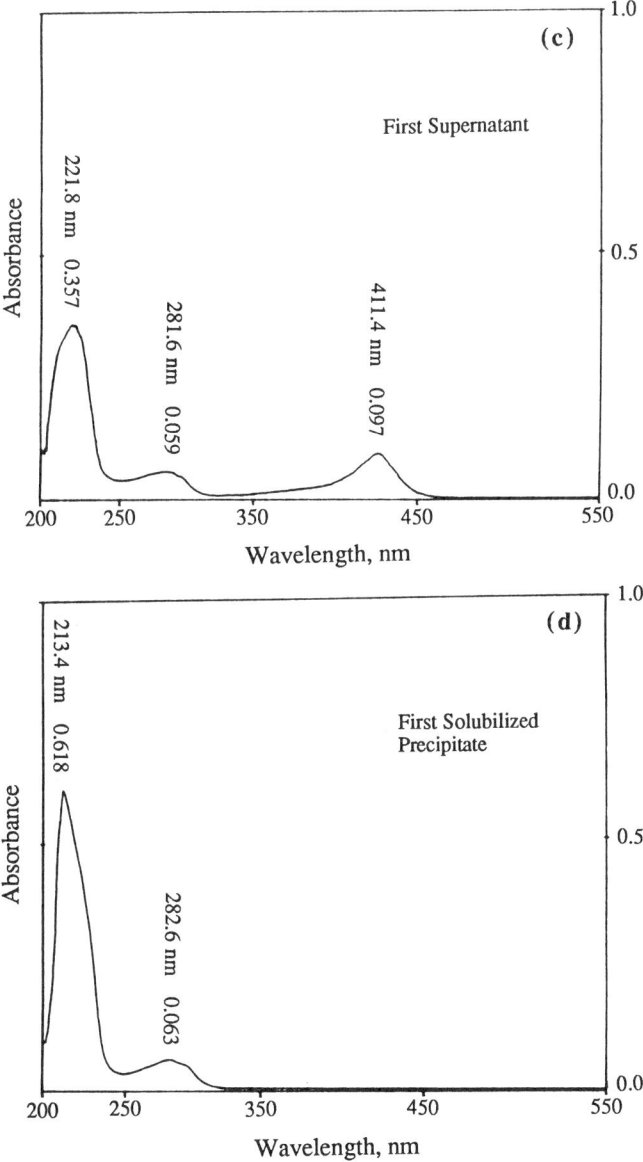

Figure 9. *Continued.* (c) Ultraviolet–visible spectrum of the supernatant after affinity precipitation of avidin from an avidin–myoglobin solution (both 10^{-6} M in 0.2 M ammonium carbonate buffer at pH 8.9) by a DMPE-B-C12E8 solution. The supernatant was sampled after precipitation for two hours and centrifugation at 5000 rpm for twenty minutes. (d) Ultraviolet–visible spectrum of avidin–DMPE-B precipitate described in text and caption 9(c) after resuspension in 10^{-3} M C12E8 solution in 0.2 M ammonium carbonate buffer at pH 8.9.

DMPE-B-C12E8 solutions to effect affinity precipitation. The disappearance of free avidin from the solutions after aggregation and precipitation was characterized by reverse phase HPLC, as described in the experimental section. Figure 10a is a reverse phase chromatogram of a standard avidin solution (10^{-6} M avidin in buffer). The peak eluting at a retention time of 10.2 minutes is avidin while the other smaller peaks are impurities in the solvent used for the gradient elution. A reverse phase chromatogram of a sample from the fraction (Fraction B) eluting between 0.6 and 0.8% ammonium actetate in eluting buffer is shown in Figure 10b. Again, the peak at retention time 10.3 minutes is avidin while now there is a large impurity protein peak at 11.2 minutes in addition to the solvent impurities seen in the standard avidin chromatogram. After affinity precipitation and centrifugation of this fraction, the chromatogram of the supernatant (Figure 10c) is obtained. In Figures 10b and 10c, the injection volume was 50 μl. The avidin peak eluting at 10.3 minutes in Figure 10c has been reduced in area by 91% relative to Figure 10b, indicating that virtually all of the avidin has precipitated. Fractions of the affinity precipitate were resolubilized in 5 ml of 10^{-3} M C12E8 solution, avidin was denatured and reconstituted as described in the Methods section above, and proteins were analyzed by reverse phase HPLC (Figure 10d). The peak at retention time 10.1 minutes corresponds to reconstituted avidin while the other peaks are the solvent impurities observed upon application of the gradient described in the Methods section above.

A standard biotin titration indicated that at least 80% of the reconstituted avidin from Fraction B was active. The UV spectrum of this reconstituted avidin was identical in form to that of native standard avidin (Figure 9a). In order to unequivocally establish the purity of this reconstituted avidin recovered from egg whites, SDS-PAGE analysis was performed on a standard avidin solution, on an aliquot of Fraction B before and after affinity precipitation, and on the purified, reconstituted avidin after affinity precipitation. Representative SDS-PAGE analyses of these three samples are shown in Figure 11. The SDS-PAGE pattern of the standard Vector Labs avidin is very similar to that of the avidin purified from egg whites by affinity precipitation. Both show darkly stained low molecular weight (ca. 15,000 daltons) lines corresponding to avidin subunits, as well as higher molecular weight fractions which move little under the applied field which are probably avidin oligomers. It is clear from the SDS-PAGE pattern of Fraction B from the egg whites that there are many other proteins present in this fraction (presumably globulins of similar chemical composition but without affinity for biotin) which do not coprecipitate with avidin.

CONCLUSIONS

The use of a novel affinity precipitation scheme, applicable to multi-binding site biomolecules, has been illustrated by the purification of avidin from model protein solutions and from partially purified egg whites. The

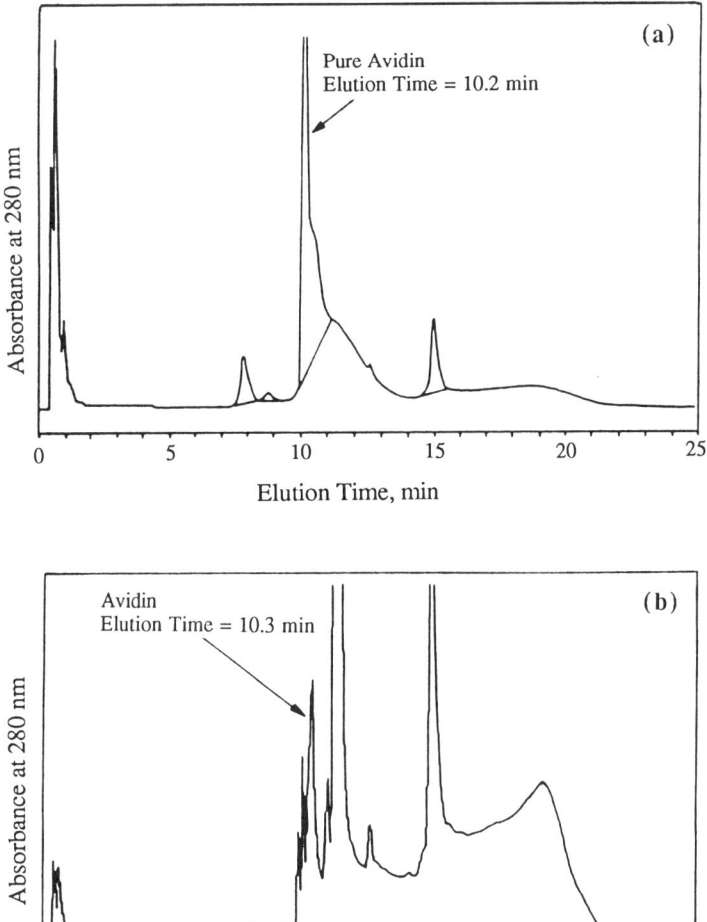

Figure 10. Reverse phase HPLC chromatograms of (a) a standard avidin solution and (b) Fraction B from ion-exchanged partially purified egg whites (described in experimental section). *Continued on next page.*

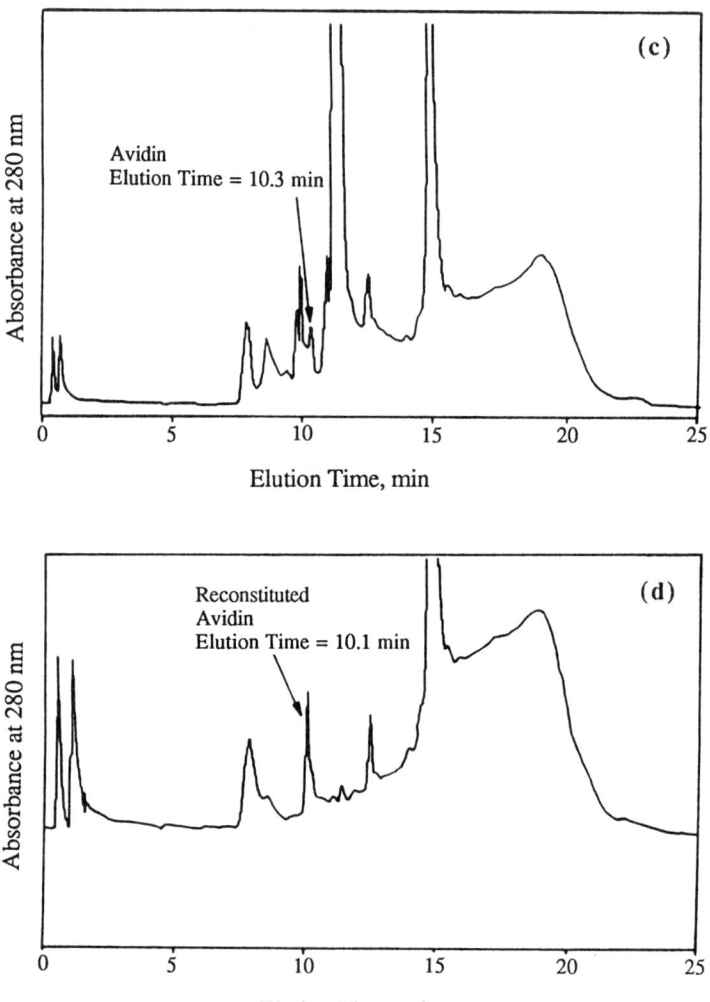

Figure 10. *Continued.* Reverse phase HPLC chromatograms of (c) supernatant of Fraction B following affinity precipitation and (d) reconstituted avidin from egg whites following affinity precipitation.

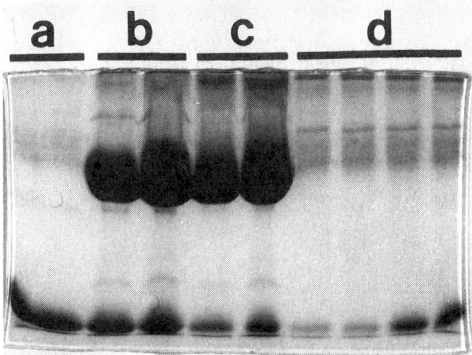

Figure 11. SDS-PAGE patterns from (a) standard Vector Labs avidin solution, (b) Fraction B from ion-exchanged partially purified egg whites before affinity precipitation (described in experimental section), (c) same as (b) but after affinity precipitation, and (d) purified avidin from Fraction B after affinity precipitation and reconstitution.

technique relies on the solubilization of a double-tailed surfactant, whose polar head group has been derivatized with an affinity ligand, in an aqueous solution of nonionic surfactant. The desired biomolecule is then contacted with this mixed surfactant solution and affinity binding between protein and surfactant-ligand occurs. The resulting protein-surfactant complexes then apparently crosslink by hydrophobic interactions between surfactant tail groups on adjacent complexes. This three-dimensional network ultimately achieves a sufficiently large size to precipitate from solution. The precipitation can be monitored by spectrophotometric turbidity. Avidin was purified from model protein mixtures containing myoglobin, bovine serum albumin, and lysozyme. In each case, no measureable coprecipitation of the undesired proteins was observed while the avidin was recovered in yields of 80-90%. It was conclusively shown that the precipitation only occurred as a result of the specific binding of avidin to biotin on the phospholipids, and it was not the result of non-specific hydrophobic aggregation of proteins with the derivatized surfactant. The avidin could be reconstituted in active form from the resulting solid precipitate by redissolving in nonionic surfactant solution, denaturing the avidin with guanidine hydrochloride, and renaturing the avidin after separation of the surfactant by ultrafiltration. Activity recoveries of more than 80% were obtained. Finally, avidin was also purified from one of its naturally occurring sources: hen egg whites. A partial purification of the egg whites by ion exchange chromatography to remove hydrophobic and negatively charged contaminants was first performed. Then avidin was recovered in pure active form by affinity precipitation and reconstitution. The purity and activity of this resulting avidin was verified by HPLC, UV spectroscopy, biotin titration, and SDS-PAGE analysis. It is likely that this technique will be generally applicable to the purification of proteins with multiple binding sites that are capable of strong interactions with small ligands that can be readily attached to hydrophobic surfactants, such as phospholipids.

ACKNOWLEDGMENT

The authors gratefully acknowledge the financial support of the National Science Foundation under grants EET-8514790 and CBT-8705562.

LITERATURE CITED

1. Bell, D. J., M. Hoare, and P. Dunnill, Adv. Biochem. Eng. Biotech., 26, 1 (1983).
2. Fisher, R. R., C. E. Glatz, and P. A. Murphy, Biotech. and Bioeng., 28, 1056 (1986).
3. Schneider, M., C. Guillot, and B. Lamy, Ann. N. Y. Acad. Sci. USA, 369, 257 (1981).
4. Larsson, P.-O., and K. Mosbach, FEBS Letters, 98, 333 (1979).
5. Flygare, S., T. Griffin, P.-O. Larsson, and K. Mosbach, Anal. Biochem., 133, 409 (1983).
6. Torres, J. L., R. Z. Guzman, R. G. Carbonell, and P. K. Kilpatrick, Anal. Biochem., 171, 411 (1988).
7. Powers, J. D., P. K. Kilpatrick, and R. G. Carbonell, Biotech. and Bioeng., 33, 173 (1989).
8. Bayer, E. A., and M. Wilchek, Methods in Biochemical Analysis, 26, 1 (1979 a).
9. Bayer, E. A., B. Rivnay, and E. Skutelsky, Biochim. Biophys. Acta, 550, 464 (1979 b).
10. Green, N. M., Biochem. J., 89, 585 (1963).
11. Melamed, M. D., and N. M. Green, Biochem. J., 89, 591 (1963).
12. Fraenkel-Conrat, H., N. S. Snell, and E. D. Ducay, Arch. Biochem. Biophys., 39, 80 (1952).
13. Smith, B. J., "SDS Polyacrylamide Gel Electrophoresis of Proteins" in Methods in Molecular Biology, Vol. 1: Proteins, Chap. 6, J. M. Walker, ed., Humana Press, 1984.
14. Ueno, M., Y. Takasawa, H. Miyashige, Y. Tabata, and K. Meguro, Colloid Polymer Sci., 259, 761 (1981).

RECEIVED October 24, 1989

NOVEL ISOLATION AND PURIFICATION PROCESSES

Chapter 12

Ultracentrifugation as a Means for the Separation and Identification of Lipopolysaccharides

Marshall Phillips and Kim A. Brogden

Agricultural Research Service, U.S. Department of Agriculture, National Animal Disease Center, P.O. Box 70, Ames, IA 50010

> Ultracentrifugation can be used as a method to separate and identify different types of lipopolysaccharides (LPS). To illustrate this, LPS from Pasteurella multocida, Brucella abortus, and other gram-negative microorganisms were fractionated on 34% to 45% CsCl field-formed gradients. After centrifugation, their apparent buoyant densities ranged from 1.34 to 1.44 gm/ml. The buoyant density positions in the gradients are thought to be not only a function of their chemical composition, but also of the hydration and conformation of the LPS molecules in their respective solutions. These differences can be capitalized on for the isolation of contaminant-free LPS as well as the isolation of LPS with slightly differing molecular composition. This may prove to be an ideal way to separate and recover recombinant LPS for biotechnology purposes. The LPS can then be used directly for the development of immunizing and diagnostic reagents.

Lipopolysaccharide (LPS) is an integral component of endotoxin found at the outer surface of gram-negative bacteria. This moiety is responsible for much of the toxicity ascribed to endotoxin and is an important determinant involved in the pathogenesis of gram-negative bacterial infections. Chemically, monomeric molecules of LPS have a molecular weight of about 2 kilodaltons, and are comprised of a hydrophilic region of polysaccharide (somatic polysaccharide of repeating units up to 100 monosaccharides long) and a hydrophobic region of long chain (n-14, n-20) fatty acids linked to N-acetylglucosamine (Figure 1). The two regions are joined by a "core" polysaccharide containing 2-keto-3-deoxyoctulosonic acid (KDO), heptose, and amino sugars. Ethanolamine and phosphate groups are covalently attached. Mutant bacterial strains exist that lack components of the core polysaccharide (Ra through Re). Excellent reviews have been written on the composition of LPS (1, 2) including its components such as lipid A (3) and KDO (4). In this paper, we present ultracentrifugation both as a preparative technique to separate and purify different types of LPS and

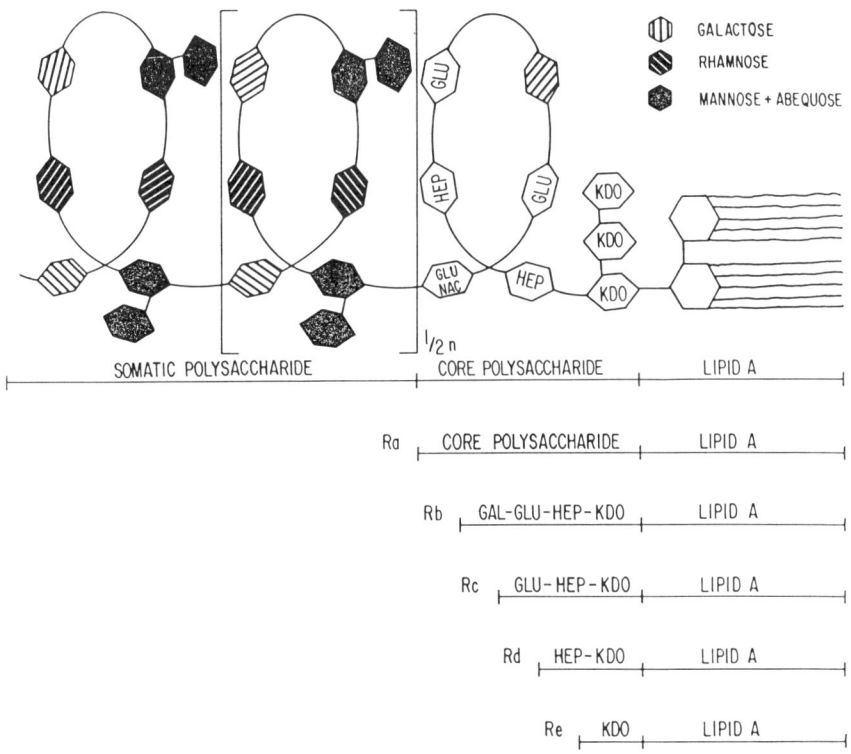

Figure 1. A schematic diagram illustrating a basic lipopolysaccharide molecule from <u>Salmonella</u> <u>typhimurium</u> with the polysaccharide and lipid regions. Also shown are core polysaccharide mutants (Ra through Re).

as an analytical technique to assess the purity of LPS preparations and to distinguish LPS of slightly differing chemical composition. To illustrate this, LPS from different gram-negative bacteria as well as LPS from Salmonella minnesota Ra, Rc, and Re mutants were fractionated on 34% to 45% CsCl gradients. Differences in density of these LPS can be capitalized on for their isolation, purification, and characterization.

Equilibrium (isopycnic) centrifugation is well established (5-8). The basic concept involves the generation of a gradient during centrifugal force: the density near the bottom of the centrifuge tube is greater than that of the most dense particle in the mixture and, at the meniscus, the density is less than that of the lightest particle. The initial application of the technique was applied to the characterization of nucleic acids, and is of historical importance in the discovery of DNA replication (9). Proteins and glycoproteins have also been characterized with this technique (10). However, there is a paucity of reports on the use of the technique for the study of LPS. Morrison and Leive (11) first used the technique to separate LPSs with slightly differing chemical composition extracted from Escherichia coli. They observed that the chemical composition of the LPS was related to its apparent buoyant density and was due to the length of the somatic polysaccharide. A nearly direct correlation was observed between the amount of carbohydrate in the LPS preparation and its buoyant density. Later, the biological properties of the LPS was also shown to correlate with the buoyant density (12-14). In 1978, Phillips and Rebers (15) published the first analytical equilibrium centrifugation patterns of LPS in CsCl gradients. This report and subsequent reports (16-18) indicated, from the schlieren patterns, that well-defined molecular distributions of LPS molecules exist at equilibrium in CsCl gradients.

A number of media may be employed for the preparation of gradients (5, 8). These include simple sugars and analogous polyhydroxyl compounds, polysaccharides, proteins, iodinated organic compounds (i.e. metrizamide), colloidal silica (Ludox), and inorganic salts (8). Inorganic salts, such as CsCl, are the most commonly used media for the preparation of gradients (8). The advantages of CsCl are numerous. First, CsCl is well characterized and its thermodynamic properties known. Second, CsCl is very soluble in water resulting in very dense solutions with a relative viscosity close to unity. Third, pure CsCl solutions do not absorb ultraviolet light. Finally, high concentrations of CsCl (34% to 50%) are very chaotropic and provide conditions for the disruption of hydrogen bonding. Thus, under the conditions of CsCl equilibrium centrifugation, the apparent buoyant density positions of LPS will be an equilibrium balance among the forces of hydrophobic interactions, solvation, and chaotropic-mediated actions. However, CsCl diffuses rapidly and a preformed gradient is stable for only a short time.

METHODS FOR EXTRACTION OF LPS

Since LPS varies in chemical composition from one bacterial species to another, different methods are used for its extraction. LPS is generally extracted from smooth strains of bacteria in a mixture of phenol and water (PW) at 68°C as described by Westphal and Jann (19). After extraction, the solution is allowed to cool and partition. Nearly all proteins remain in the phenol phase or interphase, while polysaccharides, LPS, and nucleic acids remain in the aqueous phase. In some instances, the LPS from certain organisms (i.e. Brucella abortus) remain in the phenol phase. Lipopolysaccharide is extracted from rough strains of bacteria in a mixture of phenol, chloroform, and petroleum ether (PCP) at room temperature as described by Galanos, et al. (20). After extraction, the chloroform and petroleum ether are removed by rotary evaporation and the LPS is then precipitated from the phenol with water. With some bacteria, PW or PCP will not extract sufficient quantities of LPS (21) and EDTA (22), chloroform-methanol (23), or butanol (11) must be used.

PREPARATIVE CsCL GRADIENT ISOLATION OF LPS

After extraction, LPS solutions usually contain protein and nucleic acid contamination. These can be removed by a variety of procedures (24). The most commonly used procedures treat LPS solutions with nucleases and proteolytic enzymes. Proteinase K or proteinase K solutions containing sodium dodecyl sulfate (SDS) can also be used. Alternately, contaminant-free LPS can be obtained by preparative equilibrium centrifugation. In our work, crude preparations of LPS were suspended in 0.05 M Tris buffer, pH 7.8, and added to solutions of CsCl in the same buffer. The concentration of CsCl was adjusted such that the final concentration was 34% after adding 1 ml of LPS in Tris buffer. The gradient was attained in 38 hours at 39,000 x g in the Beckman SW 50.1 rotor. Recovery of LPS from the gradient is accomplished by sequentially emptying the tube for recovery of the LPS band with a Pasteur pipette. The CsCl is then removed by dialysis. In this technique, LPS buoy in a range from 1.34 to 1.45 gm/cm^3 in CsCl gradients. Centrifugation was performed at 25°C to limit diffusion of the CsCl in the tubes when manipulated at room temperature. Contaminants will buoy at different densities. For example, the lipids will buoy on the surface of the gradient, most proteins will buoy above 1.34 gm/cm^3 (usually 1.25 to 1.32 gm/cm^3), extraneous complex carbohydrates will buoy from 1.49 to 1.58 gm/cm^3 or higher, and nucleic acids will band at even higher values (Figure 2). The buoyant density of these molecules in CsCl is not only a function of their chemical content but also of their degree of hydration, conformation, orientation, and extent of hydrogen bonding. Procedures and calculations for obtaining equilibrium for other density regions are also described (6-8).

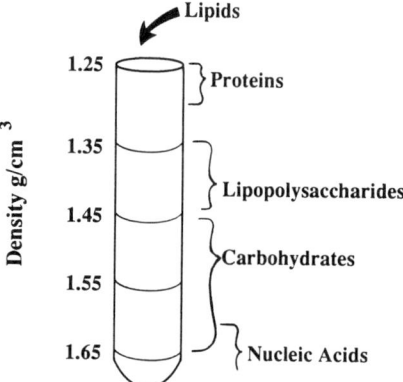

Figure 2. Predicted buoyant position of hydrated constituents in CsCl gradients.

The amount of carbohydrate in the LPS molecule influences its buoyant density position. This can be shown by mixing Ra, Rc, and Re LPS forms from Salmonella minnesota and seperating them in a 39% CsCl gradient (Figure 3). The respective forms buoy at 1.38, 1.42, and 1.50 gm/cm^3. Each LPS in the Ra, Rc, and Re mixture separated at each density as clearly as each individual form alone. Similarly, representative patterns of Brucella LPS are shown in Figure 4. The LPS reaches equilibrium in a narrow region at one position, and other contaminants reach equilibrium at other positions in the tubes. Minor contaminants may not be identifiable as a band but only detected in isolated fractions by chemical means (i.e. protein determination, ultraviolet absorption, etc.). It is also possible that LPS molecules may bind with components and thus be found in other regions in the gradient. Preparations of LPS isolated from Brucella abortus strains S-19 and 2308S band at nearly identical positions. The LPS from B. meletensis can be found in two bands. The B. abortus rough strains contain a band at 1.35 gm/cm^3 and, under conditions of the gradient, also can be found in more than one band. Table I contains a listing of the apparent buoyant densities of LPS isolated from several sources.

ANALYTICAL CsCL GRADIENT SEPERATION OF LPS

Analytical ultracentrifugation of LPS on CsCl gradients reveals a number of properties about the LPS molecule that cannot be easily determined by typical chemical analysis (i.e. protein, total carbohydrate, KDO content, etc.). Analytical ultracentrifugation is performed in the Beckman Model E analytical ultracentrifuge. During centrifugation, a schlieren peak is generated. The shape and number of peaks present determine not only the purity of the LPS preparation but the number of LPS molecules with slightly differing chemical composition. Therefore very sharp, narrow schlieren peaks indicate a monodisperse, highly uniform LPS, free of contamination whereas a wide, less defined peak indicates the presence of a population of differing LPS molecules of questionable purity. Also, the use of the analytical centrifuge can be used to characterize the LPS obtained by the preparative procedure. Methods for determining the apparent density values from the schlieren peaks are based on the general equation of equilibrium centrifugation (9, 25-27). For example, a monodisperse schlieren peak was observed when B. abortus LPS was recovered from the discrete bands as shown in preparative tube A, Figure 5. Monodisperse peaks are not always observed from the Brucella LPS (Figure 6). Monodisperse peaks were also obtained with LPS from Pasteurella multocida (Figure 6c). Figure 7 shows the computer-assisted conversion of components expected from schlieren patterns as described by Johnson (28).

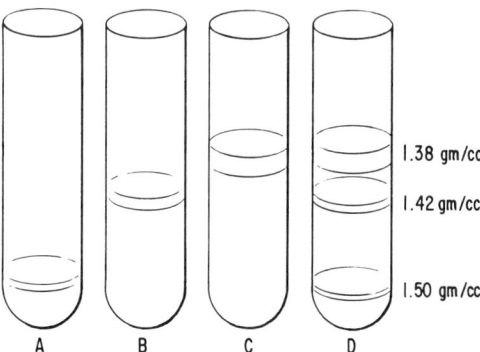

Figure 3. Buoyant density separation of <u>Salmonella</u> <u>minnesota</u> Ra, Rc, and Re lipopolysaccharides. Centrifugation is 39% CsCl for 40 hours at 25°C. A: Ra; B: Rc; C: Re; and D: Mixture of Ra, Rc, and Re.

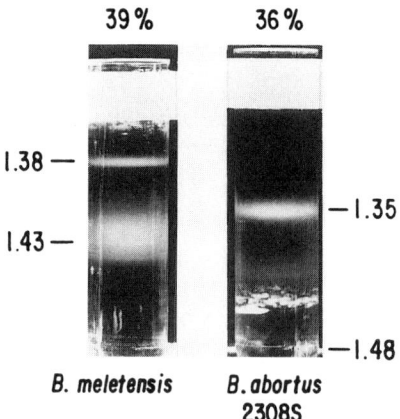

Figure 4. Preparative 36% and 39% CsCl gradients of <u>Brucella</u> LPS. Shows separation of other components. Left: 39% gradient showing <u>B</u>. <u>meletensis</u> LPS. Right: 36% gradient showing <u>B</u>. <u>abortus</u> LPS.

Table I. Apparent Buoyant Densities of LPS

Origin	Apparent Density (g/cm³)	Reference
Escherichia coli 0811 (phenol extraction)	1.45	11
Escherichia coli 08111 (butanol extraction)	1.35	11
Salmonella minnesota R595	1.38	12
Pasteurella multocida X-73	1.39	16
Pasteurella multocida	1.38-1.44	16
Serratia marcescens	1.50	13
Brucella abortus 2308	1.35	17
Brucella abortus S-19	1.35	17
Brucella abortus 2308R	1.35-1.38	-
Brucella meletensis	1.38	-

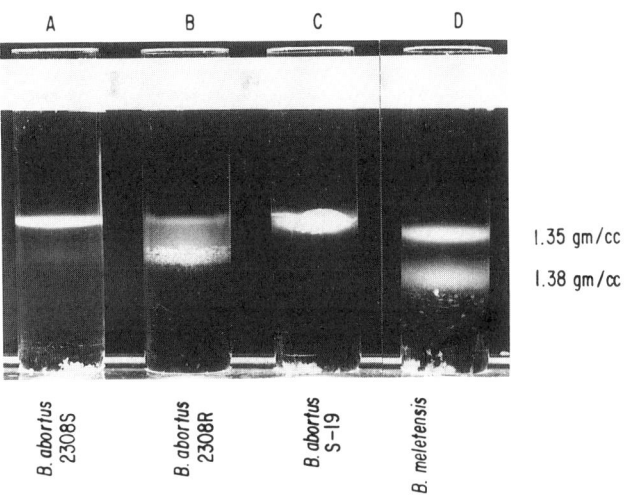

Figure 5. 34% CsCl gradients of Brucella LPS. The LPS from B. abortus strain indicated under tube.

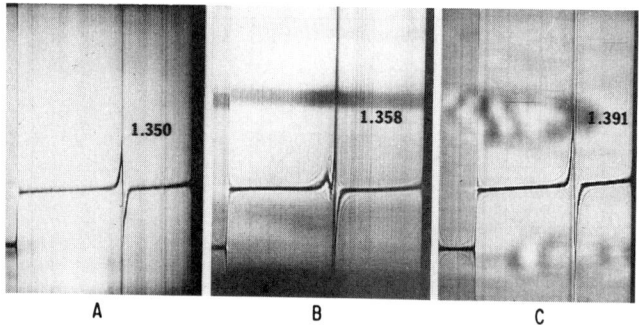

Figure 6. Model E analytical schlieren patterns of purified LPS. Equilibrium attained at 36,000 rpm at 25°C for 18 hours in 2°C sector cells. A: 34% CsCl pattern of B. abortus LPS, sample from preparation run in Figure 5A. B: 34% CsCl pattern of B. abortus LPS. C: 39% CsCl pattern of P. multocida LPS.

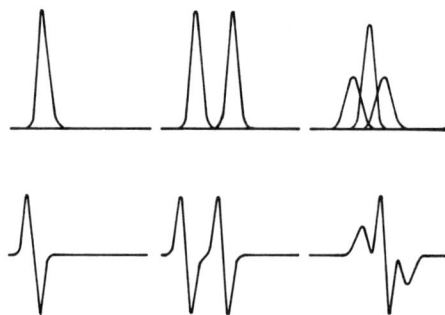

Figure 7. Computer-generated density gradient distributions.
Note first row: concentration of components; second row: schlieren pattern.

EFFECT OF CHEMICAL TREATMENT OF LPS ON SEPARATION

The buoyant density of LPS has been shown to differ between that extracted from smooth and rough bacteria (for example, B. abortus strain 2308, Figure 5), as well as differ within a single bacterial species depending upon its method of extraction (11), ionic environment, and association with other molecules. The effect of chemical treatment on LPS was shown by Amano and Fukushi (29). Treatment of S. minnesota wild and R mutants (Ra to Re) with alkali caused partial removal of somatic polysaccharide linked fatty acids. Thus, the monolayer formation in alkali-treated LPS resulted from reduction of the steric hindrance by fatty acids and added to the reduction of its hydrophobicity. At acidic pH, changes in aggregation of LPS have also been observed. Following treatment with acetic acid (pH 2.8) for 3 hours at room temperature, Acker and Wartenberg (30) observed LPS strands with regularly spaced, "loosened" regions. After treatment for several days, the LPS strains were disconfigured.

Divalent cations are important to the physical-chemical properties of LPS (31). The LPS contains inorganic cations such as Na^+ K^+, Ca^{++}, and Mg^{++}, as well as low molecular weight amines (1). These cations neutralize the negative charges on LPS molecules necessary for ordered assembly (32).

Treatment of LPS with detergents result in a reversible dissociation of LPS into subunits (33). Both Ribi, et al. (34) and Horisberger and Dentam (35) observed that LPS treated with sodium deocycholate, followed by dialysis, did not alter the size of the particles, yet resulted in a more uniform preparation. Polymyxin B will dissociate and degrade LPS components. These effects have been suggested to be caused by the interactions of the lipophilic and lipophobic groups of the antibiotic with those of the LPS (36).

All of these treatments may influence the density of LPS preparations by removing contamination components that interact with the LPS molecules.

EFFECTIVENESS OF CsCL GRADIENT SEPARATION

The preparative CsCl gradients effectively separate LPS from contaminants and the analytical CsCl gradients effectively characterize the LPS in them. The hypersharp peak patterns indicate large aggregates (or highly associated molecules) of LPS under the conditions of the CsCl. This is the result of the association of the hydrophobic portions of the molecules into the formation of a highly oriented conformation. Based on the expected patterns of molecules in CsCl (Figure 6), the schlieren patterns indicate only a limited population distribution of molecules in the preparations. In the gradients, it is thought that the LPS form a continuum of different size micelles. Since LPS are amphipathic molecules, they undergo association - dissociation phenomena, and enter into micellular formation. The LPS molecules combine by hydrophobic - hydrophilic interactions; lipid A regions align toward each other internally, and their carbohydrate regions orientate toward the aqueous interphase. The nature of such aggregate formation is based on pH, temperature, and other solution conditions.

FUTURE APPLICATION

Ultracentrifugation can be used as a method to separate and identify different types of lipopolysaccharides (LPS). The buoyant density positions of LPS in the gradients not only reveal something about their state of purity but also about their chemical composition. These differences can be capitalized on for the isolation of contaminant-free LPS as well as the isolation of LPS with slightly differing molecular composition. This may prove to be an ideal way to separate and recover recombinant LPS for biotechnology purposes. The LPS can then be used directly for a variety of biomedical purposes including immunomodulators, vaccines, and diagnostic reagents.

LITERATURE CITED:

1. Luderitz, O.; Freudenberg, M. A.; Galanos, C.; Lehmann, V.; Rietschez, E. T.; Shaw, D. E.; Current Topics in Membranes and Transport; Membrane Lipids of Prokaryotes; Razin, S.; Rottem, S., Eds.; Academic: New York, 1982; Vol. 17, pp. 79-151.
2. Westphal, O.; Jann, K.; Himmelspach, K.; Prog. Allergy 1983, 33, pp. 9-39.
3. Rietschel, E. T.; Brade, H.; Brade, L.; Brandenburg, K.; Schade, U.; Seydel V.; Zahringer, U.; Galanos, C.; Luderitz, O.; Westphal, O.; Labischinski, H.; Kusumoto, S.; Shiba, T.; Progr. Clin. Biol. Res. 1987, 231, pp. 25-53.
4. Unger, F. M.; Advances in Carbohydrate Chemistry and Biochemistry; Tipson, R. S.; Horton, D., Eds.; Academic: New York, 1981; pp. 323-88.
5. Osterman, L. A.; Equilibrium (Isopycnic) Centrifugation; Springer-Verlag: New York, 1984; pp. 275-307.
6. Ifft, J. B.; A Laboratory Manual of Analytical Methods of Protein Chemistry; Alexander, P.; Lundgren, H. P.; Eds.; Pergamon: Oxford, 1969; Vol. 5, pp. 151.
7. Hearst, J. E.; Schmid, C. W.; Methods in Enzymology; Hirs, C. H.; Timsahoff, S. N., Eds.; Academic: New York, 1973; Vol. 27, pp. 111-27.
8. Birnie, G. D.; Rickwood, D.; Centrifugal Separations in Molecular and Cell Biology; Butterworths: London, Boston, 1978; pp. 169-217.
9. Meselson, M.; Stahl, F. W.; Vinograd, J. ; Proc. Nat. Acad. Sci. U.S.A. 1957, 43, pp. 581-588.
10. Dunstone, J. R.; Prep. Sci. 1969, 4, pp. 267-85.
11. Morrison, D. C.; Leive, L.; J. Biol. Chem. 1975, 250, pp. 2911-19.
12. Ulevitch, R. J.; Johnston, A. R.; Clin. Invest. 1978, 62, 1313-24.
13. Wilson, M. R.; Morrison, D. C.; Eur. J. Biochem. 1982, 128, pp. 137-41.
14. Vukajlovich, S. W.; Morrison, D. C. ; J. Immunol. 1985, 135, pp. 2546-50.
15. Phillips, M.; Rebers, P. A. ; Anal. Biochem. 1978, 85, pp. 265-70.
16. Rimler, R. B.; Rebers, P. A.; Phillips, M.; Am. J. Vet. Res. 1984, 45, pp. 759-63.
17. Phillips, M.; Pugh Jr, G. W.; Deyoe, B. L.; Am. J. Vet. Res. 1989, 50, pp. 311-317.
18. Phillips, M.; Brogden, K. A.; Infect. Immun. 1987, 55, pp. 2047-51.

19. Westphal, O.; Jann, K.; Methods in Carbohydrate Chemistry; Whistler, R. L., Ed.; Academic: New York, 1965; pp. 83-92.
20. Galanos, C.; Luderitz, O.; Westphal, O.; Eur. J. Biochem. 1969, 9, pp. 245-49.
21. Wu, L.; Tsai, C-M.; Fresch, D. E.; Anal. Biochem. 1987, 160, pp. 281-89.
22. Leive, L.; Methods in Enzymology; Ginsburg, V., Ed.; Academic Press: New York, 1972; Vol. 28, pp. 254-62.
23. Amano, K.; Fukushi, F. ; Microbiol. Immunol. 1984, 28, pp. 135-148.
24. Rietschel, E. T.; Handbook of Endotoxin; Chemistry of Endotoxin; Elsevier Science: The Netherlands, 1984; Vol. 1.
25. Svedberg, T.; Pederson, K. O.; The Ultracentrifuge; Oxford Press: Oxford, 1940.
26. Vinograd, J.; Hearst, J. E.; Prog. Chem. Organic. Natural Products 1962, 20, pp. 371-405.
27. Chervenka, C. H.; A Manual of Methods for the Analytical Ultracentrifuge; Spinco Div. Beckman Instruments, California, 1969.
28. Johnson, N. W.; J. Gen. Virol. 1973, 18, pp. 207-09.
29. Amano, K.; Fukushi, F.; Microbiol. Immunol. 1984, 28, pp. 161-168.
30. Acker, G.; Wartenberg, K.; Zbl. Bakt. Hyg. I Abt. Orig. A 1976, 235, pp. 439-52.
31. Galanos, C.; Luderitz, O.; Eur. J. Biochem. 1975, 54, pp. 603-10.
32. Graham, G. S.; Treick, R. W.; Brunner, D. P.; Curr. Microbiol. 1979, 2, pp. 339-43.
33. Hurlbert, R. E.; Hurlbert, I. M.; Infect. Immun. 1977, 16, pp. 983-94.
34. Ribi, E.; Anacker, R. L.; Brown, R.; Haskins, W. T.; Malmgren, B.; Milnder, K. C.; Rudbach, J. A. ; J. Bacteriol. 1966, 92, pp. 1493-1509.
35. Horisberger, M.; Dentam, E.; Arch. Microbiol. 1980, 128, pp. 12-18.
36. Weber, D. A.; Nadakavukaren, J.; Tsang, J. C.; Microbios 1976, 17, pp. 149-61.

RECEIVED October 2, 1989

Chapter 13

Removal and Inactivation of Viruses by a Surface-Bonded Quaternary Ammonium Chloride

I-Fu Tsao and Henry Y. Wang

Department of Chemical Engineering, University of Michigan, Ann Arbor, MI 48109−2136

A quaternary ammonium chloride (QAC) covalently bound to alginate beads was shown to be capable of removing viruses from protein solutions. Bacteriophage T2 and herpes simplex virus type 1 (HSV-1) were used as model viruses in this investigation. The enveloped HSV-1 showed a higher susceptibility to surface-bonded QAC due to hydrophobic interaction. The elution experiments using 1% tryptone solution demonstrated that inactivation, instead of non-specific adsorption, is the main factor for virus titer reduction. Both equilibrium and kinetic studies were conducted in distilled water and protein (BSA, 0.5%) solution to observe the effect of protein molecules on reducing the capacity and the rate of virus adsorption/inactivation process. Heating thermolabile proteins at 60°C for 10 hours with various stabilizers has historically been used to inactivate any possible viral contaminants. However, the protecting effect of these stabilizers may minimize the destruction of the virus. Thermal inactivation and immobilized QAC inactivation procedures were compared using β-lactamase, a thermolabile enzyme, as a model protein. The activity of β-lactamase dropped down to 40% of the initial value after heating at 60°C for 10 hours using sucrose as a stabilizer. The virus titer diminished from 10^6 to 10^3 (PFU/ml) during the first four hours without further reduction. The titer of T2, a more hydrophilic virus, decreased only one order of magnitude and the recovery of β-lactamase activity was 70%. Computer simulation results demonstrating the effects of various process variables are also presented.

Quaternary ammonium chlorides (QAC) are cationic surface-active agents with antimicrobial activity (1). The major mode of action of QAC was identified as the effects on cell permeability and cytolytic damage (2). Klein and Deforest (3) summarized the virucidal capacities of Zephiran (alkyldimethylbenzylammonium chloride) against various types of viruses. They reported that Zephiran can effectively inactivate lipid-containing viruses, some non-lipid viruses and bacteriophages but is not effective against smaller but non-lipid viruses, such as picornaviruses.

Quaternary ammonium chlorides display their antimicrobial activity even being immobilized on inert supports because they can act on the membranes of various cells. The effects of surface-bonded organosilicon QAC on bacteria, yeasts, fungi, algaes have been extensively investigated (4, 5, 6, 7, 8). However, the activity against viruses of these immobilized compounds has not been demonstrated.

The presence of low level infectious viral contaminants has been one of the main hurdles for wider application of genetically engineered therapeutic proteins such as blood-clotting factors, interferon, insulin, growth hormones, antibodies, and various clinically significant proteins such as tissue plasminogen activator. Heat treatment is a widely accepted method for sterilizing these complex protein solutions. It is well known that albumin can be pasteurized by heating at 60°C for 10 hours in the presence of certain stabilizers (9). Unfortunately, similar attempts at other blood proteins (e.g. clotting factors) resulted in marked reduction or elimination of their functional activities. Schwinn et al., (10) and Fernandes and Lundblad (11), modified the pasteurization procedure using high concentrations (>30% w/v) of polysaccharides as stabilizers. Unfortunately, the protein stabilizing effect of these polysaccharides likewise provides increased thermo-resistance to the viruses (12). Heat treatment of lyophilized clotting factor concentrates was also investigated (13).

The phenomenon of viral adsorption to various surfaces was extensively studied from an environmental standpoint as reviewed by Daniels (14) and Gerba (15) for prevention of various waterborne viral transmissions. The problem of virus removal from complex protein solutions is very different from that of sewage and drinking water treatment processes because most protein molecules compete for the active sites of the adsorbents. Hence, both the adsorption rate and capacity diminish in the presence of protein molecules (16). It is the intention of this paper to demonstrate and to compare the antiviral activity of a surface-bonded QAC in aqueous solutions against 2 model viruses with and without the presence of proteins. The efficacy of the accepted antiviral thermo-inactivation was compared with the viral inactivation method by the surface-bonded QAC treatment. Beta-lactamase was used as a thermolabile model protein (17), and bacteriophage T2 and herpes simplex virus type 1 (HSV-1, an enveloped animal virus) were used as model hydrophilic and hydrophobic viruses to test these chemical inactivation methods.

MATERIALS AND METHODS

Chemicals. Three-(Trimethoxysilyl)-propyldimethyloctadecyl ammonium chloride (Si-QAC), known as Dow Corning 5700 Antimicrobial

Agent, was provided by W. Curtis White (Dow Corning Corp., Midland, Mich.). It is a methanolic solution containing 42 wt% of this active ingredient. Beta-lactamase was obtained from Sigma Chemical Co. (St. Louis, Mo.). Other chemicals were of reagent grade and were purchased from various commercial sources.

Organisms. Escherichia coli B and bacteriophage T2 are regularly maintained in our laboratories. BSC-1 cells and HSV-1 strain 148 were obtained from Dr. Charles Shipman, Jr. (Dental School, U. of Michigan, Ann Arbor, Mich.). The cultures are regularly passaged to maintain their viability.

Preparation of Beads. Dried alginate/magnetite beads were prepared by a method similar to that described by Burns et al., (18). Barium chloride was used as a gel-inducing agent for better stability (19). The beads were further stabilized by treating with glutaraldehyde in the presence of polyethyleneimine to avoid the dissolution problem (20). Beads with diameters between 0.15 and 0.25 mm were obtained by crushing and then systematically sieving the original spherical beads. Various concentrations of Si-QAC solution were prepared by diluting the 42% active material in distilled water at pH 5. After the beads were added to the Si-QAC solution, the reaction temperature was raised to about 50°C for 10 minutes. Then, the pH was adjusted to 10.5 for another 10 minutes. The beads were then dried in an oven (100°C), rinsed several times with sterile deionized water (pH 7.0) and stored at 4°C.

Cell Culture. BSC-1 cells were grown in minimal essential medium (MEM) with Earle salts supplemented with 10% fetal bovine serum (FBS) and 1.1 g/l sodium bicarbonate. Cells were passaged according to conventional procedures by using 0.05% trypsin plus 0.02 wt% ethylenediaminetetraacetic acid (EDTA) in a HEPES-buffered balanced salt solution. Tissue culture flasks were incubated at 37°C in a humidified 3% CO_2 - 97% air atmosphere. Total cell counts were made using a Coulter counter equipped with a 100-μm orifice and microscopic cell count.

Titration of viruses. HSV-1 was assayed by using monolayer cultures of BSC-1 cells grown in 6-well multidishes. The cells were plated 3×10^5 cells/well in MEM(E) with 10% FBS and 1.1 g/l sodium bicarbonate. After 24 hours, the cell sheet was about 75% confluent and was inoculated with 0.2 ml of the virus suspension to be assayed and incubated for 1 hour to permit viral adsorption. The cell sheets were then overlaid with 3 ml medium containing 0.5% methocel and incubated for another two days.

After aspiration of the overlay, the cells were stained with crystal violet, and macroscopic plaques were enumerated.

The assay procedures for T2 used here were described by Rovozzo and Burke (21).

Assays. Samples collected in all experiments were cooled and stored at 4°C, then the concentration of total protein in the solution was assayed by the Bradford method (22), and the concentration of β–lactamase was assayed according to Sykes and Nordstrom (23).

Batch Experiments. During these experiments, adsorbents and viruses were continuously mixed in Erlenmeyer flasks by a gyratory shaker at 22°C. Reaction mixtures of known composition were made by adding stock solution to 0.01 M Tris/HCl buffer at pH 7.0. All stock chemical solutions were autoclaved and stored at 4°C.

In the equilibrium studies, tests were conducted with various initial concentrations of adsorbents and viruses to determine the amount of virus adsorbed per unit gram of adsorbent and the virus concentration remaining in the solution at equilibrium. The time required to reach equilibrium was determined by periodically sampling over a 24-hour period. In the kinetic studies, samples were withdrawn at predetermined time intervals and assayed for virus titer.

RESULTS AND DISCUSSION

Inactivation of T2 phage by QAC's in Free Solutions.

Susceptibility of bacteriophage T2 to QAC is shown in Table 1. Survivors could not be found in solutions without the bovine serum albumin (BSA). These results demonstrated that bacteriophage T2 can be inactivated by QAC as well as Si-QAC solutions. The presence of protein molecules inhibited the activities of these antimicrobial agents. In fact, BSA was even coagulated in the presence of high concentration (>0.05%) of Si-QAC.

Inactivation of Viruses by Surface-Bonded QAC.

The attachment of this Si-QAC to surfaces involves a rapid ion-exchange process which coats as a monolayer on the bead surface. Then, the immobilization is further strengthened by the polymerization reactions (24). Table 2 shows the effects of dried alginate beads treated by various Si-QAC concentrations. Zero percent means untreated beads and served as controls. When the titer was very low (4.7×10^2), viruses were eliminated completely in all cases including the control. This was due to non-specific adsorption. When the titer was raised to 4.0×10^4, the adsorption capacities of treated beads were distinctly better than the control. For a titer as high as 2.0×10^8, it seems that the beads were nearly saturated with viruses in all cases.

Table 1. Antiviral Activity of QAC Against T2 Phage

Disinfectant	Solution	Initial (PFU/ml)	Survivors (PFU/ml)
QAC*	D.W.#	7.4×10^5	0
QAC	0.5% BSA+	7.4×10^5	7.0×10^2
Si-QAC ^	D.W.	2.0×10^4	0
Si-QAC	D.W.	1.5×10^3	0
Control	D.W.	7.4×10^5	7.1×10^5

* 0.5% hexadecyltrimethyl ammonium chloride.
^ 0.5% Dow Corning 5700 antimicrobial agent.
# distilled water buffered by 0.01 M Tris/HCl, pH 7.0.
+ bovine serum albumin buffered by 0.01 M Tris/HCl, pH 7.0.
No disinfectant was added to the control.

Table 2. Effects of dried alginate beads treated by various concentrations of Si-QAC

Initial (PFU/ml)	Survivors (PFU/ml)				
	10%	1%	0.1%	0.01%	0%
4.7×10^2	0	0	0	0	0
4.0×10^4	0	0	2.0×10	2.0×10	1.4×10^4
2.0×10^8	1.5×10^7	1.3×10^7	1.6×10^7	1.3×10^7	2.8×10^7

bead preparation: 2g of dried alginate beads in 20 ml of Si-QAC solution.
inactivation reaction: 2g of treated beads in 10 ml of 0.01 M Tris/HCl buffer solution, pH 7.0.

The titer reduction and adsorption capacities of the T2 phage and HSV-1 are compared in Table 3. For similar initial titers (10^6 PFU/ml), the survivor titer of HSV-1 was at least 2 orders of magnitude lower than that of T2. For similar equilibrium titer remaining in the solution (10^2 PFU/ml), the adsorption capacity (PFU/ml) of HSV-1 was 2 orders of magnitude higher than that of T2. Evidently, HSV-1 is much more susceptible to the surface-bonded QAC than T2. Since HSV-1 is an enveloped virus, the lipid bilayer surrounding the capsid binds strongly to the QAC-treated surface due to additional hydrophobic interaction. It should be noted that the adsorption experiments of T2 were carried out in buffer solutions without proteins, while those of HSV-1 were in buffered 1 vol% FBS solution.

Viruses can be considered as biocolloids with surface charges that result from ionization of carboxyl and amino groups of proteins localized on the surface. At a characteristic pH, defined as isoelectric point (pI), the virions exist in a state of zero net charge. Isoelectric point of a virus may vary by the type and the strain of the virus (25). The phage T2 (pI=4.2) possesses a net negative surface charge in solutions of pH 7.0. On the other hand, the QAC treated bead renders a positively-charged surface. This suggests that electrostatic force may play an important role in the adsorption process. However, the electrostatic force may not be the sole mechanism. Besides Brownian motion, the electrical double-layer (26), which is influenced by ionic strength and pH of the medium, may also facilitate the virus adsorption to the solid surface. Reduction of this double layer allows the van der Waals and hydrophobic to effect the adsorption of viruses to the immobilized QAC surface. Quantification of these effects is generally difficult in these complex protein solutions.

Elution Experiments. In addition to reversible adsorption, inactivation or degradation of viruses by various types of surfaces such as metal oxides (27), aluminum metal (28), magnetite (29), clays (30) and soils (31) have been reported. The mechanisms were identified to be either degradation of the capsid and/or the nucleic acid. However, such inactivation may be only specific to certain types of viruses.

Bacteriophage T4 attached on activated carbon can be reversibly eluted by 1% tryptone solution (32). In this case, the majority of the adsorbed viruses could not be recovered by the tryptone elution (Table 4). The results suggest that the viruses were eluted off of the surfaces but in an inactivated form (33).

Adsorption Isotherms. Removal of T2 onto QAC-treated surfaces with and without the presence of BSA can be correlated using the Freundlich isotherms:

$$q = K C_e^n \qquad (1)$$

Table 3. Comparison of Titer Reduction and Adsorption Capacity Between HSV-1 and T2 Phage Using Surface-Bonded QAC

Virus	Initial (PFU/ml)	Survivors (PFU/ml)	Titer in Solution (PFU/ml)	Viruses Adsorbed (PFU/g)
T2*	2.12×10^6	1.10×10^4	1.50×10^2	3.70×10^5
HSV-1#	2.03×10^6	6.67×10^2	6.67×10^2	2.03×10^7

* distilled water buffered by 0.01 M Tris/HCl, pH 7.0.
\# 1% FBS buffered by 0.01 M HEPES, pH 7.0.

Table 4. Elution of phage T2 after the adsorption/inactivation process using 1% tryptone solution

Initial (PFU/ml)	Survivors (PFU/ml)	After elution (PFU/ml)	Inactivated T2/Total titer reduction (%)
2.1×10^6	7.0×10^3	1.1×10^4	99.5
1.8×10^5	2.0×10^2	3.9×10^3	97.8
2.0×10^4	4.2×10^1	1.9×10^2	99.0
1.9×10^3	0	0	100

inactivation reactions: 0.5 g of Si-QAC treated beads in 5 ml 0.01 M Tris/HCl buffer solution, pH 7.0.

where q is the amount of viruses removed and C_e is the virus titer in equilibrium remaining in the solution, K and n are coefficients which can be determined by linear regression. Typical isotherms for removal of viruses by QAC-treated beads are shown in Figure 1. The value of n is close to one in both cases. A significant reduction in the adsorption capacity is observed in 0.5% BSA solution because the BSA molecules interfere with the adsorption of the viruses.

In Figure 2, kinetics of T2 removal using QAC treated beads is presented. It is obvious that the competitive adsorption between viruses and BSA molecules also reduced the adsorption rate. In both cases, viruses were inactivated rapidly at the initial 2 hour mark and titer reduction slowed down after that. This inconsistency with the first-order inactivation model may be due to various interfering mechanisms such as displacement, molecular orientation, multilayer effects, surface heterogeneity, and virion clumping.

Adsorption/Inactivation of T2 Phage in a β-Lactamase Solution.

The denaturation or unfolding of a protein leads to loss of its functional activity. The activation energy of the protein unfolding process can be increased in the presence of sucrose (34). The activity of β-lactamase, a model protein, dropped down to 40% of the initial value after heating the mixture at 60°C for 10 hours using sucrose (0.8 g/ml) as a stabilizing agent (Figure 3). The total amount of soluble proteins decreased because of coagulation. The decline of β-lactamase activity agreed with that of the total protein. These experimental results compared favorably with various sucrose stabilization studies of thermolabile proteins using blood clotting factors (10, 11). It was assumed in those studies that the treatment can render the protein solutions free of hepatitis infection. However, Figure 3 shows the T2 titer also diminished from 10^6 to 10^3 during the first four hours without further reduction.

Inactivation of phage T2 in β-lactamase solution by QAC-treated beads is shown in Figure 4. The initial T2 titer was 3.0×10^6 PFU/ml. A quantity of 0.8 g of beads were mixed with 10 ml of β-lactamase solution. Fifteen percent of the viruses survived this treatment. The amount of total protein in the solution was 80% of the initial value after the adsorption process, while the recovery of β-lactamase activity was at least 70%. It was the purpose of this experiment to demonstrate that QAC-treated beads can effectively remove viruses from a protein solution without significantly losing the activity of the protein. Optimal adsorption condition and mode of operation ought to be determined by studying the interactive effects of pH, ionic strength, and temperature of the solution, with the specific types of virus and protein of interest.

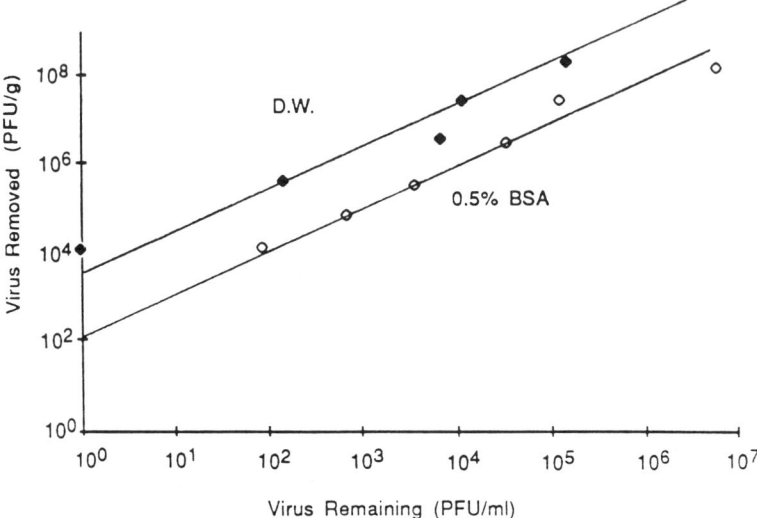

Figure 1. Equilibrium isotherms of phage T2 inactivation using surface-bonded QAC

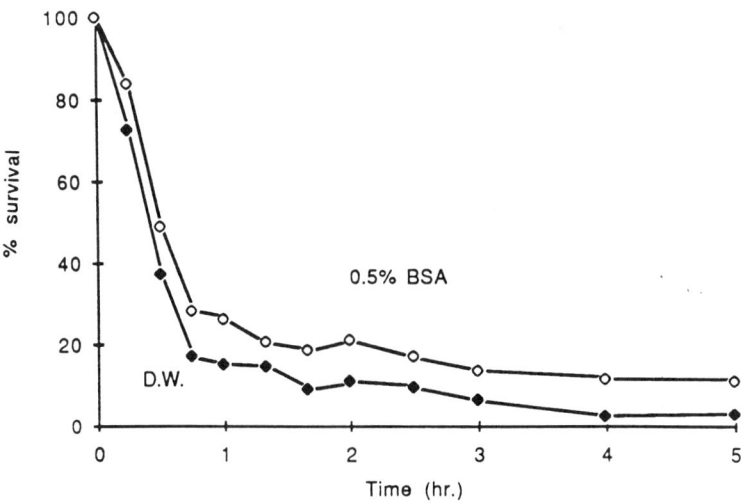

* initial T2 titer : 5.5×10^6 PFU/ml

Figure 2. Kinetic study of phage T2 removal using surface-bonded QAC

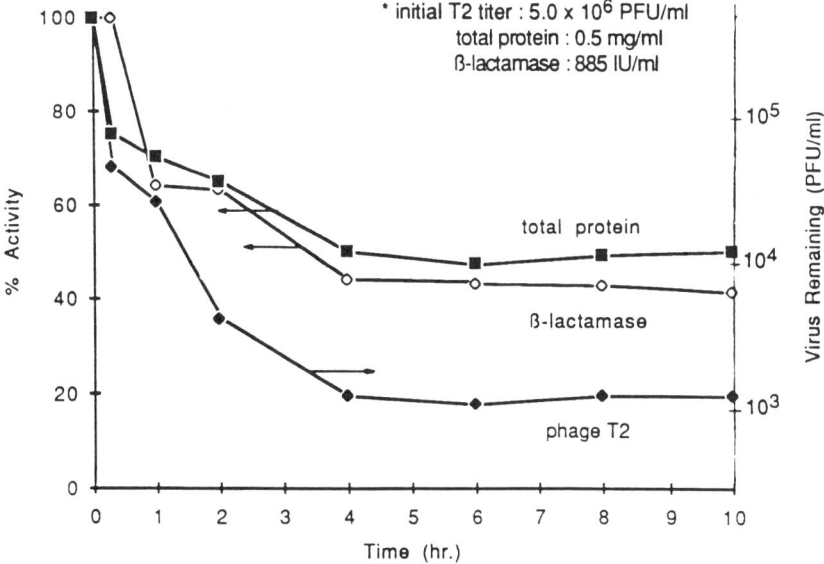

Figure 3. Thermal inactivation of phage T2 in β-lactamase solution using sucrose as stabilizer

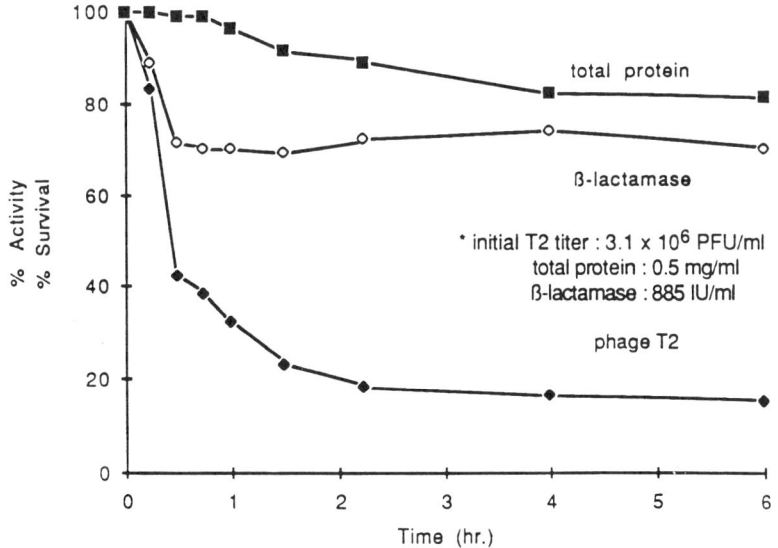

Figure 4. Removal of phage T2 from β-lactamase solution using surface-bonded QAC

Simulation Studies of Virus Removal Using Adsorption Column.

A fixed-bed adsorption has several advantages over batch and continuous stirred tank reactor (CSTR) because the rates of adsorption depend on the concentration of viruses in solution. This point is especially important for virus removal because of the low concentration of viral contaminants. The design of a fixed-bed adsorption column involves estimation of the shape of the breakthrough curve and the appearance of the breakpoint. Computer simulation studies were done here to demonstrate the performance of a virus adsorber using the surface-bonded QAC beads which have a higher binding affinity for viruses over other proteins.

The diffusion model of the system can be described mathematically by sets of material balance equations together with appropriate boundary and initial conditions:

Equation of Continuity

$$\frac{\partial C_i}{\partial t} + v \frac{\partial C_i}{\partial z} - D_L \frac{\partial^2 C_i}{\partial z^2} = \frac{1-\varepsilon}{\varepsilon} R_{Ai} \qquad (2)$$

Adsorption Rate

$$R_{Ai} = \frac{3k_{ci}}{R} (C_i - C_{ei}) \qquad (3)$$

Adsorption Equilibrium

$$q_i = K_i C_{ei}^n \qquad (4)$$

Initial Conditions

$$C_i(z,0) = 0 \qquad (5)$$

$$q_i(z,0) = 0 \qquad (6)$$

Boundary Conditions

$$C_i(0,t) = C_{i0} \qquad (7)$$

$$\frac{\partial C_i(L,t)}{\partial z} = 0 \qquad (8)$$

where C_i, C_{ei}, and C_{i0} are the concentrations of ith adsorbate in the bulk solution, at the interface, and of the influent, v is the linear velocity, L is the bed length. Linear adsorption isotherms (n=1) are assumed for both virus and total protein. The equilibrium constants K were obtained from batch experiments. It was also assumed that the complex proteins can be considered as a single component, no radial concentration gradient, and diffusion coefficients, fluid viscosity and density remained constant.

The quantity D_L is the longitudinal dispersion coefficient of viruses and can be determined by the empirical correlation given by Chung and Wen (35) as a function of Reynolds number (Re), density and viscosity of the fluid.

$$\frac{\rho D_L}{\mu} = \frac{Re}{0.2 + 0.11\, Re^{0.48}} \qquad (9)$$

The mass transfer coefficient k_c is estimated by the correlation of dimensionless j, or Colburn factor, with Sherwood number (Sh), Schmidt number (Sc), and void fraction as described by Cookson (36).

$$j = \frac{Sh}{Re\, Sc^{1/3}} = \frac{k_c}{v}\left(\frac{v}{D}\right)^{2/3} \qquad (10)$$

and

$$j = Be\, Re^{-2/3} \qquad (11)$$

These equations were solved numerically using finite difference method with double precision.[1]

Figures 5(a) and 5(b) show the simulated breakthrough curves of both total protein and HSV-1 respectively. It should be noticed that the dimensionless time scales in these two figures differ by four orders of magnitude. The breakpoint of HSV-1 is the operating endpoint at which the effluent from the adsorption column can no longer meet the desired sterilization criterion. Since the HSV-1 has a much higher affinity to the bead surface, the breakpoint of HSV-1 appears much later than that of the total protein. To optimize the protein recovery, one should improve the design of the bead surface (better selectivity, higher loading capacity), size, and operating parameters of the filter to further delay the breakpoint of the virus elution. A stochastic approach to model the removal process may be more appropriate in low concentrations of viruses.

The effects of desired sterilization criterion on total protein recovery and the amount of adsorbent required are demonstrated in Figure 6. Stringent sterilization criterion (10^{-3}) can only be achieved with reduced protein recovery based on our current design of the beads.

CONCLUSIONS

The surface-bonded QAC can effectively adsorb and inactivate viruses based on our initial experimental results. HSV-1, an enveloped virus, is

---

[1] The physical parameters used for simulation are listed in Table 5.

Table 5. Physical Parameters Used for Simulation Studies

Adsorbates parameters:	HSV-1	FBS
Equilibrium constant K(ml solution/ml adsorbent)	5.0×10^4	5.25
Diffusion coefficient $D(cm^2/sec.)$	8.0×10^{-8}	6.0×10^{-7}
Influent titer (PFU/ml) or concentration (mg/ml)	1.0×10^5	500

Column parameters:

column diameter (cm)	: 3
column porosity (-)	: 0.5
bead diameter (cm)	: 0.02
bead density (g/ml)	: 2.11

Fluid parameters:

viscosity (centipoise)	: 1.20
density (g/ml)	: 1.01
flow rate (ml/min)	: 1.0

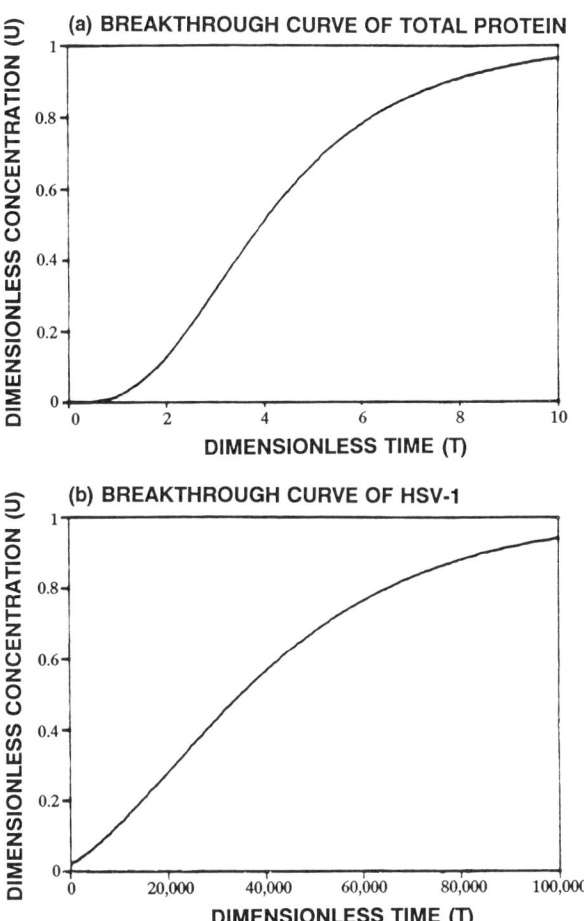

Figure 5. Simulated adsorption breakthrough curves of total protein and HSV-1

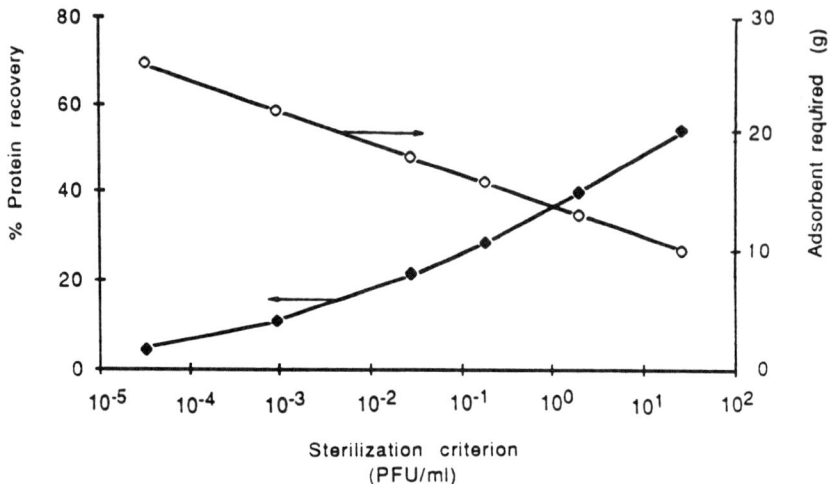

Figure 6. The effect of sterilization criteria on the protein recovery and the required amount of adsorbent

more susceptible than the non-enveloped bacteriophage T2 to the QAC treatment. However, as a non-specific adsorption process, both the rate and capacity were reduced due to the competitive binding of the protein molecules. Thermo-inactivation and surface-bonded QAC treatment were compared in terms of titer reduction and remaining functional activity of a model protein, β–lactamase. Process modeling and computer simulation enable us to predict the breakthrough curves of a virus adsorption column. Choosing a specific sterilization criterion has to be compromised with reduced protein recovery if adsorption has to be used for the removal of viruses in protein solutions.

ACKNOWLEDGMENT

The partial financial support of the National Science Foundation is acknowledged.

Notation

Be	=	constant depending on void fraction.
C	=	fluid phase-concentration, virions/cm^3 or µg/cm^3.
C_0	=	fluid phase inlet concentration, virions/cm^3 or µg/cm^3.
D	=	diffusion coefficient, cm^2/sec.
D_L	=	longitudinal dispersion coefficient, cm^2/sec.
D_p	=	mean diameter of adsorbent, cm.
j	=	dimensionless Colburn factor.

K	=	volume equilibrium constant, cm^3/cm^3.
k_c	=	mass transfer coefficient, cm/s.
L	=	length of column, cm.
n	=	parameter in Freundlich isotherm.
pe	=	vL/D_L, Peclet number.
q	=	solid phage concentration, virions/cm^3 or μg/cm^3.
R	=	mean radius of adsorbent, cm.
Re	=	vD_p/υ, Reynolds number.
R_A	=	adsorption rate, virions/cm^3/sec or μg/cm^3/sec.
Sc	=	υ/D, Schmidt number.
Sh	=	k_cD_p/D, Sherwood number.
T	=	vt/L, dimensionless time.
t	=	time, sec.
U	=	C/C_0, dimensionless fluid-phase concentration.
v	=	average linear velocity, cm/sec.

Greek Letters

ε	=	void fraction, cm^3/cm^3
ρ	=	fluid bulk density, g/cm^3
μ	=	absolute viscosity, g/cm/sec
υ	=	kinematic viscosity, cm^2/sec

LITERATURE CITED

1. Petrocci, A. N. Surface-active agents: quaternary ammonium compounds, In S.S. Block (ed.), Disinfection, Sterilization and Preservation, 3rd ed. Lea and Febiger, Philadelphia, PA. 1983. p. 309-334.
2. Hugo, W. B. "The mode of action of antimicrobial agents" J. Appl. Bacteriol. 1967. 30: 27-60.
3. Klein, M. and Deforest, A. Principles of viral inactivation, In S.S. Block (ed.), Disinfection, Sterilization and Preservation, 3rd ed. Lea and Febiger, Philadelphia, PA. 1983. p. 422-434.
4. Walters, P. A., Abbott, E. A., and Isquith, A. J. "Algicidal activity of a surface-bonded organosilicon quaternary ammonium chloride" Appl. Microbiol. 1972. 25: 253-256.
5. Isquith, A. J., Abbott, E. A. and Waters, P. A. "Surface-bonded antimicrobial activity of an organosilicon quaternary ammonium chloride" Appl. Environ. Microbiol. 1972. 24: 859-863.
6. Isquith, A. J. and McCollum, C. J. "Surface kinetic test method for determining rate of kill by an antimicrobial solid" Appl. Environ. Microbiol. 1978. 36: 700-704.
7. Speier, J. L. and Malek, J. R. "Destruction of microorganisms by contact with solid surfaces" J. Colloid Interface Sci. 1981. 89: 68-76.
8. Nakagawa, Y., Hayashi, H., Tawaratani, T., Kourai, H., Horie, T. and Shibasaki, I. "Disinfection of water with quaternary ammonium salts insolubilized on a porous glass surface" Appl. Environ. Microbiol. 1983. 47: 513-518.
9. Gellis, S. S.; Neefe, J. R.; Stokes, J.; Stong, L. E.; Janeway, C. A. and Scatchard, G. "Chemical, clinical, and immunological studies on the virus of homologous serum hepatitis in solutions of normal human serum albumin by means of heat" J. Clin. Invest. 1948. 27: 239-244.

10. Schwinn, H., Heimburger, N., Kumpe, G. and Herchenhan, B., "Blood coagulation factors and process for their manufacture" U.S. Patent, 4,297,344. 1981.
11. Fernandes, P., Lundblad, J. L. "Pasteurized therapeutically active protein composition" U.S. patent, 4,440,679. 1984.
12. Ng, P. K. and Dobkin, M. B., "Pasteurization of antihemophilic factor and model virus inactivation studies" Thromb. Res. 1985. 39: 439-447.
13. Rubinstein, A., "Heat treatment of lyophilized blood clotting factor VIII concentrate" U.S. patent, 4,456,590. 1984.
14. Daniels, S. L. "Mechanisms involved in sorption of microorganisms to surfaces" In G. Bitton and N.C. Marshall (ed.), Adsorption of Microorganisms to Surfaces. John Wiley and Sons, Inc., New York. 1980. p. 7-58.
15. Gerba, C. P. "Applied and theoretical aspects of virus adsorption to surfaces" Adv. Appl. Microbiol. 1984. 30: 133-168.
16. Lipson, S. M. and Stotzky, G. "Effect of proteins on reovirus adsorption to clay minerals" Appl. Environ. Microbiol. 1984. 48: 525-530.
17. Smith, J. T. "R-factor gene expression in gram-negative bacteria" J. Gen. Microbiol. 1969. 55: 109-120.
18. Burns, M., Kvesitadze, G. I., and Graves, D., "Dried alginate/magnetite spheres: a new support for chromatographic separations and enzyme immobilization" Biotech. Bioeng. 1985. 27: 137-145.
19. Paul, F. and Vignais, P. M. "Photophosphorylation in bacterial chromatophores entrapped in alginate gel: improvement of the physical and biochemical properties of gel beads with barium chloride as gel-inducing agent" Enzyme Microb. Technol. 1980. 2: 281-287.
20. Birnbaum, S., Pendleton, R., Larsson, P-O., Mosbach, K., "Covalent stabilization of alginate gel for the entrapment of living whole cells" Biotech. Lett. 1982. 3: 393-400.
21. Rovozzo, G. C. and Burke, C. N. "A Manual of Basic Virological Techniques" Prentice-Hall Inc., Englewood Cliffs, New Jersey. 1973.
22. Bradford, M. "A rapid and sensitive method for the quantitation of microgram quantities of protein utilizing the principle of protein-dye binding" Anal. Biochem. 1985. 72: 248-254.
23. Sykes, R. B. and Nordstrom, K. "Microiodometric determination of β-lactamase activity" Antimicrob. Agents Chemotheraphy. 1972. 1: 94-99.
24. Malek, J. R. and Speier, J. L. "Development of an organosilicon antimicrobial agent for the treatment of surfaces" J. Coated Fabrics. 1982. 12: 38-45.
25. Zerda, K. S. Ph.D. Dissertation. Baylor College of Medicine, Houston, Texas. 1982.
26. Verwey, E. J. W. and Overbeck, J. G. "Theory of the stability of lyophobic colloids" Elsevier, Amsterdam. 1984.
27. Murray, J. P. and Laband, S. J. "Degradation of poliovirus by adsorption on inorganic surfaces" Appl. Environ. Microbiol. 1979. 37: 480-486.
28. Murray, J. P. "Physical chemistry of virus adsorption and degradation on inorganic surfaces" U.S. Environmental Protection Agency, Cincinnati, Ohio. 1980.
29. Atherton, J. G., Bell, S. S., "Adsorption of viruses on magnetic particles" Water Res. 1983. 17: 943-953.
30. Tayler, D. H., Bellamy, A. R. and Wilson, A. T. "Inactivation of bacteriophage R-17 and reovirus type III with the clay mineral allophane" Water Res. 1980. 14: 339-346.
31. Yeager, J. G. and O'Brien, R. T. "Structural changes associated with poliovirus inactivation" Appl. Environ. Microbiol. 1979. 38: 702-709.
32. Cookson, J. T., North, W. J. "Adsorption of viruses on activated carbon" Environ. Sci. Technol. 1967. 1: 46-52.
33. Tsao, I-F., Wang, H. Y., Shipman, Jr., C. "Interaction of infectious viral particles with a quaternary ammonium chloride (QAC) surface" Biotech. Bioeng. 1989. 34, 5: 639-646.

34. Lee, J. C. and Timasheff, S. N. "The stabilization of proteins by sucrose" J. Biol. Chem. 1981. 256: 7193-7201.
35. Chung, S. F. and Wen, C. Y. "Longitudinal dispersion of liquid flowing through fixed and fluidized bed" AIChE J. 1968. 14: 857-866.
36. Cookson, J. T. "Removal of submicron particles in packed beds" Environ. Sci. Technol. 1970. 4: 128-134.

RECEIVED October 27, 1989

Chapter 14

Mathematical Model of a Rotating Annular Continuous Size Exclusion Chromatograph

Sandeep K. Dalvie, Ketan S. Gajiwala, and Ruth E. Baltus

Department of Chemical Engineering, Clarkson University, Potsdam, NY 13676

A mathematical model of a rotating annular continuous size exclusion chromatograph has been developed. In this process, the conventional packed chromatography column is replaced by rotating concentric cylinders with the packing in the annular region. The displacement of components with time which occurs in conventional chromatography is transformed into a displacement with angular position. The steady state continuity equations for solute in the mobile phase and in the stationary phase results in a coupled set of partial differential equations. Mass transfer effects external to the particles as well as within the porous packing are included in the model. Axial dispersion is accounted for by using a term analogous to Fickian diffusion in the continuity equation in the mobile phase. The differential equations were solved using two different numerical methods. The concentration profiles predicted using each method were in close agreement and showed solute concentration versus angular position profiles which were close to Gaussian distributions. The moments of the concentration distribution in the angular direction were determined and expressions for the peak variance and the resolution in terms of system and solute parameters were derived. These expressions were used to evaluate the effect of various operating parameters and column dimensions on the separation efficiency of this device. The properties of a mixture containing three proteins representing the molecular weight range typical of many proteins was used in this simulation.

Size exclusion chromatography (SEC), also called gel permeation or gel filtration chromatography, is a technique used to separate a mixture of macromolecules according to size. The principle of separation involves the distribution of molecules between the solution contained within the porous packing (stationary phase) and the solution surrounding the porous particles (mobile phase). The extent of permeation into the stationary phase depends on the size of the solute molecules relative to the pores in the packing. A schematic diagram of a continuous annular chromatograph is shown in Figure 1. Both cylinders are rotated at a

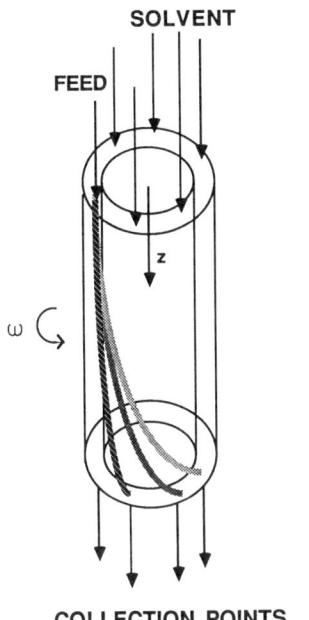

Figure 1: Schematic of a rotating annular continuous chromatograph

constant angular velocity, ω. At the top of the bed, a multicomponent feed solution is continuously introduced over an angular segment at a fixed position and the solvent is introduced at all other positions. As the sample passes through the bed, separation occurs because the extent to which the components are able to penetrate the porous packing differs depending upon their molecular size. Helical bands result, each of which contains similarly sized molecules. If each of the components in the mixture has a unique size, then each will have a unique, fixed elution position at the base of the unit. Therefore, the separated components can be easily collected.

Chromatography is a common analytical technique used in the laboratory. There has been considerable interest in recent years in developing the technology to utilize the separation principles of chromatography on a process scale. The need for low temperature, non-destructive separations which are provided by chromatography is particularly important in the biotechnology industry. At the present time, preparative scale chromatography as a unit operation is performed in large diameter batch columns (1-3). In recent years, there has been an effort to develop continuously operating bioreactors (1, 4, 5). In order to take full advantage of a continuously operating reactor, a continuously operating product recovery unit is needed.

The concept of a continuous annular chromatograph was first proposed by Martin in 1949 (6). More recently, workers at Oak Ridge National Laboratory have investigated a rotating annular ion exchange chromatograph (7 - 13). The theoretical efforts began with Scott et al. (7) who applied the plate theory approach of chromatography to the rotating annular ion exchange unit. An analytical solution to the governing differential equations was presented by Bratzler and Begovich (8). This model neglected any axial or angular dispersion effects. To improve the model, modifications were made by imposing a Gaussian distribution on the predicted elution profiles (9 - 10). Experiments were carried out by the Oak Ridge group on a rotating ion exchange chromatograph (10 - 13). Data was obtained for the separation of nickel and cobalt salts and the experimental results were in excellent agreement with the extended analytical model cited above. The size exclusion capabilities of a rotating annular chromatograph were qualitatively investigated by separating $CoCl_2$ and dextran (a polysaccharide of glucose with molecular weight 2000) using Sephadex gel as the stationary phase.

A mathematical analysis of a crossflow magnetically stabilized fluidized bed chromatograph has been presented (14). The geometry of this system is similar to the rotating annular chromatograph and therefore the modeling approach is quite similar to that reported here. A parametric sensitivity study was conducted and the results indicated that the extent of band broadening was most sensitive to two factors. These factors were the external resistance to mass transfer and the width of the feed band.

Two approaches have been used in theoretically describing conventional liquid chromatographic separations. These have been called the plate theory and the rate theory. The important parameter which results from the plate theory is the number of theoretical plates or the height equivalent to a theoretical plate. This parameter is important in characterizing band spreading and is useful when interpreting experimental data when the objective is to evaluate the effect of various parameters on peak spreading and resolution (15 - 16). However, this approach does not provide the fundamental relationships necessary to predict the capabilities of a unit with given operating conditions and geometric characteristics. The rate theory approach provides a more rigorous description of the mass transfer phenomena occurring in a chromatographic column and is therefore more useful for predictive studies (17 - 20).

In general, the governing equations for conventional size exclusion chromatography are similar to those for other forms of liquid chromatography. However, since the solute size is large, mass transfer rates between the mobile and stationary phases and within the stationary phase can be relatively slow. Therefore, the assumption of equilibrium between mobile and stationary phases, which is generally valid for other forms of liquid chromatography, is questionable when applied to size exclusion chromatography. The model we have developed incorporates finite mass transfer rates to and within the stationary phase.

MATHEMATICAL MODEL

The following assumptions were made in formulating this model: 1) there is no solute adsorption to the stationary phase, 2) the porous particles which form the stationary phase are of uniform size and contain pores of identical size, 3) there are no interactions between solute molecules, 4) the mobile phase is treated as a continuous phase, 5) the intrapore diffusivity, the dispersion coefficient and the equilibrium partition coefficient are independent of concentration. The mobile phase concentration, C_m, is defined as the mass (or moles) per interstitial volume and is a function of the axial coordinate z and the angular coordinate θ. The stationary phase concentration, C_s, is defined as the mass per pore volume and depends on z, θ and the radial coordinate, r, of a spherical coordinate system whose origin is at the center of one of the particles.

The steady state continuity equations for solute in the mobile phase and in the stationary phase are:

$$\omega \frac{\partial C_m}{\partial \theta} + \upsilon \frac{\partial C_m}{\partial z} = D_m \frac{\partial^2 C_m}{\partial z^2} + \frac{D_\theta}{r^2} \frac{\partial^2 C_m}{\partial \theta^2} + \frac{k_m S \beta (1-\varepsilon)}{\varepsilon} [C_s |_{r=R} - C_m K_{eq}] \quad (1)$$

$$\omega \frac{\partial C_s}{\partial \theta} = D_s [\frac{1}{r^2} \frac{\partial}{\partial r} r^2 \frac{\partial C_s}{\partial r}] \qquad (2)$$

where ω is the rotation rate; v is the interstitial flow velocity; D_m is the solute dispersion coefficient in the axial direction; D_θ is the solute dispersion coefficient in the angular direction; k_m is the mass transfer coefficient between the mobile and stationary phases; S is the particle area per unit column volume; β is the internal porosity; ε is the external porosity and K_{eq} is the equilibrium partition coefficient ($[\beta\, C_s / C_m]_{eq}$). The boundary conditions for this problem are:

$$z = 0 \quad \text{all } \theta \quad C_m = C_f [\, H(\theta) - H(\theta - \theta_f)\,] \qquad (3)$$

$$z = \infty \quad \text{all } \theta \quad C_m = 0 \qquad (4)$$

$$\text{all } z \quad \theta = 0 \quad C_m = C_s = 0 \qquad (5)$$

$$r = 0 \quad \partial C_s / \partial r = 0 \qquad (6)$$

$$r = R \quad \frac{-D_s}{r} (\partial C_s / \partial r) = k_m\, S\, \beta\, [C_s |_{r=R} - C_m K_{eq}] \qquad (7)$$

where C_f is the feed concentration and $H(x)$ is the Heaviside step function. The radius of the particles is R. In solving the problem numerically using the finite difference algorithm, the Danckwerts condition ($\partial C_m / \partial z = 0$) was used for the boundary condition at the bottom of the bed. If one assumes that angular dispersion occurs because of molecular diffusion in the angular direction, a comparison of axial to angular dispersion time scales can be made. This comparison indicates that the time scale for angular diffusion is several orders of magnitude larger than that for axial dispersion because axial dispersion is governed by convection. Therefore, it is reasonable to neglect the angular diffusion term in equation 1.

These equations were solved using Laplace transforms where the transformation was made with respect to the angular variable, θ:

$$\overline{C_m}(z) = \int_0^\infty e^{-p\theta}\, C_m(\theta, z)\, d\theta \qquad (8)$$

The solution in the Laplace domain is:

$$\overline{C_m/C_f} = \left[\frac{1 - \exp(-p\,\theta_f)}{p}\right] \exp\left[\frac{vL}{D_m}\left(1 - \sqrt{1 - \frac{4\phi D_m}{vL}}\right)\right] \qquad (9)$$

where

$$\phi = \frac{k_m S \beta (1-\varepsilon) L^2 K_{eq}}{\varepsilon D_m} \left[\left(\frac{1}{k_m S \beta} \sqrt{\frac{p\omega D_s}{R^2}} \coth \sqrt{\frac{p\omega R^2}{D_s}} + 1 - \frac{D_s}{k_m S \beta R^2} \right)^{-1} - 1 \right]$$
$$- \frac{p\omega L^2}{D_m} \tag{10}$$

An expression for solute concentration versus angular displacement at the column outlet requires inversion of this solution back to the θ domain, a procedure which cannot be performed analytically. A fast Fourier transform algorithm was used to perform the inversion numerically (21). Equations 1 and 2 were also solved using a finite difference algorithm.

The moments of the concentration profile in the θ domain can be determined directly from the solution in the Laplace domain (equation 9). The nth moment, m_n of the concentration distribution $C_m(\theta, L)$ (at the base of the unit) is defined by:

$$m_n = \int_0^\infty \theta^n C_m(\theta, L) \, d\theta = (-1)^n \frac{\partial^n \overline{C_m}}{\partial p} \bigg|_{p=0, \, z=L} \tag{11}$$

The zeroth, first and second moment were determined by evaluating successive derivatives of m_n with respect to the Laplace variable, p:

$$m_0 = \theta_f \tag{12}$$

$$m_1 = \frac{\theta_f^2}{2} + \frac{\omega \theta_f L}{\upsilon} \left[\frac{K_{eq}(1-\varepsilon)}{3\varepsilon} - 1 \right] \tag{13}$$

$$m_2 = \frac{\theta_f^3}{3} + \frac{\theta_f^2 \omega L}{\upsilon} \left(\frac{K_{eq}(1-\varepsilon)}{3\varepsilon} + 1 \right) + \frac{\theta_f \omega^2 L^2}{\upsilon^2} \left(\frac{K_{eq}(1-\varepsilon)}{3\varepsilon} + 1 \right)^2 \left[1 + \frac{2 D_m}{\upsilon L} \right]$$
$$+ \frac{2}{9} \frac{(1-\varepsilon)}{\varepsilon} \frac{K_{eq} \theta_f \omega^2 L}{\upsilon k_m S \beta} + \frac{2}{45} \frac{(1-\varepsilon)}{\varepsilon} \frac{K_{eq} \theta_f \omega^2 L R^2}{\upsilon D_s} \tag{14}$$

The variance of the distribution is related to these moments by:

$$\sigma^2 = \frac{m_2}{m_0} - \left(\frac{m_1}{m_0}\right)^2 \tag{15}$$

$$\sigma^2 = \frac{\theta_f^2}{12} + \frac{2\omega^2 L D_m}{\upsilon^3}\left[\frac{K_{eq}(1-\varepsilon)}{3\varepsilon}+1\right]^2 + \frac{2\varepsilon L K_{eq} \omega^2}{9(1-\varepsilon)\upsilon k_m S}$$
$$+ \frac{2\varepsilon\omega^2 L R^2 K_{eq}}{45\beta(1-\varepsilon)\upsilon D_s} \tag{16}$$

Equation 16 shows that the peak variance or band broadening is comprised of individual contributions from different aspects of the separation process. The first term in equation 16 represents the contribution of the width of the feed band to the peak variance. The second term represents the contribution to band broadening from dispersion due to eddy diffusion. The third term represents the contribution of mass transfer effects external to the particles while the fourth term represents the contribution of diffusional resistances within the stationary phase. The significance of each term relative to the total variance depends upon the operating parameters, the column and packing dimensions and the size of the solute.

If one assumes the concentration profile is represented by a normal or Gaussian distribution, then the analytical expression for $C_m(\theta, L)$ is

$$C_m(\theta, L) = C_f \frac{m_0}{\sqrt{2\pi\sigma^2}} \exp\left[-\frac{(\theta - m_1/m_0)^2}{2\sigma^2}\right] \tag{17}$$

RESULTS

The properties of three proteins, cytochrome-c, carboxypeptidase and bovine albumin were used to evaluate the model. The molecular weight and diffusion coefficient of each protein is listed in Table 1 (22). These molecular weight values represent the molecular weight range typical of many proteins. The effect of various operating and column parameters on the band broadening for each of these proteins as well as the resolution between proteins was investigated.

The transport properties, D_m and D_s, as well as the equilibrium property, K_{eq}, were assumed to be independent of solute concentration. Therefore, the elution profile of each component was calculated independently of the others and the elution profiles were superimposed when determining the peak resolution.

A correlation developed by Chung and Wen (23) for dispersion in fixed beds was used to estimate the dispersion coefficient D_m:

$$\frac{D_m \rho}{\mu} = \frac{Re}{0.20 + 0.011 \, Re^{0.48}} \qquad (18)$$

The Reynolds number is based on the particle diameter and the superficial velocity (ϵv).

A correlation by Ohashi et al. (24) was used to estimate the mass transfer coefficient, k_m:

$$Sh = \frac{2 k_m R}{D_\infty} = 2.0 + 0.51 \left[\frac{\kappa^{1/3} (2R)^{4/3}}{\mu} \right]^{0.60} Sc^{1/3} \qquad (19)$$

where the constant κ is given by

$$\kappa = \frac{1200 \, (1 - \epsilon) \, \epsilon^3 \, v^3 / Re}{2R} \qquad (20)$$

The solvent viscosity is represented by μ and the solvent kinematic viscosity is represented by v. This correlation was developed for mass transfer to solid particles. Correlations for mass transfer to porous particles are not available and it is difficult to estimate the influence of the porous structure on k_m. Therefore, equation 19 was used as a reasonable estimate.

In order to evaluate the solute diffusion coefficient in the stationary phase, D_s, and the equilibrium partition coefficient, K_{eq}, a model for the pore is required. A simple model where the pore is considered to be an infinitely long cylinder and the solute is a rigid sphere has been shown to be adequate in describing the elution process (25). The intrapore diffusivity, D_s, was estimated from the hydrodynamic theory of hindered diffusion for spherical solutes in cylindrical pores (19):

$$\frac{D_s}{D_\infty} = 1 - 2.104 \left(\frac{a}{r_o} \right) + 2.089 \left(\frac{a}{r_o} \right)^3 - 0.948 \left(\frac{a}{r_o} \right)^5 \qquad (21)$$

where D_∞ is the solute bulk phase diffusivity and 'a' is the solute radius which can be determined from D_∞ using the Stokes-Einstein equation. For spherical solutes limited to steric interactions within the pore:

$$K_{eq} = \beta \left(1 - \frac{a}{r_o}\right)^2 \qquad (22)$$

where the internal porosity, β, is included because C_s is based on pore volume while C_m is based on mobile phase volume. Although equations 21 and 22 can yield values for D_s and K_{eq} when $a > r_o$, these expressions are only valid when $a < r_o$.

A comparison of the predicted concentration profiles (C_m versus θ) at the base of the bed obtained using the finite difference algorithm, the numerical inversion of the analytical Laplace solution and the results predicted by assuming a normal distribution (equation 17) is shown in Figure 2. The properties of bovine albumin were used in these simulations. This comparison reveals a very close agreement between the solution obtained using the fast Fourier transform inversion and a normal distribution. The small difference between the finite difference results and the other two solutions is likely attributable to the fact that a different axial boundary condition was used in obtaining the numerical solution. In our model we have neglected any solute adsorption on the stationary phase and we have neglected any concentration dependence for D_m, D_s and K_{eq}. Therefore, peak tailing is not expected to be significant and the close agreement between the numerical results and the normal distribution shown in Figure 2 is not surprising.

The extent of separation between two proteins, A and B, is characterized by the resolution, R_s. For a normal distribution, resolution can be defined by:

$$R_s = \frac{(m_1/m_0)_B - (m_1/m_0)_A}{2\left(\sqrt{\sigma_A^2} + \sqrt{\sigma_B^2}\right)} \qquad (23)$$

The agreement between the concentration profiles predicted using the more exact numerical schemes and that obtained by assuming a normal distribution indicates that this definition of resolution is consistent with the other assumptions made in this model.

In defining the resolution using equation 23, one must note that in an annular chromatograph, the effluent streams are constrained to elute in 2π radians. In our model we consider the range of θ to be $(0, \infty)$ and therefore it was necessary to modify this definition of R_s for instances where the front of one peak containing smaller solutes overlaps the opposite front of another peak containing larger solutes. In each case, the resolution between peaks was calculated for the separation between the closest fronts.

We have used this mathematical model to investigate the effect of various operating parameters and column and packing dimensions on peak variance and resolution. We begin with a set of parameters which provides an excellent separation between the three proteins investigated and these values are listed in Table 2. This set of parameters is termed the

Table 1: Properties of Proteins Used in the Model Evaluation

Protein	Molecular Weight	D_∞ (cm^2/sec)
cytochrome c equine	12,000	13.0×10^{-7}
carboxypeptidase bovine	35,000	9.2×10^{-7}
albumin bovine	67,000	5.9×10^{-7}

SOURCE: Data are from ref. 22.

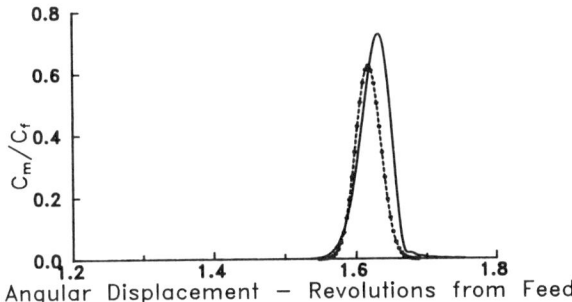

Figure 2: Comparison of results for bovine albumin obtained using different methods to solve equations 1 and 2.

Parameters used for the simulation were: rotation rate, 1 rph; feed angle, 10 deg.; interstitial velocity, 2.0×10^{-4} m/sec; column length, 1 m; particle size, 10 μm; pore radius, 100 Å.

—————— Finite difference
● ● ● ● ● ● Numerical inversion of equation 9 with fast Fourier transform
---------- Normal distribution (equation 17)

'base case'. The predicted elution profiles for the three proteins under these operating conditions is shown in Figure 3. The resolution between these proteins is listed in Table 3. These results show that albumin is easily separated from cytochrome-c and carboxypeptidase but the cytochrome-c is just resolved from the carboxypeptidase. Operating conditions and column dimensions could be different if the only product of interest was albumin.

The value of each contribution to the peak variance (the individual terms in equation 16) was determined for each protein under the base case conditions and these values are presented in Table 3. These results show that the contribution of the width of the feed band to peak variance is independent of solute molecular size, as expected. The influence of dispersion decreases with molecular size. This is because the solute residence time in the column decreases with molecular size. As expected, the contributions of external mass transfer and internal diffusion increase with molecular size because transport rates from the mobile phase to the stationary phase and within the stationary phase decrease with increasing solute size. Under the conditions of this simulation, the total variance decreases with molecular size because the dispersion term is dominant over the other terms.

An investigation of the influence of each parameter listed in Table 2 was performed by changing each value while holding the other parameters at their base case values. Each parameter was changed by a factor of 4 with the exception of the flow velocity and the column length which were each changed by a factor of 2. The rotation rate and column length were decreased for this analysis while the other parameters were increased in value. All changes investigated resulted in a decrease in the separation efficiency over that achieved with the base case values. The changed value for each parameter is listed in the second column in Table 2.

The objective was to examine the sensitivity of the separation efficiency of this unit to each of the parameters in Table 2. In order to optimize a system used for preparative chromatography, one must balance the desired separation capability with the desired product throughput. By investigating the influence of various operating parameters and column dimensions on solute resolution, a strategy for increasing throughput without an unacceptable decrease in separation efficiency will be provided.

The observed effect of changing each parameter is dependent upon the values for the other parameters. For example, if the width of the feed band (θ_f) is large, then the contribution of the feed band to the variance and the angular displacement (m_1/m_0) will dominate over the other effects. In choosing the base case values and the subsequent changed values, we have attempted to maintain some balance between the various contributors to the peak resolution. The result of each change on band broadening is summarized in Table 4. The result of each change on peak resolution is summarized in Table 5. The results are reported as the ratio

Table 2: Operating Parameters and Column and Packing Dimensions

Parameter	Value for Base Case	Changed Value
Rotation rate, rph	1.0	0.25
Feed angle, degrees	5.0	20.0
Interstitial velocity, m/sec	2.2×10^{-4}	4.4×10^{-4}
Column length, m	1.0	0.5
Particle Size, μm	10	40
Pore Radius, Å	75	300

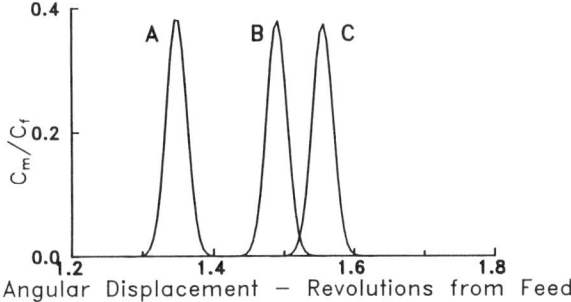

Figure 3: Elution Profile for 'Base Case' Parameters assuming a normal distribution (equation 17).
  Peak A - albumin
  Peak B - carboxypeptidase
  Peak C - cytochrome c

Table 3: Contributions to the Peak Variance for each Protein using Base Case Parameter Values

	Cytochrome c	Carboxypeptidase	Albumin
Feed Band Width	0.000635	0.000635	0.000635
Dispersion	0.00686	0.00630	0.00546
External Mass Transfer	0.00129	0.00142	0.00122
Internal Diffusion	0.000593	0.000903	0.00166
Total Variance	0.00948	0.00926	0.00898

Resolution between Cytochrome c and Carboxypeptidase = 1.062
Resolution between Carboxypeptidase and Albumin = 1.675
Resolution between Cytochrome c and Albumin = 2.739

Table 4: Effect of a Change in One Parameter on Peak Variance

The ratio listed is the variance with the parameter change relative to the variance with the base case parameters (Table 3)

Parameter	Cytochrome c Variance	Ratio of Carboxypeptidase Variance	Albumin Variance
Feed Band Width	2.012	2.019	2.060
Flow Velocity	0.350	0.364	0.384
Rotation Rate	0.125	0.126	0.129
Column Length	0.534	0.534	0.536
Particle Size	6.209	6.803	7.636
Pore Radius	1.194	1.296	1.460

Table 5: Effect of a Change in One Parameter on Resolution

The ratio listed is the resolution with the parameter change relative to the resolution with the base case parameters (Table 3)

Parameter	Cytochrome c Carboxypeptidase Resolution	Ratio of Carboxypeptidase Albumin Resolution	Albumin Cytochrome c Resolution
Feed Band Width	0.703	0.700	0.700
Flow Velocity	0.836	0.818	0.826
Rotation Rate	0.703	0.699	0.698
Column Length	0.681	0.684	0.684
Particle Size	0.392	0.372	0.381
Pore Radius	0.286	0.322	0.308

of the peak variance or resolution with the changed parameter value relative to the variance or resolution determined with the base case value.

Effect of feed band width. An increase in the width of the feed band will increase the product throughput in the unit in proportion to the increase in θ_f (for constant C_f). The payoff for this increased throughput comes in a resulting decrease in resolution. The decrease in resolution arises predominantly because of an increase in the width of the solute band in the column. For the conditions of this simulation, the payoff for a factor of 4 increase in throughput is a decrease in peak resolution by a factor of only about 0.7.

Effect of flow velocity. An increase in flow rate through the column also increases the product throughput with a resulting payoff in decreased separation efficiency. A change in the flow velocity influences all contributions to the variance except the contribution of the feed band because the time scale for flow relative to time scales for dispersion, external mass transfer and internal diffusion is altered. Because the time scales for external mass transfer and internal diffusion are dependent upon solute size, a change in the fluid velocity influences the peak variance for each protein to a different extent. Both the solute displacement and the peak variance decrease when v is increased; therefore the change in the peak resolution is less than the resulting change in either the displacement or the peak variance.

Effect of rotation rate. This is one parameter which does not have an analogous counterpart in batch column operation. Therefore, this parameter provides additional flexibility when optimizing a separation performed in an annular chromatograph. The rotation rate does not influence the solute residence time in the unit nor does it influence the relative time scales for flow, internal diffusion or external mass transfer. However, the rotation rate does influence the angular distance that a solute traverses in a given time. The result of a slower rotation rate is to decrease both the peak variance and the displacement for each solute and therefore again the resolution between proteins changes less than either m_1/m_0 or σ^2.

A change in the rotation rate does not affect the product throughput for a given unit. However, because peak resolution is influenced by ω, it is possible to enhance the separation efficiency by increasing the rotation rate.

Effect of column length. The length of the column influences the solute residence time in the column without affecting the product throughput (because this is a continuous process). However, the size of the unit and the amount of packing needed to fill it will influence the initial cost of the chromatograph. A decrease in column length decreases each contribution to the variance with the exception of the contribution of the feed band. The solute angular displacement also decreases resulting in a significant decrease in the peak resolution. A comparison of the effect of the column length and flow velocity shows that a factor of 2 increase in v and a factor of 2 decrease in L have the same effect on the peak displacement. However, the peak variance decreases more by changing velocity than by changing column length; therefore, the change in resolution is greater when L is changed than when v is changed by the same factor.

Effect of particle size. The size of the packing influences the Reynolds number and therefore influences the contributions of dispersion and external mass transfer. The contribution from internal diffusion to the total variance is also affected by the particle size. An increase in particle size from 10 μm to 40 μm increases the contribution from external mass transfer more than the other factors investigated here and this influence is dependent upon solute size. Solute displacement is not affected by particle size and therefore the increased band broadening results in a significant decrease in peak resolution.

The product throughput in the unit is not influenced by the size of the packing. However, the particle size affects the pressure drop in the column and therefore influences the operating costs. Also, because peak resolution is influenced by particle size, it is possible to enhance the separation efficiency by decreasing the particle size.

Effect of pore size. The choice of pore size in the packing influences the solute intrapore diffusion coefficient, D_s. The equilibrium partition coefficient, K_{eq}, which signifies the magnitude of the driving force for transport between the mobile and stationary phases is also dependent on the pore size in the packing. Therefore the peak displacement as well as the peak variance are influenced by pore size and the influence increases with increasing solute size.

The pore size in the packing does not influence the product throughput nor will it influence substantially the initial cost of a unit of given size. However, because pore size does have a strong influence on separation efficiency, a smaller pore size can be used to obtain good resolution with a smaller unit or with larger throughput.

CONCLUSIONS

A mathematical analysis of a rotating annular continuous size exclusion chromatograph has been performed. The mathematical model incorporates finite mass transfer rates between the mobile and stationary phases and within the pores of the stationary phase. The model predicts that this continuous chromatograph can provide an excellent separation between three proteins with molecular weights ranging from 12,000 to 67,000 under reasonable operating conditions. The results presented in Tables 4 and 5 provide a basis for determining the optimal parameters which provide sufficient separation capabilities with maximum product throughput. Depending on the nature of the mixture to be separated, several units in series, each with different dimensions and operating conditions, may provide the optimal separation for a given objective.

An investigation of the effects of various parameters on the separation capabilities of this chromatograph indicates the following. The peak resolution was found to be most sensitive to changes in the particle size of the packing and the pore radius in the packing. Changes in the two parameters which control product throughput, the width of the feed band and the flow velocity, were found to cause a smaller change in peak resolution when compared to the other parameters investigated. This is an encouraging observation for preparative chromatography.

These conclusions differ somewhat from those of Pirkle and Siegell in their analysis of adsorption chromatography in a crossflow magnetically fluidized bed (14). They found the dominant effects to be the width of the feed band and the external mass transfer resistance. It is not surprising that the effect of internal diffusion would be more important in size exclusion chromatography with macromolecular solutes.

LITERATURE CITED

1. Bailey, J.E. and Ollis, D.F., Biochemical Engineering Fundamentals, Second edition, McGraw-Hill, New York (1986).
2. Mascone, C.F., Chemical Engineering, Jan. 19, 1987.
3. Wankat, P.C., Large-Scale Adsorption and Chromatography, CRC Press, Boca-Raton, FL (1986).
4. Feder, J. and W.R. Tobert, Scien. Amer., 248, 36 (1983).
5. Ku, K., Kuo, M.J., Delente, J., Wildi, B.S. and J. Feder, Biotechnol. Bioeng., 23, 79 (1981).
6. Martin, A.J.P., Disc. Faraday Soc., 7, 332 (1949).
7. Scott, C.D., Spence, R.D. and W.G. Sisson, J. Chromatogr., 126, 655 (1976).
8. Bratzler, R.L. and J.M. Begovich, ORNL/TM - 6706 (1980).
9. Torres, R.J., Chang, C.S. and H.A. Epstein, ORNL/MIT - 329 (1981).
10. Begovich, J.M. and W.G. Sisson, A.I.Ch.E. J., 30, 705 (1984).
11. Canon, R.M. and W.G. Sisson, J. Liq. Chromatogr., 1, 427 (1978).
12. Canon, R.M., Begovich, J.M. and W.G. Sisson, Sep. Sci. Technol., 15, 655 (1980).
13. Begovich, J.M., Byers, C.H. and W.G. Sisson, Sep. Sci. Technol., 18, 1167 (1983).
14. Pirkle, J.C., Jr. and J.H. Siegell, Ind. Eng. Chem. Res., 27, 823 (1988).
15. Horvath, C. and H.-J. Lin, J. Chromatogr., 126, 401 (1976).
16. DeLigney, C.L. and W.E. Hammers, J. Chromatogr., 141, 91 (1977).
17. Kucera, E. J. Chromatogr., 19, 237 (1965).
18. Van Krevald, M.E. and N. Van den Hoed, J. Chromatogr., 149, 71 (1978).
19. Ouano, A.C. and J.A. Barker, Sep. Sci., 8, 673 (1973).
20. Lenhoff, A.M., J. Chromatogr., 384, 285 (1987).
21. Hsu, J.T. and J.S. Dranoff, Comput. Chem. Eng., 11, 101 (1987).
22. Creighton, T.E. Proteins - Structures and Molecular Properties, W.H. Freeman and Co., p. 268 (1983).
23. Chung, S.F. and C.Y. Wen, A.I.Ch.E. J., 14, 857 (1968).
24. Ohashi, H., Sugawara, T., Kikuchi, K. and H. Konno, J. Chem. Eng. Japan, 14, 433 (1981).
25. Knox, J.H. and J.P. Scott, J. Chromatogr., 316, 311 (1984).
26. Anderson, J.L. and J.A. Quinn, Biophys. J., 14, 130 (1974).

RECEIVED September 28, 1989

Chapter 15

The Continuous Rotating Annular Electrophoresis Column

A Novel Approach to Large-Scale Electrophoresis

Randall A. Yoshisato[1], Ravindra Datta, Janusz P. Gorowicz[2], Robert A. Beardsley, and Gregory R. Carmichael

Department of Chemical and Materials Engineering, University of Iowa, Iowa City, IA 52242

> The continuous rotating annular electrophoresis (CRAE) column design is capable of processing relatively large flowrates and can be scaled-up for industrial use. Previous modelling studies indicate that this column design offers considerable flexibility in meeting process objectives through the decoupling of several important design parameters. The column geometry, voltage gradient, residence time, angular velocity, and packing material/size can be adjusted in order to achieve the desired separation while controlling the peak temperature rise in the bed. However, actual specification of the operating parameters requires careful consideration of buoyancy effects, dispersion, and electrophoretic mobilities in order to achieve optimum results. A laboratory scale CRAE column has been constructed to verify these findings. This paper summarizes the work that has been done so far in developing the CRAE column.

Electrophoresis is one of the most sensitive methods for the separation and purification of charged chemicals available. Most species acquire a charge in a polar or ionic solution through ionization or ion adsorption. These species can be separated from one another based on the relative differences between their electrophoretic migration velocities. Proteins, ions, colloids, cellular materials, organelles and whole cells (1-5) have been separated by electrophoresis on an analytical scale. This ability to separate a wide range of compounds with high selectivity suggests that large-scale electrophoretic methods may be a useful adjunct to current techniques used in downstream bioprocessing (6-8).

Many novel electrophoretic devices and techniques have been proposed for continuous electrophoretic separations such as the velocity-stabilized Biostream/Harwell device (9-11), the recycle continuous-flow electrophoresis device (12-14), Bier's isoelectric focusing technique (15),

---

[1]Current address: Dow Chemical U.S.A., 2800 Mitchell Drive, P.O. Box 9002, Walnut Creek, CA 94598-0902
[2]Current address: Johnson Controls, Automation Systems Group, Saline, MI 48176

Gidding's electrical field-flow fractionation technique (16), and the rotating annular electrochromatograph (17-18). The rotating annular electrochromatography column developed by Scott (17) has bulk flow in the axial direction. This device is actually a rotating continuous chromatograph, with a radial electric field providing a radial electrophoretic separation, similar to the Biostream/Harwell device.

Despite these advances and the success of electrophoresis in analytical applications, the scale-up of electrophoresis for industrial separations has been hampered variously by low throughput, substantial Joule heating, significant dispersion phenomena, and the inability to handle multicomponent separations. The recently developed CRAE column is a design that utilizes an axial electric field in an annular column (19-21). Unlike other electrophoretic separators, the CRAE column operates with the electric field imposed in the same direction as the elutant flow. The bed is rotated slowly about its axis such that each component leaves the column at a different angular position. Products form helical bands as they traverse from the stationary feed point, down the column to stationary product collection points at the bottom of the column, as shown in Figure 1. Similar rotating annular separators have been developed previously for use in gas and liquid chromatography (21-24); however, the CRAE column is the first attempt to utilize this principle in electrophoresis. A key advantage of this configuration for electrophoresis is that it decouples the directions of separation (angular), electric field gradient (axial) and heat removal (radial), thus offering greater design flexibility.

BASIC ELECTROPHORESIS THEORY

For a charged species i carried by elutant flow and under the influence of an electric field, the net species velocity, $<v_i>$, is the sum of the convective and electrophoretic migration velocities,

$$<v_i> = <v> + w_i \quad (1)$$

The convective velocity is the bulk average velocity of the elutant given by

$$<v> = \frac{L}{\bar{t}} \quad (2)$$

where $\bar{t}$ is the mean residence time of elutant. The electrophoretic migration velocity for species i can be written as

$$w_i = - u_i \frac{dE}{dz} \quad (3)$$

where E is the electric potential and u_i is the electrophoretic mobility of species i which can be positive, negative, or zero depending upon whether

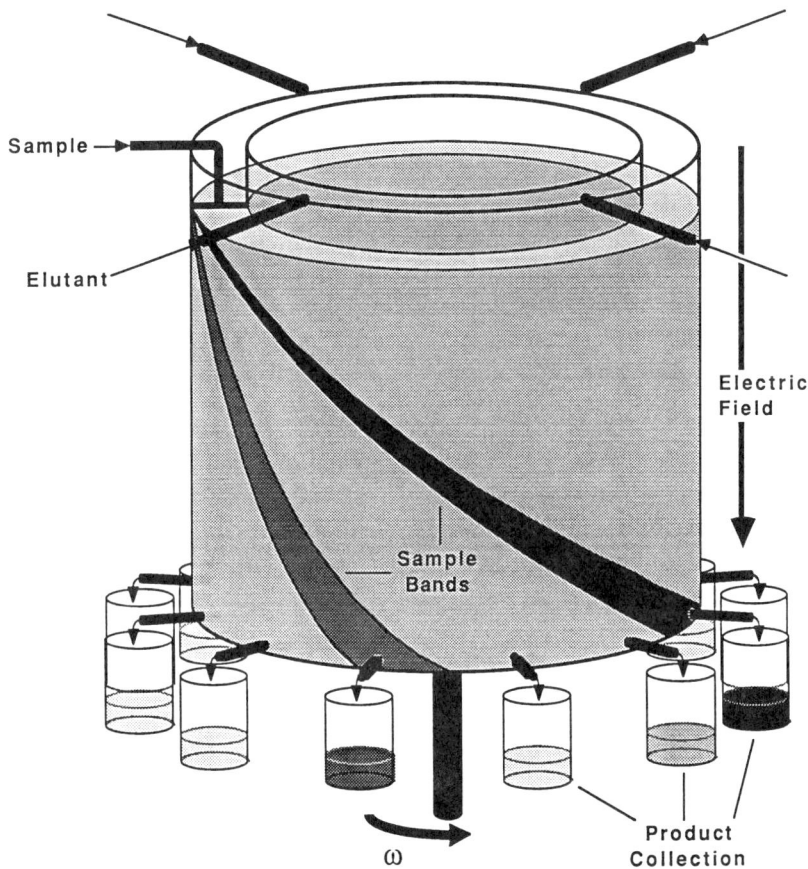

Figure 1. CRAE Helical Product Bands

the species has a net charge that is positive, negative, or zero, respectively. For a spherical particle, the electrophoretic mobility is related to its total net charge, particle radius, and fluid viscosity by

$$u_i = \frac{Q_i}{6\pi r_p \mu} \tag{4}$$

However, the electrophoretic mobility calculated using Eq. (4) is usually not accurate and, in general, the electrophoretic mobility must be measured experimentally. An alternate expression for u_i in terms of the zeta potential is given by Henry

$$u_i = \frac{\varepsilon \zeta}{6\pi \mu} f(\phi) \tag{5}$$

where ϕ is the ratio of particle radius and double layer thickness, and $f(\phi)$ varies between 1.0 for small ϕ and 1.5 for large ϕ.

The mean residence time for species i in the column is given by

$$\bar{t}_i = \frac{L}{<v_i>} = \frac{L}{<v> + w_i} \tag{6}$$

If the species residence times are sufficiently different, the various components will be well-resolved. The resolution between two exiting bands is defined by

$$R_s = \frac{\bar{t}_2 - \bar{t}_1}{4\sigma_2} \tag{7}$$

assuming that the baseline band width is given by four times the standard deviation.

MATHEMATICAL MODEL FOR THE CRAE COLUMN

A comprehensive mathematical model has been formulated for the CRAE column which considers temperature and velocity gradients, dispersion, electroosmosis and adsorption (20). For the sake of completeness, the governing equations are summarized below. Conservation of momentum is expressed by

$$\frac{1}{r}\frac{\partial}{\partial r}(r\mu \frac{\partial v_z}{\partial r}) = \frac{dP}{dz} - \bar{\rho}g + \bar{\rho}\bar{\beta}g(T - \bar{T}) \tag{8}$$

with boundary conditions

$$P = P_0 \quad \text{at } z = 0 \tag{9}$$
$$P = P_L \quad \text{at } z = L \tag{10}$$
$$v_z = v_{eo} \quad \text{at } r = r_i \tag{11}$$
$$v_z = v_{eo} \quad \text{at } r = r_o \tag{12}$$

A slip boundary condition is assumed to exist at the walls due to electroosmotic flow given by the Helmholtz-Smoluchowski equation

$$v_{eo} = -u_{eo} \frac{dE}{dz} \tag{13}$$

where the electroosmotic mobility u_{eo} is given by

$$u_{eo} = -\frac{\varepsilon \zeta}{4\pi\mu} \tag{14}$$

The conservation of energy is expressed by the heat equation

$$\varepsilon_B \langle v_i \rangle \rho C_p \frac{\partial T}{\partial z} = k_h \frac{1}{r}\frac{\partial}{\partial r}\left(r\frac{\partial T}{\partial r}\right) + k_m \left(\frac{E_o}{L}\right)^2 \tag{15}$$

with boundary conditions

$$T = T_0 \quad \text{at} \quad z = 0 \tag{16}$$
$$\frac{\partial T}{\partial z} = 0 \quad \text{at} \quad z = L \tag{17}$$
$$T = T_0 \quad \text{at} \quad r = r_i \tag{18}$$
$$T = T_0 \quad \text{at} \quad r = r_o \tag{19}$$

The conservation of species i in a packed CRAE column is given by

$$\varepsilon_B \omega \frac{\partial C_i}{\partial \theta} + (1-\varepsilon_B)\omega \frac{\partial n_i}{\partial \theta} + \varepsilon_B \langle \bar{v}_i \rangle \frac{\partial C_i}{\partial z} = \frac{K_{i\theta}^{eff}}{r^2}\frac{\partial^2 C_i}{\partial \theta^2} + K_{iz}^{eff}\frac{\partial^2 C_i}{\partial z^2} \tag{20}$$

The second term on the left accounts for possible adsorption of species i onto the surface of the packing material. For a single feed inlet, the boundary conditions are

$C_i = 0$	at	$\theta = 0$	For all z	(21)
$C_i = C_{if}$	at	$0 < \theta < \theta_f$	$z \leq 0$	(22)
$C_i = 0$	at	$\theta_f < \theta \leq 2\pi$	$z \leq 0$	(23)
$C_i = 0$	at	$z = \infty$	For all θ	(24)

The coordinate system is oriented such that the feed stream enters between $\theta = 0$ and $\theta = \theta_f$.

DESIGN OF THE CRAE COLUMN

The CRAE column constructed for laboratory testing is designed to process approximately 0.5 liters of feed per hour. As shown in Figure 2, a 4 cm by 45.75 cm annulus is formed between a 35.5 cm (14 in.) OD inner cylinder and a 43.5 cm (17.25 in.) ID outer cylinder. The annulus is cooled by inner and outer cooling jackets. The polycarbonate cylinders which form the annulus are held between nylon/teflon bearings. These bearings also seal the annulus from the inner and outer cooling jackets. Fiberglass rods connect the inner and outer cylinders. The annulus is rotated by a drive shaft attached to the inner cylinder and driven by a 1/3 hp variable speed motor through a 60,000:1 gear reducer. The speed of rotation is controlled by an electronic controller with tachometer feedback, which is capable of controlling the motor speed to within 1 rpm.

The feed stream enters the annulus through a stationary "feed comb" which distributes the feed stream across a thin radial section of the annulus. The feed comb is constructed from three teflon-sheathed stainless steel needles. Elutant enters the annulus through five ports spaced at 60° intervals (along with the feed comb) around the annulus. Product is collected continuously through 72 outlet ports spaced every 5 degrees around the bottom perimeter of the annulus. The stationary electrodes are annular stainless steel plates attached at the top and bottom of the annular space. Nylon membranes separate the electrodes from the bulk fluid and create top and bottom electrode compartments. These compartments are flushed with buffer to remove electrolysis gases as they form. This flush solution is degassed and recycled. A 0-300 VDC, 0-10 A power supply generates the electric field. An anti-convective packing material is used in the annulus. This packing may be inert or may have adsorption properties such that electrochromatography can be accomplished.

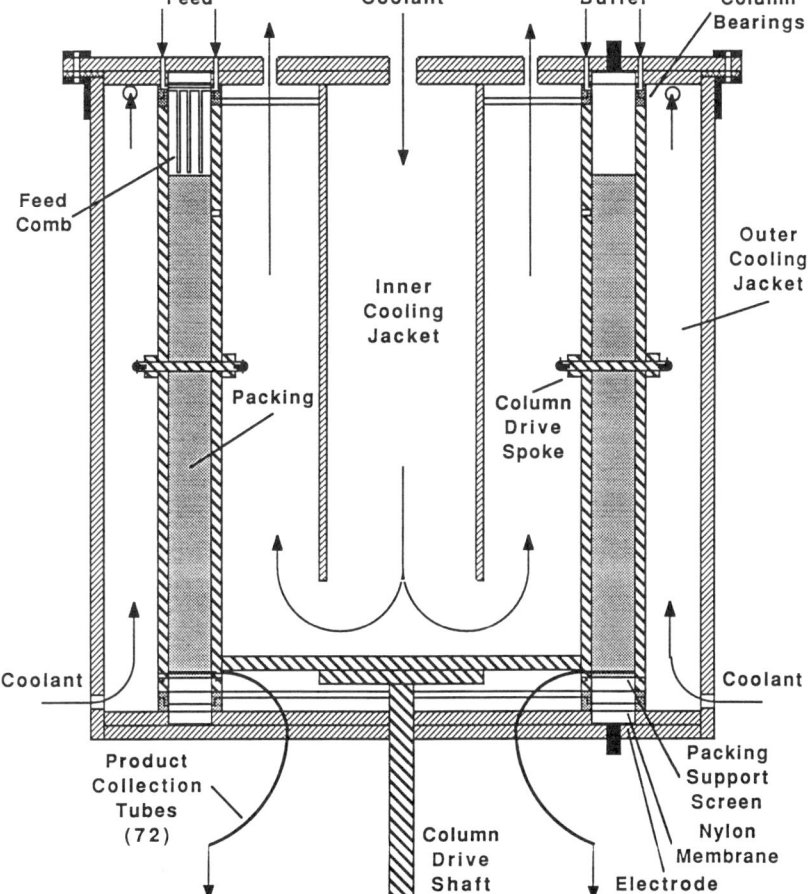

Figure 2. CRAE Column Schematic (cross-section)

REVIEW OF CURRENT STATUS

Theoretical studies of the CRAE column have shown that it has the potential for achieving large-scale, multi-component separation of charged substances. Before this potential can be fully realized, however, the mechanical design must be perfected so that the CRAE column is able to operate within the prescribed manner. Moreover, it is recognized that the flow pattern through the CRAE column for a particular species will be greatly influenced by thermal convection, electroosmosis, surface adsorption, and other dispersive phenomena. The research up to this point has centered on two principal areas: the design and construction of a prototype column, and the study of the mechanisms of dispersion in electrophoresis.

The CRAE column's design attempts to minimize the effect of Joule heating in the column while maintaining a relatively high throughput. Korndorf (25) solved Eq. (14) to obtain the temperature profile in the CRAE column as a function of Fourier number. She showed that the CRAE design offers improved performance in terms of minimizing the peak bed temperature and power consumption as compared to other thick-bed devices. She also developed equations for estimating the location and magnitude of the peak bed temperature.

Zingher (26) investigated the effect of the temperature field on the flow field in packed and unpacked CRAE columns. Velocity profiles in the CRAE column were obtained for packed and unpacked operating conditions. The packed-bed velocities were calculated using a parallel annular channel model for the bed. The analysis involved numerically solving the momentum balance equation using the temperature profile obtained as above (by assuming plug flow) and temperature dependent physical properties. These simulations confirmed that in unpacked wide-gap annular columns buoyancy effects significantly degrade column performance even at small temperature gradients. Packing the column with appropriate size beads, however, was shown to effectively control the effects of buoyancy. The calculations for various operating parameters indicated that the temperature effects on the physical properties, particularly density and viscosity, can have a significant effect on the velocity profiles.

The major factor limiting resolution in free-flow, packed-bed or gel electrophoresis is dispersion; yet, until recently, electrokinetic dispersion has not been extensively studied. There are several factors which affect the extent of dispersion in a CRAE column. These include: elutant flowrate and ionic strength; electric field strength; size, shape, zeta potential and sorption characteristics of packing used; non-uniformity of electric field; and thermally-generated flow instabilities. Such factors will assume increased importance in attempts to scale the CRAE for various industrial applications. Gorowicz (27) studied electrokinetic dispersion in the simpler geometry of a batch packed-bed electrophoresis column. The dependence of the dispersion coefficients on the electric field strength and direction, flowrate, species charge and size, and packing size was established, using water-soluble dyes as tracers.

Gorowicz (27) assumed that the concentration of charged species i being carried by the elutant through a packed-bed with an applied electrical potential is described by

$$\frac{\partial <C_i>}{\partial t} + <v_i> \frac{\partial <C_i>}{\partial z} = K_i \frac{\partial^2 <C_i>}{\partial z^2} \qquad (25)$$

where K_i is the experimentally determined bulk dispersion coefficient. With experimentally determined residence time distributions, the solution of Eq. (25) may be used to estimate the dispersion in an electrophoretic column. The bulk average concentration is given by

$$<Ci> = \frac{\frac{m_i}{\pi r^2}}{\sqrt{4\pi K_i t}} \exp\left(-\frac{Z_i^2}{4 K_i t}\right) \qquad (26)$$

where m_i is the mass of the impulse and $Z_i = z - <v_i> t$.

The bulk average concentration and the bulk average velocity can be measured as a function of time for a charged species in an electrophoretic column. These experimental electrokinetic dispersion response curves can be analyzed using the method of moments. This technique has been used for the estimation of dispersion coefficients in packed-bed columns in the absence of an electric field (28-31). Expressions for moments in the time domain are given by (32):

(nth moment) $\qquad M_n = \int_0^\infty <C_i>(t, x=L) \, t^n \, dt \qquad (27)$

(normalized nth moment) $\mu_n = \dfrac{M_n}{M_o} = \dfrac{\int_0^\infty <C_i>(t, x=L) \, t^n \, dt}{\int_0^\infty <C_i>(t, x=L) \, dt} \qquad (28)$

(nth Central moment) $\mu_n' = \dfrac{\int_0^\infty <C_i>(t, x=L)(t-\mu_1)^n \, dt}{\int_0^\infty <C_i>(t, x=L) \, t^n \, dt} \qquad (29)$

Note that M_o is the total mass of species i, μ_1 is the mean residence time of the peak ($\bar{t_i}$), and μ'_2 is the variance of the residence time distribution (σ^2). The higher order moments μ'_3 and μ'_4 are related to the skewness:

$$\gamma_1 = \frac{\mu_3}{\sigma^3}$$

and kurtosis:

$$\gamma_2 = \frac{\mu_4}{\sigma^4}$$

respectively.

For an impulse of a tracer dye, the mean and variance of the experimentally measured residence time distribution are given by the following moment equations

$$\mu_1 = \frac{L}{<v_i>} + \frac{t_o}{2} + t_d \qquad (30)$$

$$\mu_2' = \frac{2 L K_i}{<v_i>^3} + \frac{t_o}{12} \qquad (31)$$

where L is the length of the column, t_o is the duration of the input pulse and t_d is the empirical time delay introduced into Eq. (30) to account for dead spaces in equipment outside the test section of the packed-bed.

Estimates of the dispersion coefficient and time lag can be obtained by integrating the experimental data to obtain the mean and variance of the residence time distribution. These values may be substituted into the moment expressions (Eqs. 30, 31) to obtain K_i and t_d. In this study, it was convenient to integrate the experimental data using the rectangular rule. The accuracy of the moment analysis can be improved by the use of optimum truncation points (33) or by the use of weighted moments (32, 34). The disadvantage of these methods is that the expressions relating moments and model parameters become more complicated.

The experimental apparatus used in this study is shown in Figure 3. The elutant flow was caused by the difference in the fluid level between the buffer in tank 2 and the column sample collection outlet. The level in tank 2 was maintained constant during the experiments by setting the buffer flowrate from tank 1 to tank 2 equal to the column throughput. The outlet flowrate was controlled by an adjustable metering valve. The electrophoretic column was cooled by water circulating through an external cooling jacket. A regulated power supply with a range of 0-400 VDC and 0-125 mA was utilized.

The elutant was a pH 7 KH_2PO_4-NaOH buffer. Seven dyes of different charge and molecular weight were used as listed in Table 1. The dyes were in the form of the sodium salt. After dissociation in water, the dyes formed anions with a charge equivalent to the number of dissociating sodium cations. The sample at a concentration of 2.5×10^{-4} g dye/g buffer was injected through a rubber septum using a glass chromatography syringe. Blue dextran (Reactive Blue 2 - C.I. 61211) with an average molecular weight of 2.0×10^6 Daltons was used as a neutral tracer. The blue dextran was injected at a concentration of 9.0×10^{-4} g dye/g buffer.

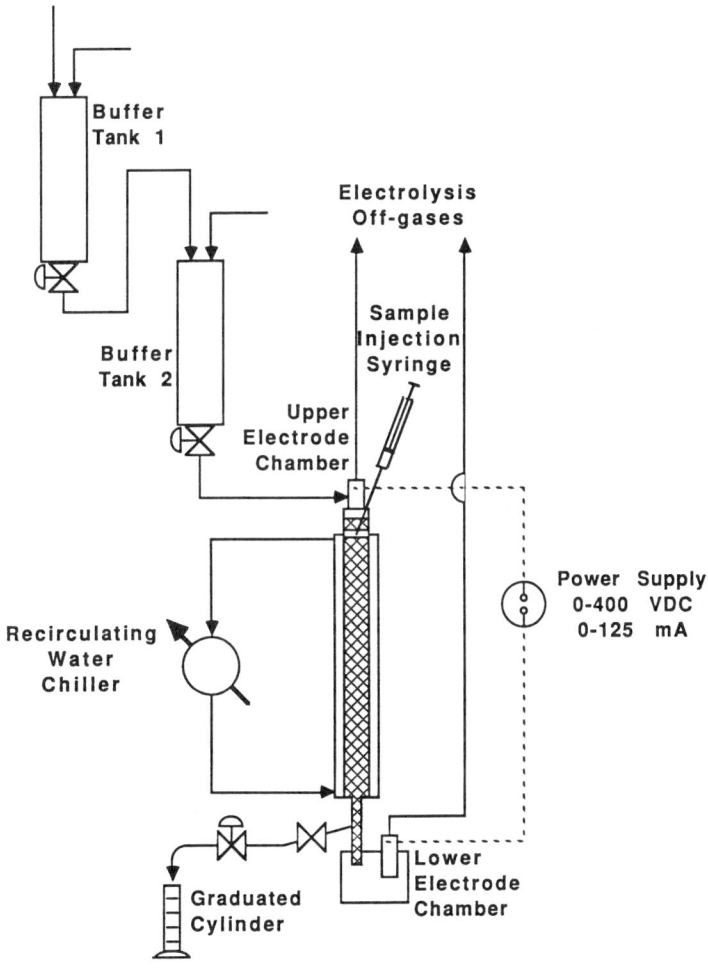

Figure 3. Packed-Bed Electrophoresis Column (PBEC) Apparatus

Table 1. Dyes Used

Dye	Charge	M.W.
Direct Yellow 50	-4	956.8
Tartrazine	-3	534.0
New Coccine	-3	604.5
Napthol Yellow S	-2	358.2
Fast Green FCF	-2	808.9
Crocein Orange G	-1	350.0
Lissamine Green B	-1	576.6

Samples of the outlet flow were collected as a function of time until all the dye passed through the column. Since the dyes were colored, the duration of the peak could be estimated visually; however, sampling was continued until analysis indicated the outlet dye concentration was negligible. The concentration of the samples was determined spectrophotometrically on a scanning UV-Vis spectrophotometer using each dye's maximum absorbance wavelength.

Experiments were conducted at flowrates of 0.5, 1.0, 1.5, and 2.0 ml/min and applied voltages of 0, 100, 200, 300, and 400 V across the column. Moment analysis of the residence time distribution of the dyes was used to estimate the electrokinetic dispersion coefficients. The effects of the direction of the electric field and the packing size on K_i were also investigated.

Figure 4 shows typical results of the normalized outlet concentration leaving the packed-bed column. In this example, the results are from a set of experiments using Direct Yellow 50 (charge = - 4) as the sample with an elutant flowrate of 1 ml/min. Curve A is the residence time distribution with no applied voltage. Curve B is the residence time distribution for an applied voltage of 400 V with the anode at the top ("downward" electric field). Curve C is the residence time distribution for an applied voltage of 400 V with the anode at the bottom ("upward" electric field). The current drawn was 28-31 mA for all runs. Each curve represents the average of at least three experimental runs. The distributions are essentially Gaussian; however, there is some tailing of the peaks at the longer times. The mean sample residence time, $\bar{t}_i$, ranges from 57.5 minutes, when the electrophoretic velocity is in the same direction as the elutant velocity, to 71 minutes, when the electrophoretic velocity is in the direction opposite of the elutant velocity. The mean sample residence time is 64 minutes in the absence of an electric field, which is the same as the elutant residence time, $\bar{t}$, which was held constant.

At first glance, Figure 4 would not indicate anything unusual, with the band width increasing with residence time. However, when K_i was calculated using Eq. (31) it was found not to be a constant, contrary to expectations. The results of K_i thus calculated for the seven dyes are summarized in Table 2. The electrophoretic mobilities for the seven dyes were also calculated from the difference $(\bar{t}_i - \bar{t})$ and varied linearly with the ratio Q_i/r_p in accord with Eq. (4), although the coefficient was not $1/6\pi\mu$. In the absence of an electric field, the dyes have a similar mean residence time of about 63.5 minutes indicating that adsorption of dyes was not significant. As might also be expected for the case when convective dispersion dominates diffusion, it was observed that the dispersion coefficient increased somewhat with increasing molecular weight.

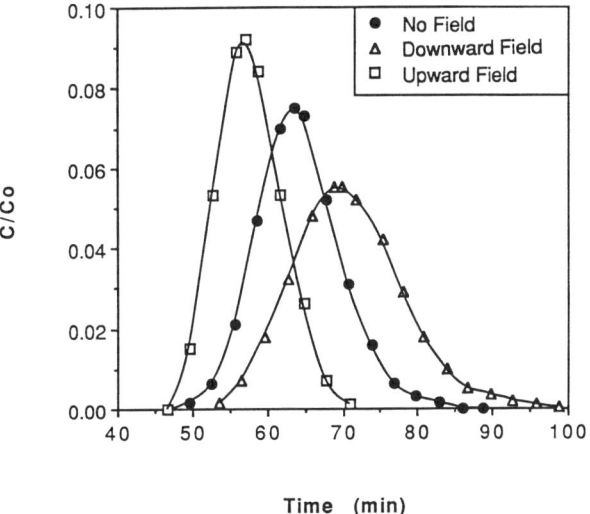

Figure 4. PBEC Residence Time Distributions for Direct Yellow 50

Table 2. Mean Residence Times and Variance

Dye	Electric Field	t (min)	σ^2 (min^2)	K_i (cm^2/min)	$\dfrac{K_i}{K_{io}}$
Crocein Orange G	none	63.49	27.7	0.077	1.00
	down	67.67	55.3	0.127	1.65
	up	60.56	20.5	0.062	0.81
Direct Yellow 50	none	64.12	32.6	0.091	1.00
	down	70.95	60.7	0.121	1.33
	up	57.52	17.6	0.066	0.73
Fast Green FCF	none	63.42	29.2	0.081	1.00
	down	68.01	58.6	0.133	1.64
	up	59.61	21.8	0.073	0.90
Lissamine Green B	none	63.17	28.4	0.079	1.00
	down	66.28	55.7	0.136	1.72
	up	60.04	23.1	0.076	0.96
Napthol Yellow S	none	63.85	26.3	0.073	1.00
	down	71.54	67.8	0.132	1.81
	up	58.99	18.2	0.064	0.87
New Coccine	none	63.29	26.1	0.073	1.00
	down	70.87	83.0	0.166	2.27
	up	57.83	19.3	0.071	0.97
Tartrazine	none	62.92	26.6	0.074	1.00
	down	71.18	58.7	0.116	1.57
	up	58.45	18.3	0.065	0.88

The last column in Table 2 clearly demonstrates that the direction of the applied electric field has a pronounced effect on the dispersion coefficient. We believe that this is a result of the interaction of the electroosmotic and hydrodynamic flow components within the interstitial channels between the packing particles, and we are investigating this further.

In addition to the effect of direction, the influence of the electric field strength on the electrokinetic dispersion coefficient in the packed-bed column was also evaluated. The ratio of the electrokinetic dispersion coefficient to the ordinary dispersion coefficient (without an external electric field) was calculated for the dye tartrazine and is shown in Figure 5 as a

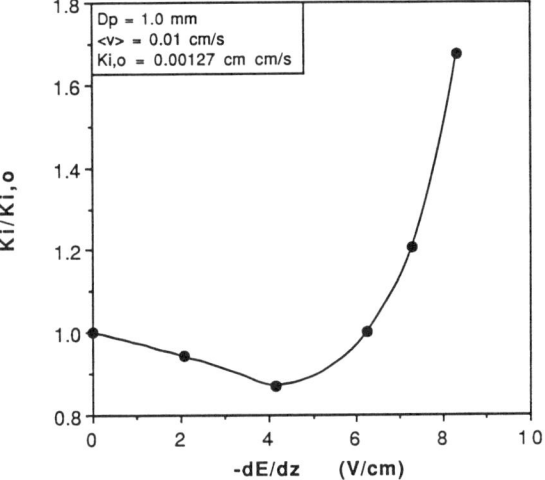

Figure 5. Ratio of the Electrophoretic Dispersion Coefficient to the Ordinary Dispersion Coefficient as a Function of Voltage Gradient for Tartrazine in the PBEC

function of the electric field strength. Under these experimental conditions, the electrokinetic dispersion coefficient decreases with increasing field strength to a value somewhat smaller than the ordinary dispersion coefficient. However, for electric field strengths greater than about 4 V/cm the electrokinetic dispersion increases rapidly and exceeds the ordinary dispersion for electric fields greater than about 6 V/cm. This result was unexpected and needs to be confirmed with further experimentation. These studies are currently under way.

It, thus, appears that under certain circumstances the dispersion might be controlled in the CRAE column and other packed electrophoresis devices. However, even if conditions of voltage gradient and direction exist that result in a substantial reduction of K_i, these may not necessarily be optimum operating conditions because of the trade-off between throughput and resolution.

SUMMARY AND FUTURE WORK

The CRAE column is a new device capable of continuous electrophoretic separation of charged species. A comprehensive model has been developed which predicts the temperature, velocity, and concentration profile in the column. The model accounts for electroosmosis at the walls of the column and adsorption onto the surface of the packing material (if any). Model calculations indicate that the CRAE column design is capable of minimizing large temperature increases in the bed, and that when properly packed dispersion due to buoyant effects can be controlled. The low power requirements and high throughput indicate that the CRAE column is competitive with other large-scale electrophoretic separators. The CRAE column prototype has been constructed and is currently undergoing testing and characterization. Preliminary experimental results should be available shortly.

Recent experiments in a small-scale packed-bed electrophoretic column suggest that dispersion in a packed CRAE column is a function of electric field strength and direction. The results indicate that under certain conditions it may be possible to obtain electrophoretic dispersion coefficients which are less than the ordinary dispersion coefficient. An explanation of this phenomena is the subject of continuing investigation.

ACKNOWLEDGMENTS

The authors wish to acknowledge James Jones and Michael Wilson for their assistance in the dispersion studies. Some of this work was funded by a grant from the State of Iowa High Technology Council.

SYMBOLS

C_i	concentration of species i
$<C_i>$	area average concentration of species i
C_{if}	feed concentration of species i
C_p	heat capacity
E	electric potential
E_o	applied electric field
g	acceleration due to gravity
K_i	axial dispersion coefficient of species i
K_{io}	axial dispersion coefficient of species i in the absence of electric field
K_{iz}^{eff}	effective axial dispersion coefficient of species i, including adsorption
$K_{i\theta}^{eff}$	effective angular dispersion coefficient of species i, including adsorption
k_h	effective thermal conductivity of bed
k_m	effective electrical conductivity of bed
L	length of electrophoresis chamber
M_n	nth moment
M_o	zeroth moment
m_i	mass of feed impulse
n_i	concentration of adsorbed species i
P	pressure
P_L	Pressure at $z = L$
P_0	Pressure at $z = 0$
Q_i	net charge on species i
r	radial coordinate
r_i	inner radius of annular space
r_o	outer radius of annular space
r_p	particle radius
T	temperature
T_0	temperature of walls, entering elutant
$\bar{T}$	reference temperature
t	time of elution
$\bar{t}$	mean residence time of elutant
t_d	lag time
$\bar{t}_i$	mean residence time of species i
t_o	feed pulse time
u_i	electrophoretic mobility of species i

u_{eo}	electroosmotic mobility of elutant
v_{eo}	electroosmotic flow velocity of elutant
v_i	velocity of species i
v_z	axial velocity
$<v_i>$	mean velocity of species i
$<v>$	mean velocity of elutant
w_i	electrophoretic migration velocity
Z	adjusted axial coordinate
z	axial coordinate

Greek

β	coefficient of thermal expansion
γ_1	skewness
γ_2	kurtosis
ε	dielectric constant of elutant
ε_B	bed porosity
ζ	zeta potential of surface
θ	angular coordinate
θ_f	angle of feed inlet
λ	double layer thickness
μ	elutant viscosity
μ_n	nth normalized moment
μ_1	first normalized moment (mean)
μ'_2	second central moment (variance)
ρ	density of elutant
$\bar{\rho}$	elutant density at the reference temperature
σ^2	variance
ω	angular velocity
ϕ	r_p/λ

LITERATURE CITED

1. Hannig, K. J. Chromatogr., 1978, 159, 183-191.
2. Hannig, K. Electrophoresis, 1982, 3, 235-243.
3. Deyl, Z.; Everaerts, F. M.; Prusik, Z.; Svendsen, P. J. Electrophoresis: A Survey of Techniques and Applications, Pt. A. Techniques, Pt. B. Applications, Elsevier: Amsterdam, 1979.
4. Righetti, P. G.; van Oss, C. J.; Vanderhoff, J. W. Electrokinetic Separation Methods, Elsevier/New Holland Biomedical Press: Amsterdam, 1979.
5. Simpson, C. F.; Whittaker, M. Electrophoretic Techniques, Academic Press: New York, 1983.

6. Huebner, V. R.; Lawson, R. H. Sep. Sci. Technol., 1968, 3, 265-277.
7. Vermeulen, T.; Nady, L.; Krochta, J. M.; Ravoo, E.; Howery, D. Ind. Eng. Process Des. Develop., 1972, 10, 91-102.
8. Arcus, A. C.; McKinnon, A. E.; Livesey, J. H.; Metcalf, W. S.; Vaughn, S.; Keey, R. B. J. Chromatogr., 1980, 202, 157-177.
9. Mattock, P.; Aitchison, G. F.; Thomson, A. R. Sep. Pur. Methods, 1980, 9, 1-68.
10. Thompson, A. R. In Electrophoretic Techniques; Simpson, C. F.; Whittaker, M., Eds.; Academic Press: New York, 1983; pp. 253-274.
11. Noble, P. T. Biotechnol. Prog., 1985, 1, 237-249.
12. Gobie, W. A.; Beckwith, J. B.; Ivory, C. F. Biotech. Prog., 1985, 1, 60-68.
13. Gobie, W. A.; Ivory, C. F. In Separation, Recovery, and Purification in Biotechnology; Asenjo, J. A.; Hong, J., Eds.; ACS Symposium Series No. 314, American Chemical Society: Washington, D. C., 1986; pp. 169-184.
14. Gobie, W. A.; Ivory, C. F. A. I. Ch. E. J., 1988, 34, 474-482.
15. Bier, M. In Separation, Recovery, and Purification in Biotechnology; Asenjo, J. A.; Hong, J., Eds.; ACS Symposium Series No. 314, American Chemical Society: Washington, D. C., 1986; pp. 185-192.
16. Giddings, J. C. Sep. Sci. Technol., 1984-85, 19, 831-847.
17. Scott, C. D. Sep. Sci. Technol., 1986, 21, 905-917.
18. Rudge, S. R.; Ladisch, M. R. Biotechnol. Prog., 1988, 4, 123-133.
19. Yoshisato, R. A.; Korndorf, L. M.; Carmichael, G. R.; Datta, R. Sep. Sci. Technol., 1986, 21, 727-753.
20. Datta, R.; Yoshisato, R. A.; Carmichael, G. R. In Recent Advances in Separation Techniques III; Li, N. N.; Hatton, T. A.; Hwang, S.-T.; LaRue, D. M.; Leeper, S. A.; Roberts, D. L., Eds.; A. I. Ch. E. Symposium Series No. 250; American Institute of Chemical Engineers: New York, 1986, pp. 179-192.
21. Carmichael, G. R.; Zingher, H. L.; Datta, R.; Yoshisato, R. A. J. Chem. Tech. Biotechnol., 1988, 41, 207-221.
22. Giddings, J. C. Anal. Chem., 1962, 34, 37-39.
23. Fox, J. B.; Calhoun, R. C.; Eglinton, W. J. J. Chromatogr., 1969, 43, 48-54.
24. Begovich, J. M.; Sisson, W. G. A. I. Ch. E. J., 1984, 30, 705-709.
25. Korndorf, L. M. M.S. Thesis, The University of Iowa, Iowa City, IA, 1984.
26. Zingher, H. L. M.S. Thesis, The University of Iowa, Iowa City, IA, 1985.
27. Gorowicz, J. P. M.S. Thesis, The University of Iowa, Iowa City, IA, 1986.
28. Horn, F. J. M., A. I. Ch. E. J., 1971, 17, 613-620.
29. Edwards, M. F.; Richardson, J. F. Chem. Eng. Sci., 1968, 23, 109-123.
30. Suzuki, M.; Smith, J. M. Chem. Eng. J., 1972, 3, 256-265.
31. Wolf, H. J.; Radeke, K. H.; Gelbin, D. Chem. Eng. Sci., 1979, 34, 101-107.
32. Furusawa, T.; Suzuki, M.; Smith, J. M. Catal. Rev. Sci. Eng., 1976, 13, 43-76.
33. Skopp, J., A. I. Ch. E. J., 1984, 20, 151-155.
34. Ramachandran, P. A.; Suzuki, M; Smith, J. M. Ind. Eng. Chem. Fund., 1978, 17, 148-160.

RECEIVED October 24, 1989

Author Index

Agrawal, Pradeep K., 108
Baltus, Ruth E., 268
Beardsley, Robert A., 285
Brogden, Kim A., 257
Cabezas, Heriberto, Jr., 38
Carbonell, Ruben G., 232
Carmichael, Gregory R., 285
Clark, Kathleen M., 188
Dall-Bauman, Liese, 206
Dalvie, Sandeep K., 268
Datta, Ravindra, 285
Dubin, Paul L., 176
Evans, Janis D., 38
Flinta, C. Daniel, 176
Forciniti, Daniel, 53
Gajiwala, Ketan S., 268
Giuliano, Kenneth A., 71
Glatz, Charles E., 188
Gorowicz, Janusz P., 285
Guzman, Roberto Z., 232
Hall, Carol K., 53
Hamel, Jean-François P., 1
Hamilton, Bruce K., 130
Hunter, Jean B., 1
Ivory, Cornelius F., 206
Jolivalt, Claude, 87
Kilpatrick, Peter K., 232
Levy, Peter F., 130
May, Sheldon W., 108
Minier, Michel, 87
Phillips, Marshall, 257
Renon, Henri, 87
Sheehan, John J., 130
Simmons, Donald K., 108
Snyder, Steven M., 71
Strege, Mark A., 176
Szlag, David C., 38,71
Tsao, I-Fu, 158
Wang, Henry Y., 158
West, Jeffrey S., 176
Yoshisato, Randall A., 285

Affiliation Index

Clarkson University, 268
Cornell University, 1
Ecole Nationale Supérieure des Mines de Paris, 87
Georgia Institute of Technology, 108
Indiana University—Purdue University at Indianapolis, 176
Iowa State University, 188
Lockheed Engineering and Sciences Company, 206
Massachusetts Institute of Technology, 1
National Institute of Standards and Technology, 38,71
North Carolina State University, 53,232
U.S. Department of Agriculture, 257
University of Arizona, 38
University of Iowa, 285
University of Michigan, 158
W. R. Grace & Company–Conn., 130
Washington State University, 206

Subject Index

A

Affinity chromatography, protein purification technique, 3–4
Affinity-mediated membrane transport process, description, 27–28
Affinity partitioning
  extraction of YADH and total protein from yeast enzyme concentrate, 80,83f
  preparation of polyethylene glycol–dye ligands, 80,81f
  recovery of YADH activity vs. pH, 80,84,85f
  triazine dye ligand screening, 80,82f
Affinity precipitation of avidin by using ligand-modified surfactants
  advantages, 215,217
  binding studies of dimyristoyl-phosphatidylethanolamine B to avidin, 219,221,222f
  determination of avidin activity, 217,218f,219,220f
  dissolution of biotinylated phospholipid in nonionic surfactant solution, 219
  experimental materials, 217
  from partially purified egg whites, 229,232,233–235f
  precipitation and recovery of avidin from model protein solutions, 221,227,229,230–231f
  purification of avidin from egg whites, 221,223,224f,225
  reversed-phase high-pressure liquid chromatograms, 229,232,233–234f
  schematic illustration of three-dimensional network, 215,216f
  sodium dodecyl sulfate–polyacrylamide gel electrophoretic patterns, 232,235f
  structural formulas of surfactants, 215,216f
  synthesis of biotinylated phospholipid, 217
  titration of avidin with solubilized dimyristoylphosphatidylethanolamine B, 225,226f,227,228f
Affinity precipitation of proteins
  advantages, 213–214
  disadvantages, 214
  schemes, 213–214
Affinity separation
  advantages, 25–26
  description, 24
  example, 25
  isolation by affinity interaction, 26–27
  recovery strategies, 27–28
  steps, 25
  supports, 25
Anion carriers for liquid membrane transport
  bidirectional transport of L-tyrosine, 116,117f,118t
  examples, 114
  transport of L-tyrosine, 114,115f,116t
Aqueous biphasic system for protein recovery, 4–5
Aqueous two-phase extraction system(s)
  advantages, 38–39
  applications, 71
  estimation of model parameters, 46–47
  formation, 71–72
  limitations in biotechnology, 71–72
  phase diagram calculations, 47,48–49f,50
  protein partitioning, 53
  scaling law expressions, 42–46
  statistical mechanical basis, 39–42
  theoretical treatment using virial expansions, 53–69
  See also Low-cost aqueous two-phase system for affinity extraction
Average linear velocity, definition, 131
Average transmembrane pressure, definition, 131
Avidin
  affinity precipitation from partially purified egg whites, 229,232,233–235f
  binding with dimyristoylphosphatidylethanolamine B, 219,221,222f
  precipitation and recovery from model protein solutions, 221,227,229,230–231f
  purification from egg whites, 221,223,224f,225
  titration with solubilized dimyristoylphosphatidylethanolamine B, 225,226f,227,228f

B

Bioproducts, purification scheme, 1,2f
Biotinylated phospholipids, affinity precipitation, 217,219

C

Capillary zone electrophoresis, description, 12–13
Carrier, definition and function, 110
Carrier-facilitated transport
  anion carriers, 114–118
  cation carriers, 118–119
  criteria for carriers, 111,114
  mode of action, 110

Carrier-facilitated transport—*Continued*
  rules for dissociation constant
    interpretation, 111
  transport of anions and cations in liquid
    membranes, 111,112*f*
  L-tyrosine, 111,113*f*
Carrier-mediated transport
  control of chemical environment, 190
  control of complexation–decomplexation
    reaction direction, 190
  schematic representation, 188,189*f*
Cation carriers for liquid membrane
    transport
  development, 119
  examples, 118
Cation-specific distribution between
    reversed micellar phase and bulk aqueous
    solution
  electrostatic free energy change, 99,101
  equality of electrochemical potentials, 99
  model, 98
Chemical potential prediction of
    solvent–polymer–protein system species
  dependence of polymer driving force on
    polymer molecular weight, 65,67*f*
  dependence of protein partition
    coefficient on polymer and protein
    molecular weights, 62–66
  derivation, 56–58
  K_p vs. dextran molecular weight, 66,68*f*
  K_p vs. polyethylene glycol molecular
    weight, 66,67–68*f*
  K_p vs. protein molecular weight, 66
  partition function evaluation, 59–60
  semigrand partition function, 55–57
  theory, 55–62
  virial coefficient determination, 60–61
Chromatography
  advancements, 21
  definition, 21
  novel methods in traditional geometries,
    22,23*f*,24
  process-scale applications, 270
  throughput problem, 21
  types, 21–22
Concentration polarization, definition,
    134–135
Constant pressure solution theory,
    advantages for chemical potential
    prediction, 54
Continuous annular chromatograph
  schematic diagram, 268,269*f*,270
  theoretical development, 270
Continuous annular chromatography
  description, 23–24
  limitations, 24
Continuous electrophoretic separations,
    possible devices and techniques, 285–286

Continuous rotating annular electrophoresis
  description, 18
  unit, 18*f*
Continuous rotating annular electrophoretic
    column
  advantage, 286
  design, 290,291*f*
  dyes, 294–295,296*t*
  effect of temperature field on flow
    field, 292
  experimental apparatus, 294,295*f*
  factors affecting dispersion, 292–294
  future work, 300
  helical product bands, 286,287*f*
  mathematical model, 288–290
  mean residence times and variance,
    297,298*t*
  operational procedure, 286
  ratio of electrophoretic dispersion
    coefficient to ordinary dispersion
    coefficient, 298,299*f*,300
  residence time distributions for dyes,
    296,297*f*
  review of current status, 286–300
  schematic representation, 290,291*f*
Conventional liquid chromatographic
    separations, approaches for theoretical
    description, 271
Counteracting chromatographic
    electrophoresis, description, 19*f*
Countercurrent chromatography,
    description, 24
Cross flow
  definition, 131
  generalized schematic representation,
    131,132*f*
Cross-flow filtration, use in processing
    cell suspensions, 10–11
Cross-flow magnetically stabilized fluidized
    bed chromatograph, mathematical
    analysis, 270
Crossed-field electrophoresis,
    description, 13
CsCl, use in gradient preparation in
    equilibrium centrifugation, 240
Current, effect on protein flux, 204

D

Displacement chromatography, description,
    22,23*f*
Donnan equilibrium
  electrochemical potential of permeating
    species, 193,195
  importance in ion-exchange and
    facilitated-transport membranes, 195
  occurrence, 193

INDEX

Downstream processing, role in biochemical industry, 1

E

Edmonds–Ogston expression, application, 53–54
Electrically assisted separation
 description, 20
 modeling, 20
Electrochromatography
 continuous rotating annular electrophoresis, 18f
 counteracting chromatographic electrophoresis, 19f
Electrophoresis
 advantages, 285
 basic theory, 286,288
 definition, 12
 electrically assisted separation, 20
 electrochromatography, 18–19f
 electrophoretic mobility, 286,288
 mean residence time in column, 288
 migration velocity, 286
 nanoscale separation, 12–13
 process-scale separation, 13,14–16f
 recycling free-flow methods, 16–18
 resolution between bands, 288
 scale-up problems, 286
 species velocity, 286
 types, 12
Electrostatic interactions of reversed micelles
 effect of ionic strength on solubilization of proteins, 93,94f
 effect of pH on dissociation rate of charged residues, 90–91
 effect of surfactant on solubilization of proteins, 93,95
Emulsions, characterization, 109–110
Enzyme-containing micellar solutions, preparation techniques, 88
Enzymes in liquid membranes
 anion carriers, 114–118
 carrier-facilitated transport, 110–114
 cation carriers, 118–119
 enzyme immobilization, 119–120
 model substrates, 123,125f,126
 previous studies, 120–121
 reactor configurations, 126–127
 schematic representation, 108,112f
 tyrosinase–ascorbate system, 121,122f,123,124f
Equilibrium centrifugation
 application, 240
 concept, 240
 gradient preparation media, 240

Excluded volume forces, models, 63–65
Extracellular bacterial protease recovery, pilot-scale membrane filtration process, 130–154

F

Facilitated-transport membranes, observation of Donnan equilibrium, 195
Factors influencing protein–polyelectrolyte complexes
 dosage, 171–172
 ionic strength, 171
 polyelectrolyte, 171
 protein, 170–171
 solution pH, 171
Fickian diffusion, schematic representation, 188,189f
Flory–Huggins theory of protein partitioning, advantage and disadvantage, 54

G

Gibbs–Duhem equation, 57

H

hgh–antibody system
 modeling, 195–199
 response to low current densities, 205
 schematic representation, 193,194f
High-pressure homogenization, freeing of intracellular products, 3
High-resolution purification techniques of proteins
 affinity separation, 24–28
 chromatography, 21
 electrophoresis, 12–20
Hollow-fiber chromatography, description, 22

I

Immunoaffinity chromatography
 effect of switch monoclonal antibodies on efficiency, 190
 low selectivity, 190–191
Induced field effects of hgh–antibody system
 comparisons of internal and external hgh concentrations, 199,201,202t
 concentration profile for hgh vs. salt concentration, 201,202f
 facilitation factor vs. salt concentration, 199,200f

Induced field effects of hgh–antibody system—*Continued*
  facilitation factors, 199
  hormone concentration vs. salt concentration, 199,200f
  pH dependence of antibody charge, 199
  potential profile vs. salt concentration, 201,203f
  upstream and downstream charges on hgh, antibody, and complexes, 201,203t
Instantaneous protein transmission, measurement, 131
Ion-exchange membranes, observation of Donnan equilibrium, 195
Isoelectric focusing, description, 12
Isoelectric precipitation, fractionation, and characterization of proteins, 11
Isoelectric protein partition coefficients, prediction, 54–55
Isolation by affinity interaction, description, 26–27
Isopycnic centrifugation, *See* Equilibrium centrifugation, 240

L

Lipopolysaccharides
  analytical CsCl gradient separation, 243,245–246f
  analytical schlieren patterns, 243,246f
  apparent buoyant densities, 243,245t
  buoyant density separation, 243,244f
  buoyant positions of hydrated constituents in CsCl gradients, 241,242f
  computer-generated density gradient distributions, 243,246f
  CsCl gradients, 243,245f
  description, 238
  effect of chemical treatment on separation, 247
  effectiveness of CsCl gradient separation, 247–248
  extraction methods, 241
  function, 238
  preparative CsCl gradient isolation, 241–245
  schematic diagram of basic molecule, 238,239f
Liquid membrane(s)
  carrier-facilitated transport, 110–114
  description, 108
  obstacles in development of enzyme reactor systems, 110
  properties, 109–110
  schematic representation of liquid membrane–enzyme system, 108,112f
  separation applications, 108
  transport of solutes, 110

Liquid membrane encapsulated enzymes
  enzyme immobilization, 119–120
  model substrates, 123,125f,126
  previous studies, 120–121
  reactor configurations, 126–127
  tyrosine–ascorbate system, 121,122f,123,124f
Liquid membrane system for protein recovery
  applications, 8
  diagram, 7f,8
  economic evaluations, 8
  future research, 8–9
Low-cost aqueous two-phase system for affinity extraction
  affinity partitioning, 77,80–85
  binodal curves, 74,75–76f
  comparison of maltodextrin and dextran ATPS, 84,85t
  economic considerations, 84,86
  experimental materials, 72–73
  extraction of alcohol dehydrogenase from crude yeast lysate, 73–74
  phase diagram determination, 73
  phase system preparation, 72–73
  physical properties, 74,77,78–79t
  protein assay procedure, 74
  yeast extract preparation, 73
  yeast homogenate preparation, 73

M

Mathematical model for continuous rotating annular electrophoretic column
  conservation of energy, 289
  conservation of momentum, 288–289
  conservation of species, 289–290
  electroosmotic flow, 289
  electroosmotic mobility, 289
Mathematical model of hgh–antibody system
  comparison of applied and induced fields, 205,206t,207
  effect of current on protein flux, 204–208
  facilitation factor vs. current density and salt concentration, 204,206f
  hydrogen flux vs. current density and salt concentration, 207,208f
  induced field effects, 199–203
  orthogonal collocation on finite elements, 197,199
  partial-equilibrium–combined flux technique, 196–197,198t
  unidirectional continuity equation, 195–196
Mathematical model of rotating annular continuous size exclusion chromatography
  assumptions, 268
  boundary conditions, 272

INDEX

Mathematical model of rotating annular continuous size exclusion chromatography—*Continued*
  comparison of predicted concentration profiles, 276,277f
  contributions of proteins to peak variance, 278,280t
  dispersion coefficient estimation, 275
  effect of change in parameter on peak variance, 278,280t,281
  effect of change in parameter on resolution, 278,281t
  effect of column length, 282
  effect of feed band width, 281
  effect of flow velocity, 281–282
  effect of particle size, 282–283
  effect of pore size, 283
  effect of rotation rate, 282
  elution profile for base-case parameters, 278,279f
  intrapore diffusivity estimation, 275–276
  mass transfer coefficient estimation, 275
  mobile-phase concentration, 274
  moments of concentration profile in θ domain, 273
  operating parameters and column and packing dimensions, 276,278,279t
  properties of proteins used in model evaluation, 274–275,277t
  resolution of protein separation, 276
  solution in Laplace domain, 272–273
  steady-state continuity equations for solute, 271–272
  variance of distribution, 274
Membrane(s)
  applications, 9
  use in bioprocesses, 9
Membrane flux performance for hollow fiber units
  comparison of permeate flow control vs. transmembrane pressure control, 141,142t
  effect of cross-flow velocity on membrane performance, 141
  flux vs. average transmembrane pressure profiles, 140f,141
  flux vs. cell concentration profile, 138,139f
  flux vs. time profile, 138,139f
Membrane separation techniques
  cross-flow filtration, 10–11
  microfiltration, 9–10
  ultrafiltration, 10
Membrane-strengthening agent, description, 109
Microemulsion phase, description, 96
Microfiltration, applications, 9–10
Model parameters for aqueous two-phase systems, estimation, 46–47

Multiphase systems for protein recovery
  aqueous biphasic system, 4–6
  liquid membranes, 7f,8–9
  reversed micellar systems, 6f,7

N

Nanoscale separation, description, 12–13

O

Oppositely charged polyelectrolytes, complex formation, 158
Overexpression of cloned genes, production of eukaryotic proteins, 3

P

Packing ratio, definition, 89
Partial-equilibrium–combined flux technique, description, 196–197,198t,199
Percent instantaneous transmission, calculation, 131
Permeate side of filter, function, 131
pH
  effect on α-amylase solubilization, 91,93f
  effect on extraction of amino acids or peptides, 91
  effect on solubilization of proteins, 91,92f
Phase diagrams for aqueous two-phase systems, calculations, 47,48–49f,50
Phenomenological model of solubilization in reversed micelles, description, 97–98
Plate theory, important parameter, 271
Poly(dimethyldiallylammonium chloride)–globular protein complexation
  apparent distributions of equivalent Stokes diameters, 161,164f
  phase boundaries, 161,162f
  quasi-elastic light scattering, 167
  quasi-elastic light scattering procedure, 160
  refractive index chromatogram, 161,165f,166
  size exclusion chromatographic procedure, 160–161
  size exclusion chromatography, 167–168
  surface charge density vs. ionic strength, 161,163f
  turbidimetric titration(s), 161,162f,166–167
  turbidimetric titration procedure, 159–160
  UV chromatogram, 161,165f,166
Polyelectrolyte(s), removal of oppositely charged proteins, 159

Polyelectrolyte–protein interaction mechanism, 159–168
Polymer driving force, dependence on polymer molecular weight, 65,67f
Precipitation of proteins
 additives, 213
 production, 213
 selectivity, 213
Process-scale chromatography, central issue, 21
Process-scale separation
 field-step focusing, 15,16f
 Philpot–Harwell device, 13,14f
 thin-film free-flow electrophoretic device, 14,15f
Protein(s)
 biological properties, 2
 membrane separation, 9–11
 recovery through formation of insoluble complexes with polyelectrolytes, 159
 separation by using reverse micelles, 87–104
Protein characterization, analytical and isolation techniques, 11
Protein flux, effect of current, 204–208
Protein partition coefficient, dependence on polymer and protein molecular weights, 62–66
Protein passage efficiency, definition, 131,133
Protein–polyelectrolyte complexes, factors influencing efficacy of protein precipitation, 170–172
Protein precipitation, *See* Precipitation of proteins
Protein precipitation and fractionation by (carboxymethyl)cellulose precipitation
 analytical procedures, 172
 binary solutions, 178,182–186
 critical micelle concentration dosage levels vs. pH and ionic strength, 174t,175
 effect of ionic strength, 178–181,185–186
 effect of pH, 174–175,177–178,183–184
 effect of pH on charge of ovalbumin, lysozyme, and critical micelle concentration, 172,173t
 effect of polymer dosage, 175,176f,178,182f
 pH and ionic strength conditions for precipitations, 172,173t
 precipitation procedure, 172
 single-component systems, 175–181
Protein purification
 advantages of fewer steps, 2–3
 affinity chromatography, 3–4
 high-pressure homogenization, 3
 high-resolution techniques, 12–28
 overexpression of cloned genes, 3
Protein recovery, multiphase systems, 4–9
Protein separation via affinity-mediated membrane transport
 definition of species vectors, 191,192t
 diffusion coefficients and charge profiles, 191,192t
 Donnan equilibrium, 193,195
 effect of current on protein flux, 204–208
 induced field effects, 199–203
 modeling, 195–199
 rate constants, 191
 reaction network, 191,194t
 schematic representation of hgh–antibody system, 193,194f
 system description, 191,192t,193,194f,t
Pulsed electrophoresis, description, 13

Q

Quaternary ammonium chlorides
 description, 250
 mode of action, 250
 virucidal capacities, 250–251
Quaternary ammonium salts, use for solubilization of biomolecules, 89–90

R

Radial-flow chromatography, description, 21–22
Rate theory, application, 271
Recirculation rate, definition, 131
Recombinant DNA technology, quality guidelines, 1
Recycling isoelectric focusing, procedure, 16–17
Recycling isotachophoresis, procedure, 17–18
Recycling zone electrophoresis, apparatus, 17
Resolution, definition, 276
Retentate side filter, function, 131
Reversed micellar systems for protein recovery
 advantages and applications, 6–7
 diagram, 6f
 effect of ionic strength on solubilization of proteins, 93,94f
 effect of surfactant on solubilization of proteins, 93,95
Reversed micelles
 enzyme catalysis, 88
 extraction examples, 104
 protein separation, 87–104
 purification of intracellular dehydrogenases, 103
 refolding of proteins, 104

INDEX

Reversed micelles—*Continued*
  separation of targeted protein from complex medium, 101
Rotating annular continuous size exclusion chromatograph, mathematical model, 271–283

S

Scaling law expressions of aqueous two-phase systems
  osmotic compressibility factor, 43–46
  osmotic virial coefficients, 43
Second virial coefficients, dependence on molecular weight, 63–65
Semigrand partition function, derivation, 56
Separation of proteins using reversed micelles
  applications, 101,103–104
  comparison to liquid–liquid extraction, 87
  electrostatic interaction, 90–95
  extraction procedure, 87
  parameters affecting solubilization, 89–96
  protein solubilization in organic solvent, 87–88,92f
  requirements, 88–89
  selectivity, 101,103
  steric effects, 95–96
  surfactants, 89–90
  thermodynamic models of solubilization, 96–102
Size exclusion chromatography
  principle of separation, 268
  schematic diagram of continuous annular chromatograph, 268,269f,270
Sodium bis(2-ethylhexyl)sulfosuccinate, use for solubilization of biomolecules, 89–90
Sphere–coil model of excluded volume forces, description, 63–65
Sphere–cylinder model of excluded volume forces, 63
Sphere–sphere model of excluded volume forces, description, 63–65
Statistical mechanical basis of aqueous two-phase systems
  canonical partition function, 40
  fundamental basis of model, 39–40
  model parameters, 39
  molality, 41
  osmotic pressure, 42
  osmotic virial coefficients, 41–42
Steric effects of reversed micelles
  effect of amount of alcohol, 95
  importance of size exclusion, 95–96
Structural model of spherical monodisperse droplets, description, 96–97

Switch monoclonal antibodies, improvement of immunoaffinity chromatographic efficiency, 190

T

Tangential flow, definition, 131
Therapeutic proteins, problem of presence of low-level infectious viral contaminants, 251
Thermodynamic models of solubilization in reversed micelles
  free energy change, 97
  phenomenological model, 97–98
  prediction of microscopic structure, 96–97
  reference state, 96
  schematic diagram of assumed interfacial structure, 99,100f
  schematic ternary phase diagram of oil–water–surfactant microemulsion system, 101,102f
  structural model of spherical monodisperse droplets, 96–97
Two-stage pilot-scale membrane filtration process
  background, 133–134
  capital costs for cell separation step, 151,153f
  definitions, 131,132f,133
  description of enzyme production and recovery process, 135,138
  design assumptions for cell separations, 151,152t
  development, 130–131
  economics of cell separation step, 148,151,152t,153f
  effect of cell wash, 145,146f
  effect of transmembrane pressure, 142,144f,145
  flow diagram, 135,136–137f,138
  flux recovery, 145,148,149–150f
  membrane cleaning, 145,148,149–150f
  membrane flux performance for hollow fiber units, 138–142
  protein passage efficiency, 142,143f
  spiral ultrafilter performance, 145,147f
  theory, 134–135
  total manufacturing cost for cell separation step, 151,153f
Tyrosinase–ascorbate system
  enzymatic oxidations of phenols, 121,122f
  schematic representations of complete system, 123,124f
L-Tyrosine, transport through liquid membranes, 114–118

U

Ultracentrifugation
  fractionation and characterization of proteins, 11
  future application, 248
  separation and identification of lipopolysaccharides, 238–248
Ultrafiltration, applications, 10
Unidirectional continuity, equation, 195–196

V

Virus removal and inactivation by surface-bonded quaternary ammonium chloride
  adsorption isotherms, 255,257,258f
  adsorption–inactivation of T2 phage in β-lactamase solution, 257,258–259f
  assays, 253
  batch experiments, 253
  cell culture, 252
  chemicals, 252
  comparison of titer reduction and adsorption capacity between HSV–1 and T2 phage, 255,256t
  effect of sterilization criteria on protein recovery and required amount of absorbent, 261,264f
  elution experiments, 255,256t
  equilibrium isotherms of T2 phage inactivation, 257,258f

Virus removal and inactivation by surface-bonded quaternary ammonium chloride—*Continued*
  inactivation of T2 phage in free solutions, 253,254t
  inactivation of viruses, 253,254t,255,256t
  kinetic study of T2 phage removal, 257,258f
  physical parameters used for simulation studies, 261,262t
  preparation of beads, 252
  removal of T2 phage from β-lactamase solution, 257,259f
  simulated adsorption breakthrough curves of total protein and HSV–1, 261,263t
  simulation studies with adsorption column, 260–264
  thermal inactivation of T2 phage, 257,259f
  titration of viruses, 252–253
Volume forces, role in protein partitioning, 63–65

W

Wall shear rate, function, 131

Z

Zone electrophoresis, description, 12

Production: Paula M. Befard
Indexing: Deborah H. Steiner
Acquisition: Cheryl Shanks

Elements typeset by Hot Type Ltd., Washington, DC
Printed and bound by Maple Press, York, PA

Paper meets minimum requirements of American National Standard for Information Sciences—Permanence of Paper for Printed Library Materials, ANSI Z39.48–1984 ∞